T0328222

Ecosystem and Territorial Resilience

A Geoprospective Approach

Ecosystem and Territorial Resilience

A Geoprospective Approach

Edited by

Emmanuel Garbolino

*Climpact Data Science, Nova Sophia—Regus Nova,
Sophia Antipolis Cedex, France*

Christine Voiron-Canicio

Université Côte d'Azur, CNRS, UMR ESPACE, Nice, France

ELSEVIER

Elsevier
Radarweg 29, PO Box 211, 1000 AE Amsterdam, Netherlands
The Boulevard, Langford Lane, Kidlington, Oxford OX5 1GB, United Kingdom
50 Hampshire Street, 5th Floor, Cambridge, MA 02139, United States

Copyright © 2021 Elsevier Inc. All rights reserved.

No part of this publication may be reproduced or transmitted in any form or by any means, electronic or mechanical, including photocopying, recording, or any information storage and retrieval system, without permission in writing from the publisher. Details on how to seek permission, further information about the Publisher's permissions policies and our arrangements with organizations such as the Copyright Clearance Center and the Copyright Licensing Agency, can be found at our website: www.elsevier.com/permissions.

This book and the individual contributions contained in it are protected under copyright by the Publisher (other than as may be noted herein).

Notices

Knowledge and best practice in this field are constantly changing. As new research and experience broaden our understanding, changes in research methods, professional practices, or medical treatment may become necessary.

Practitioners and researchers must always rely on their own experience and knowledge in evaluating and using any information, methods, compounds, or experiments described herein. In using such information or methods they should be mindful of their own safety and the safety of others, including parties for whom they have a professional responsibility.

To the fullest extent of the law, neither the Publisher nor the authors, contributors, or editors, assume any liability for any injury and/or damage to persons or property as a matter of products liability, negligence or otherwise, or from any use or operation of any methods, products, instructions, or ideas contained in the material herein.

British Library Cataloguing-in-Publication Data
A catalogue record for this book is available from the British Library

Library of Congress Cataloging-in-Publication Data
A catalog record for this book is available from the Library of Congress

ISBN: 978-0-12-818215-4

For Information on all Elsevier publications
visit our website at https://www.elsevier.com/books-and-journals

Publisher: Candice Janco
Acquisitions Editor: Peter J. Llewellyn
Editorial Project Manager: Mona Zhair
Production Project Manager: Anitha Sivaraj
Cover Designer: Matthew Limbert

Typeset by MPS Limited, Chennai, India

Contents

Chapter 12: Geodesign for collaborative spatial planning: three case studies at different scales 323

Matteo Caglioni and Michele Campagna

Chapter 13: How do public policies respond to spatialized environmental issues? Feedback and perspectives...347

Christine Voiron-Canicio, Emmanuel Garbolino, Nathalie Cecutti, Carlo Lavalle and José Juan Hernández Chávez

List of contributors

Géraldine Abrami G-EAU, Univ Montpellier, AgroParisTech, CIRAD, IRD, INRAE, Institut Agro, Montpellier, France

Olivier Barreteau G-EAU, Univ Montpellier, AgroParisTech, CIRAD, IRD, INRAE, Institut Agro, Montpellier, France

Jacques Baudry INRAE Centre Bretagne-Normandie, UMR 0980 BAGAP, Rennes Cedex, France

Bruno Bonté G-EAU, Univ Montpellier, AgroParisTech, CIRAD, IRD, INRAE, Institut Agro, Montpellier, France

François Bousquet CIRAD, UPR GREEN, F-34398 Montpellier, France; GREEN, Univ Montpellier, CIRAD, Montpellier, France

Matteo Caglioni University Côte d'Azur, CNRS, Laboratory ESPACE, Nice, France

Michele Campagna University of Cagliari, DICAAR, Cagliari, Italy

Nathalie Cecutti Ministry of the Ecological and Solidary Transition, Paris, France

José Juan Hernández Chávez Ministry of Environment and Natural Resources, Mexico

Warren Daniel Plant and Ecosystems (PLECO), University of Antwerp, Wilrijk, Belgium

Sandrine Dhenain G-EAU, Univ Montpellier, AgroParisTech, CIRAD, IRD, INRAE, Institut Agro, Montpellier, France; TEC Conseil, Marseille, France

Hervé Fritz Sustainability Research Unit, Nelson Mandela University-. Madiba drive, George, South Africa; REHABS International Research Lab, CNRS-Université Lyon 1-NMU, Nelson Mandela University-. Madiba drive, George, South Africa

Giovanni Fusco University Côte d'Azur, CNRS, Laboratory ESPACE, Nice, France; Université Côte d'Azur, CNRS, UMR ESPACE, Nice, France

Emmanuel Garbolino Climpact Data Science, Nova Sophia—Regus Nova, Sophia Antipolis Cedex, France

Françoise Gourmelon CNRS, UMR LETG, Institut Universitaire Européen de la Mer, Plouzané, France

Chloé Guerbois Sustainability Research Unit, Nelson Mandela University-. Madiba drive, George, South Africa; REHABS International Research Lab, CNRS-Université Lyon 1-NMU, Nelson Mandela University-. Madiba drive, George, South Africa

Thomas Lagier SoLyft, Lyon, France

Carlo Lavalle European Commission—Joint Research Center, Ispra, Italy

Romain Légé Université de Nantes, UMR LETG, Nantes, France

Jean-Christophe Loubier   University of Applied Sciences and Arts Western Switzerland, Sierre, Switzerland

Stéphanie Mahévas   Ifremer, Nantes, France

Christine Malé   MUG, Metropole of Lyon, France

Raphaël Mathevet   CEFE, CNRS, Univ. Montpellier, EPHE, IRD, Univ. Paul Valéry Montpellier 3, Montpellier, France

Guillermo Hinojos Mendoza   ASES Ecological and Sustainable Services, Pépinière d'Entreprises l'Espélidou, Parc d'Activités du Vinobre, Aubenas, France

Cédric Simi   G-EAU, Univ Montpellier, AgroParisTech, CIRAD, IRD, INRAE, Institut Agro, Montpellier, France

Clara Therville   G-EAU, Univ Montpellier, AgroParisTech, CIRAD, IRD, INRAE, Institut Agro, Montpellier, France; CEFE, CNRS, Univ. Montpellier, EPHE, IRD, Univ.
Paul Valéry Montpellier 3, Montpellier, France

Laurie Tissière   Université de Nantes, UMR LETG, Nantes, France

Brice Trouillet   Université de Nantes, UMR LETG, Nantes, France

Gilles Voiron   Université Côte d'Azur, CNRS, UMR ESPACE, Nice, France

Christine Voiron-Canicio   Université Côte d'Azur, CNRS, Laboratory ESPACE, Nice, France; Université Côte d'Azur, CNRS, UMR ESPACE, Nice, France

About the editors

Emmanuel Garbolino

Dr. Emmanuel Garbolino has a PhD in Geography as well as an HDR accreditation to supervise research in Geography, both of which he earned from the University of Nice, France. He also has a Master of Ecology and Paleoecology degree from the University of Aix-Marseille, France. After completing his PhD, for which he also worked at the Joint Research Centre (JRC) of the European Commission in Ispra (Italy), Dr. Garbolino spent 17 years in the field of public research at the Risk and Crisis Research Centre at MINES ParisTech. He is currently the Director of Climpact Data Science (CDS), a company that provides services to support decision makers in adapting their strategies, activities, and territorial management in the context of global climate and land use changes.

Christine Voiron-Canicio

Dr. Christine Voiron-Canicio is a professor of geography at Université Côte d'Azur, France. She is a recipient of bronze medal from the French National Centre for Scientific Research (CNRS) and is a co-founder of the Geoprospective research group and a co-head of the Territorial Resilience Strategy training for sustainable city stakeholders, in collaboration with Sustainable Cities France. She led the ESPACE lab in France from 2008 to 2017. She specializes in spatial analysis applied to sustainability issues, including: urban risks, slow evolutions and sudden phenomena, vulnerabilities, and resilience. She couples scientific skills in modeling and spatial simulations with geoprospective approaches, aimed at anticipating the spatial impacts of future changes. She is currently involved in research into diagnosing and improving the adaptive capacity of regions and cities.

About the authors

Géraldine Abrami

Géraldine Abrami is a research engineer at UMR G-EAU, France, and has been involved in research into the implementation of public water policies in France and Africa using participatory modeling and simulation tools. Since 2009, he has worked as a co-developer of Wat-A-Game, an open material and methodological toolkit for the participatory design of water catchment models and role-playing games.

Olivier Barreteau

Olivier Barreteau is a senior water scientist at INRAE and a director of the G-EAU Lab, France. He has 25 years experience in interdisciplinary research at the interface among hydrology, modeling, and management sciences in projects on local water governance, using social simulation with agent-based modeling and role-playing games. His recent works have focused on joint adaptive management of land and water with in the frame work of global changes with issues of interdependences between local adaptation plans.

Jacques Baudry

Jacques Baudry has been a senior research scientist at INRAE since 1986. He is a coordinator of projects fostering the design of ecological networks with stakeholders. His area of research is mainly around the landscape dynamics driven by socioecological systems at different scales. Jacques is pioneering research on connectivity using hedgerow networks as a model. He was also a scientific coordinator of the research programs of the French Ministry of Environment related to public action and biodiversity from 1987 to 2016.

Bruno Bonté

Bruno Bonté is a research fellow at INRAE and G-EAU Lab, France and has completed a PhD in computer science. His research is focused on modeling and simulation of complex systems. He has been engaged in the development of integrative tools for collective decision-making (based on the companion modeling approach). His most recent works are devoted to social experiments and multiscale modeling of social and ecological systems and the application of water resource management and adaptation of territories to global change.

François Bousquet

François Bousquet completed his PhD and is a senior researcher in the environmental field at CIRAD, France. His works are focused on biodiversity conservation and collective decision-making in natural and renewable resource management. He is also a member of the international research network Resilience Alliance.

Matteo Caglioni

Matteo Caglioni is an associate professor in geography and urban modeling, Université Côte d'Azur, France. He is also a researcher at CNRS UMR 7300 ESPACE in spatial analysis, geosimulation, artificial intelligence in geocomputation, geographical information science, volunteered geographic information, and semantic enrichment of 3D city models.

Michele Campagna

Michele Campagna is an associate professor at the Civil and Environmental Engineering and Architecture Department, University of Cagliari, Italy. His main research interests are digital technologies innovation in spatial planning and design, geodesign, metaplanning, planning support systems, and authoritative and volunteered geographic information.

Nathalie Cecutti

Nathalie Cecutti is a state architect and urban planner in chief. She has led the Prospective Mission of the Ministry in charge of Sustainable Development and the Environment from 2011 to 2017. Currently, she is an expert to the Head of the Ministry of the Ecological and Solidary Transition's research department for developing relations with non-state actors on research issues. She is also a coordinator of the European research and innovation framework programs "Horizon 2020" and the future "Horizon Europe."

Warren Daniel

Warren Daniel is an engineer in environment sciences, AgroParisTech. Currently, he is a PhD student at the Plant and Ecosystems Research Center of the University of Antwerp, Belgium. He has worked on the study of the complete greenhouse gas (GHG) balance of a tropical forest in French Guiana, and has built experiments in order to better characterize the global budget for GHG flows from a tropical forest ecosystem (at ground, tree trunk, and canopy levels).

José Juan Hernández Chávez

José Juan Hernández Chávez is a director for environmental policy in the Secretaría de Medio Ambiente y Recursos Naturales, México, where he is in charge of technical counseling and institutional support in the processes of environmental planning (ordenamiento ecológico del territorio) in the Mexican Territory. He has a degree in biology from the National School of Biological Sciences of Instituto Politécnico Nacional, México, and is a fellow of Cohort 20 from the LEAD (Leadership for Environment and Development) program México, Colegio de México.

Giovanni Fusco

Giovanni Fusco is a research fellow at the French National Centre of Scientific Research and Deputy Director of the ESPACE Laboratory, Nice, France. He is a developer of geoprospective approaches for cities and metropolitan areas using principles of urban resilience and artificial intelligence. He investigates issues of complexity and uncertain knowledge in urban research.

Françoise Gourmelon

Françoise Gourmelon is a senior research scientist in geography at CNRS, France. She supervises the LETG lab and is working in the European Institute for Marine Studies, Brest, France. She specializes in geographic information sciences (GIS) and spatial data infrastructure (SDI). After studying the interactions between nature and human activities through spatiotemporal dimensions in coastal zones, her research is currently devoted to the contribution of geographical information to public policies.

Chloé Guerbois

Chloé Guerbois is a French research fellow at the Nelson Mandela University, Sustainability Research Unit, South Africa. She is an interdisciplinary scientist interested in the socioecological processes involved in resilience to global change and particularly the contribution of protected areas to biosphere-based sustainability. She completed her PhD in the Hwange LTSER in Zimbabwe.

Guillermo Hinojos Mendoza

Guillermo Hinojos Mendoza is a ecologist, who holds a PhD in "Science and engineering of risk activities" from Mines ParisTech, France. He has been involved in the identification of risks of loss of biodiversity due to climate change and territorial transformation within Horizon 2100. Currently, he is the CEO of ASES and ASES France companies that are specialized in ecological engineering operations and the development of computer-based tools in this field of activities.

Thomas Lagier

Thomas Lagier is a founder of ForCity, a start-up on SaaS solution for urban geoprospective and systemic decision tools for cities. His is also a founder of the start-up studio SoLyft, France, for industrializing POC to professional SaaS software.

Romain Legé

Romain Legé is a PhD candidate at the University of Nantes, France. He has been involved in research projects aimed at adapting geoprospective methods to support marine spatial planning.

Jean-Christophe Loubier

Jean-Christophe Loubier is a professor at the University of Applied Sciences of Western Switzerland and a head of the GIS-LAB Laboratory, HES-SO Valais. He is also a specialist in spatial analysis and modeling in a participatory context.

Christine Male

Christine Male is an architect and urban planner in charge of the Gerland Urban Modeling project with the Head of Territorial Strategy and Urban Policies concerning Lyon Metropolitan Area.

Stéphanie Mahevas

Stéphanie Mahevas is a senior research scientist in applied mathematics for fisheries science at Ifremer (Ecology and Models for Fisheries Science, Unit EMH, Nantes, France). She is involved in the following research topics: fisheries dynamics modeling, methods for complex models, trajectories analysis, spatial stock assessment and management strategy evaluation, scenario and foresight.

Raphael Mathevet

Raphael Mathevet is a senior research scientist at CNRS, CEFE Laboratory, France. He is an ecologist and geographer working on the conservation of biodiversity, protected areas and conservation planning tools, adaptive management, and evaluation of public policies. He is an expert in simulation tools and role-playing games for resolving management conflicts. His most recent research has focused on the concepts of ecological solidarity and stewardship. He is a member of various scientific committees such as National Parks and Reserves, IHOPE, and UNESCO MAB program.

Cedric Simi

Cedric Simi is an engineer, whose areas of interests are focused on environmental issues. He has been involved in tackling climate change and is working in the MAGIC research project as part of his master's degree in biodiversity management (France).

Clara Therville

Clara Therville is asocial geographer interested in the multilevel governance of environment−society interactions. By mobilizing theoretical tools from SES resilience thinking and institutional analysis, she has worked on a diversity of social−ecological issues such as protected areas, adaptation to global change along coastlines, agroforestry systems, and locust plagues.

Laurie Tissière

Laurie Tissière is a PhD geographer, and a member of the Chaire maritime at the University of Nantes, France. She is active in the development of participatory research about policy and politics in maritime space and fisheries.

Brice Trouillet

Brice Trouillet is an associate professor in geography at the University of Nantes, France, and a member of the LETG lab, CNRS. His research uses the case of fisheries in marine spatial planning to study how power relations and data/knowledge issues intertwine in "socio-technical agencements" forming an intermediate with the environment.

Gilles Voiron

Gilles Voiron is a geographer and contract engineer at ESPACE lab, Nice, France. He has engaged in spatial analysis of mobility data and activities devoted to urban sustainability, especially focused on low-carbon mobilities and electromobility.

Preface

When this book project was conceived, at the end of the year 2010, taking up the issue of resilience from the viewpoint of the geoprospective approach was far from obvious. Research works in these two fields were very rarely combined. Therefore the option retained has been to demonstrate the relevance of bringing them closer together. Since then, the COVID-19 pandemic and its cascade of effects have illustrated the critical need to do so. Indeed, this crisis with multidimensional consequences puts into question our ability to anticipate coming changes and detect the socioecosystem's responses to these changes. The urgency of climate change, of which the risks are increasingly better understood and the impacts largely known, adds a pressing need to anticipate the appearance of yet unknown risks, and to prepare to face them. Getting prepared for the unforeseeable! Today, this oxymoron has taken on true significance. How to go about it remains a problem. There is neither a magic bullet nor a universal modus operandi. Nevertheless, there are research guidelines that help in decision-making related to public and private entities. This book contains proposals to that effect, such as broadening the field of vision by adopting a wide-angle frame, so as to integrate the subject under study—an ecosystem or territorial system—into its environment, envisaging possible events and not only those that are probable, that could affect it, and evaluating their impacts. The aim is twofold: on the one hand to assess the system's vulnerabilities and its reactivity and adaptability potential, and on the other hand, to act accordingly in order to improve its resilience over the long term.

Indeed, the resilience sought here is a general and continuous resilience that not only enables the system to face any major event, but also to hold out due to its ability to adapt to changes in its environment and generate forms of regulation to ensure its survival. When applied to regions and towns, resilience is the ability to assimilate crises by anticipating disruptions, continuously readjusting and renewing. In the long run, territorial resilience is not as dependent on a territory's ability to repel shocks, but rather on its adaptability to changing conditions. Therefore anticipating change is a major component of resilience.

To understand an ecosystem's future dynamics, it is not only indispensable to anticipate the impacts of global changes and to assess the conditions for ecosystems to adapt to their environments' spatiotemporal situations, but also to research the impact of decisions taken or not taken today on their future evolution. The same applies to territorial systems—

region, town, neighborhood— however, these have a specificity that distinguishes them from ecosystems. The dynamics of change are driven by cognition. Information, memory, and learning influence decisions and introduce some anticipation in the form of intentionality and strategy in systemic functioning. This way, anticipation intervenes in various forms in detecting future changes. Admittedly, it is the means to detect the possible transformations of socioecosystems, but it is also consubstantial to the dynamics of sociospatial systems.

Resilience is both a process and the result of this process, appraised at a given moment and in a given context. It is therefore dependent on the spatial and temporal contexts in which the system under study operates. To use the OECD wording, it is context-specific and place-based. From then on, it appears fundamental to know the characteristics of the geographic context, spatial organization, and physical and human resources.

The challenges of research into these issues lie in the development of methods integrating the prospective, ecologic, geographic, and spatial dimensions. The objective is twofold: producing new knowledge on the future dynamics of socioecosystems and a knowledge useful to stakeholders to help them in their decisions, by enabling them to weigh up, at best, the impacts of the options before them. The geoprospective movement is fully in line with this dual purpose. Unlike resilience, the geoprospective approach is still little known and seldom practiced. Like prospective, this approach does not intend to predict but to anticipate the possible evolution of a socioecosystem by considering it in its overall environment and over a long period, with the aim of informing decisions. However, the specificity of the geoprospective approach is the major place given to the spatial dimension. Whatever the nature of the phenomenon and the type of system being studied, exploring the future is carried out via a spatial entry. More precisely, the aim is to highlight the role played by space in the dynamics of the system under study. A space is first an area, but provided with a set of differentiated elements, and links between these. Some people prefer to replace the term space, which is sometimes judged too abstract, by that of the environment, which has a more physical and environmental connotation. Now, the space or environment is continuously shaped by stakeholders' interactions, and at the same time, it directs and constrains in part such interactions. It is this complex dynamics that is difficult to identify and interpret in the present and a fortiori in the future, which is the subject matter of the geoprospective approach. The guideline is to analyze the spatial change and the processes generating it differently according to the spatial context. The purpose is not only to decipher the differentiations observed in past paths, but also to assess, using conjectures and models, possible future paths and identify the existing local levers of action to move toward sustainability. Whereas the questionings and methodologies used are varied, the spatial dimension is omnipresent at every stage of the research, while being combined with the temporal dimension. Then, combining the two dimensions provides keys to enter the future of spatial systems: differentiated spatiotemporal paths, divergent dynamics, vulnerabilities versus adaptability potentials peculiar to the spatial systems being studied.

Times of crisis are conducive to prospective questionings—actually, prospective appeared in the aftermath of the 1929 Great Depression. The current period is on the whole a time of crisis, but also a time of transformation. The future is laden with uncertainties but also with challenges which must be taken up, and although the future is dependent on the hazards of the time path, there is nothing random in its spatial expression. The seeds of change to come are partly contained in the socioecosystem's vulnerabilities and adaptability, both being specific to each place and spatial context. Revealing these differentiated spatiotemporal dynamics is the core of the geoprospective approach.

The purpose of this book is to demonstrate, with the support of case studies, the contribution of the geoprospective approach in anticipating change in spatial systems. The book is made up of 13 chapters.

The first four are of a conceptual and methodological nature.

Chapter 1, "The Origins of Geoprospective", shines the light on its foundations. The construction process of this new field of research conducted at the interface of spatial modeling, land change science, and spatialized participatory approaches is expounded in this chapter. It points out the multiform character of the geoprospective approach and the way it distinguishes itself from other methods of studying the future.

Chapter 2, "Anticipating the Impacts of Future Changes on Ecosystems and Socioecosystems: Main Issues of Geoprospective", addresses the key problem of how to adapt our strategies and practices to make our territories and activities resilient. It proposes a generalized geoprospective approach to help decision-makers and stakeholders to take into account the potential impacts of global changes in their planning. This geoprospective approach is based on the main and significant concepts related to territorial vulnerability and resilience, in order to focus the attention of decision-makers on the risks that can be generated by global changes on activities and territories.

Chapter 3, "Knowledge Challenges of the Geoprospective Approach Applied to Territorial Resilience", delves into the way in which a territory—region or city—will behave when faced with disruptions. This behavior depends on the territorial resilience capacity, a multifaceted and place-based process in which the spatial context plays a major role. The proposed entry into this issue consists in going deeper into the role of the spatial dimension by examining the system's propensity to change and the adaptive potential of a system's spatial structure. Urban geoprospective is not limited to the resilience to catastrophes but integrates a broader vision of the resilience to any sociotechnical change.

Chapter 4, "Methods and Tools in Geoprospective", illustrates the diversity of methods and tools available for developing a geoprospective approach. A number of methods, such as modeling, are not specific to geoprospective; the chapter points out the specific methods entailed by their use in geoprospective. Moreover, it focuses on the following new approaches which are hitherto hardly used in geoprospective: scenarios integrating various territorial scales, modeling of the decision-making process coupled with prospective spatial modeling, geoprospective based on causal probabilistic models, graphic modeling, prospective choremes, and immersive and 3D simulation in landscapes of the future.

The following five chapters present case studies on issues as diverse as: biodiversity conservation, the spread of decarbonized mobility, the ski economy in the distant future, the Wood Energy Supply Chain sustainability, and the long-term dynamics in coastal areas.

Chapter 5, "Geoprospective Approach for Biodiversity Conservation Taking into Account Human Activities and Global Warming", demonstrates the contribution of landscape analysis for identifying the main patterns to simulate the landscape dynamics, without ancillary data. This geoprospective modeling identifies the artificialization process in the French Riviera area and its potential dispersion up to 2050. The model is based on remote sensing images and spatial data. The transition rules consider the constraints caused by the geomorphologic conditions, the natural protected areas, as well as the proximity of the transport network to urbanized areas, the closeness of the population centers, and the pattern of change between each land use category.

Chapter 6, "Assessing the Territorial Adoption Potential of Electric Mobility: Geoprospective and Scenarios", confronts the sustainable mobility issue. The aim is to assess the adoption potential of battery electric mobility in a territory. An expert system evaluates this potential in 2019 and by 2040 according to two scenarios: "toward all-electric mobility by 2040" and "toward a diversified decarbonized mobility by 2040." The approach is presented via an application concerning the communes of a French Region, the South Region. The simulations inform not only on the spatial differentiations in the adoption potential and their possible evolutions between the two dates, but also on the communes' reactivity to change and their adaptability, which is a major factor of resilience.

Chapter 7, "The Touristic Model of Valais Facing Climate Change: Geoprospective Simulations of More Environmentally Integrated Development Models", delves into the field of tourism economics in relation to the natural environment. The chapter presents a geoprospective study of the ski economy in the Canton of Valais (Switzerland). Ongoing climate change is causing a great concern in this economic model: What are the sociospatial consequences that climate change will have on the Valais tourism system? Is this system resilient enough to withstand climate change? What would be the sociospatial consequences of an imbalance in economic, ecosystem, and natural risk terms? A geoprospective simulation tool based on a multiparadigm approach has been developed to answer these questions.

Chapter 8, "Geoprospective Assessment of the Wood Energy Supply Chain Sustainability in a Context of Global Warming and Land Use Change within 2050 in the Mediterranean Area", assesses the potentiality of tree availability by using a geoprospective approach. The methodology for identifying the best territories for wood availability in 2050 integrates two models: the Climflora model of vegetation dynamics due to the climate scenarios for 2050 in order to assess the potential modifications of the vegetation structures and a model of urban dynamics, based on the use of cellular automata, for estimating the urban spread toward 2050 in order to identify the locations of the potential population increase and its energy demand.

Chapter 9, "Simulating Together Multiscale and Multisectoral Adaptations to Global Change and their Impacts: A Generic Serious Game and its Implementation in Coastal Areas in France and South Africa", proposes a new kind of participatory device designed as a serious game suitable for geoprospective workshops. The methodology is dedicated to helping in setting up multiscale and multisectoral long-term adaptation plans to global change. Two games are presented: the first concerns the George municipality in South Africa, where sessions were organized in coordination with the South African National Parks body. The second example is for a coastal area of the "Gard South," a French Mediterranean department.

The last four chapters address environmental planning and urban design. They point out the deficiencies of the foresight approaches currently used and *a contrario* the adequacy of the geoprospective approach to these questions.

Chapter 10, "Geoprospective as a Support to Marine Spatial Planning: Some French Experience-Based Assumptions and Findings", tests the advantages and limits of the geoprospective approach in terms of social learning, with a focus on the role of the different types of spatial representation and whether they helped or hindered the participatory process. The analysis is based on three experiments performed over several scales of time and space, with a variety of publics and using different objects to enable the approach of the space-participation-modeling trio from different angles. A first case combines single use and regional scale, a second links multiple uses and regional scale, and a third focuses on multiple uses and local scale.

Chapter 11, "Simulating the Interactions of Environmental and Socioeconomic Dynamics at the Scale of an Ecodistrict: Urban Modeling of Gerland (Lyon, France)", presents an interactive 3D platform. This tool allows the Métropole de Lyon decision-makers to develop and test scenarios for the future (roads, public transportation, urban project, infrastructures, buildings, etc.) based on coupled models and to visualize the results either in the form of 3D visualization or charts and tables. The city consists of systems (population, housing, transport, economy, networks of water, energy, waste production, pollution, etc.) interacting together. The geoprospective approach is based on comparing scenarios from the present day to 2040 to help decision-makers to make the best strategic decisions for their cities.

Chapter 12, "Geodesign for Collaborative Spatial Planning: Three Case Studies at Different Scales", introduces the concept of geodesign as an integrated spatial planning activity and design process informed by contextual environmental impact assessment, which includes project conceptualization, analysis, projection and forecasting, diagnosis, alternative design, and impact simulation. A parallel is drawn with geoprospective, which shares with it the aim of achieving foresight and a better place to live. Both geodesign and geoprospective evaluate the spatial changes related to different planning options or to outgoing dynamics; nevertheless, geodesign is better suited to a smaller time scale, matching with planning activities, rather than a larger time scale like geoprospective, which studies long-term changes.

Chapter 13, "How Do Public Policies Respond to Spatialized Environmental Issues? Feedback and Perspectives", gives three persons in charge of the planification and territorial and environmental prospective departments of major government bodies the opportunity to express their views on the geoprospective approach. How is the spatial dimension introduced in the prospective studies carried out by their organizations? What are the preferred themes? On what types of space, and at what scale are these prospective works conducted? How are they achieved, using what methodologies and operating modes? Backed by their experience, how do they consider the geoprospective approach, its interest, appropriation by the stakeholders concerned, and the prospects for its spread?

Christine Voiron-Canicio

May 2020

Acknowledgments

The editors of this book want to express their gratefulness to all authors of the chapters. Their involvement and determination has been crucial in order to achieve the publication of this book. We thank them not only for the very interesting results they provide but also for the reflection they bring in the field of geoprospective. Through these applications of the geoprospective concept and methods, we also want to present our gratitude to all the contributors of financial support, both public and private entities that help researchers to develop a consistent research dedicated to problem solving.

We thank Françoise Gourmelon for reviewing chapter 4. Finally, we also address our acknowledgments to the contributors of Chapter 13, How do Public Policies Respond to Spatialized Environmental Issues? Feedback and Perspectives, who are involved in public decision processes and who have dedicated to us a part of their precious time:

Nathalie Cecutti, architect and urban planner in chief at the Ministry of Ecological and Solidary Transition (France);

Carlo Lavalle, coordinator of the European Commission Knowledge Centre for Territorial Policies at Join Research Centre (European Commission);

José Juan Hernández Chávez, Director for Environmental Policy in the Secretaría de Medio Ambiente y Recursos Naturales (Mexico country).

Emmanuel Garbolino and Christine Voiron-Canicio

June 2020

CHAPTER 1

The origins of geoprospective

Christine Voiron-Canicio[1] and Emmanuel Garbolino[2]

[1]*Université Côte d'Azur, CNRS, Laboratory ESPACE, Nice, France,* [2]*Climpact Data Science, Nova Sophia—Regus Nova, Sophia Antipolis Cedex, France*

Chapter Outline

1.1 In the beginnings, the prospective approach

Geoprospective originated in the French geographers' community some 15 years ago. Research works in environmental and territorial dynamics presented at symposiums revealed a converging interest in the issue of the spatial change to come, and the models to use for anticipating it, not for predictive purposes but with a view to helping in reflecting on the future evolution of the spaces that were studied.

The purpose of this chapter is to go back over the steps which led to that convergence. The foundations of geoprospective have to be found within the discipline of geography, and more specifically in the field of spatial analysis. They also rest on the contributions of other disciplines and fields of research, the first of these being prospective.

Prospective appears at the beginning of the 20th century, as a result of the Great Depression which shook the United States, when the forecasting methods used until then were put into question (Didier, 2009). After the end of the Second World War, it developed in two centers, the United States and France, that devised anticipation methods with specific orientations for each of them. On the American side, the Rand Corporation was created, as a laboratory of prospective methods—like the Delphi method, for example—focused on technological forecasting approaches mostly conducted in a military context. On the French side, the necessity to move from a still rural economy to an economy that would be at the same time industrial and more competitive coincides with the appearance of a prospective attitude which takes a fresh look at forecasting methods and at the end goal. So, in the mid-1950s, the philosopher, Gaston Berger, initiated a new reflection on how to anticipate the future and on the methods used for making decisions and set up the Centre d'Etudes Prospectives. He contrasted a vision of the future freed from fatalism and the dependence on the past with the determinism stemmed from the positivist vision. "Tomorrow will not be like yesterday. It will be new and will depend on us" (Berger, 1957, 1958, 1967). The purpose of prospective is to prepare decisions and actions taking into account both changes in society and the potential impact of decisions on its development and, furthermore, in adopting a global approach to complex phenomena. Berger's pioneering work was continued on through the 1960s and 1970s by Bertrand de Jouvenel, the founder of the Futuribles Group (Association Internationale de Futuribles) which introduced the use of scenarios to construct positive images of the future or "scientific utopias" (De Jouvenel, 1967). Since the 1970s, the work of the French pioneers has been expanded on by Godet (1979).

The translation of the French word prospective is still a matter of debate, sometimes translated by Futures Studies, forecasting, foresight. More than a translation problem, this reveals the existence of different conceptions between the two sides of the Atlantic. As an example, Michel Godet explains in one of his books that the concept of prospective failed for a long time to find an appropriate translation. Forecasting is judged by Michel Godet as too influenced by economic modeling and technological forecasting (Godet, 2001). It was in 1996 that the relationship was established between prospective and foresight in an article about him: "The starting point of foresight, as with *la prospective* in France, is the belief that there are many possible futures" (Martin, 1996, 2010). However, the correspondence is not perfect, because Foresight expresses the image of a given future whereas prospective designates at the same time a process, the result of this process and the preparation of a plan of action for powering the change wished for. Because of this voluntarist and strategic dimension of prospective, which is one of the specificities of the French viewpoint as compared to the American approach, Michel Godet recommends the use of the expression strategic foresight or strategic scenario building (Godet, 2001).

1.1.1 Prospective and temporalities

The specificity of the prospective approach stems from the way time—past, present, future—is taken into account in anticipating possible futures. Very often, the vision of the future is formatted by the past and kept in check by the tyranny of short-termism. Gaston Berger wrote in 1957: "Our civilization is breaking away from the fascination of the past with difficulty. Of the future, it only dreams and, when it works out projects that are no longer mere dreams, it draws them on a canvas on which it is still the past that is cast. It is stubbornly retrospective. It should become prospective" (Berger, 1957). Prospective seeks to inform society on the issues that both individuals and territories will have to confront in a distant future. This concern about long-term phenomena echoes the one which, at the beginning of the 1970s, underpinned the work of the Club of Rome and the publication of The Limits to Growth, first emblematic example of global and systemic prospectives conducted on a global scale (Meadows et al., 1972).

Prospective intends to distinguish itself from extrapolation by the ambition to explore the future without extending the present. Nevertheless, the representation of possible futures is dependent on the importance given to the facts of the present. Conceptually speaking, the dynamics of the future come from the forces of change and forces of inertia which put a brake on evolutions. Then, it is important to identify them and to spot, among present events, future-oriented facts, weak signals, and bifurcation points. To do so, there is no universal technique, the prospective interpretation of the lessons of the past and facts of the present remains an art.

1.1.2 Scenarios, collective thinking, and debating: the foundations of prospective

Prospective is a discipline made up of various branches—territorial prospective, strategic prospective, and environmental prospective—developed from a common conceptual base. These branches share the common aim to make stakeholders aware of the implicit hypotheses on which their action is based and rest on the principle of freedom of action which enables individuals to have a hold on their future. Whatever the field of study, prospective consists in seeing far and wide in order to assess the consequences of decisions, carrying out an in-depth analysis so as to go beyond analogy and extrapolation, and anticipating potential breaks. On the other hand, it broadens the representations of the future to qualitative approaches and data that are not solely quantified. For example, it gives a dominant place to the points of view of stakeholders in the future.

Scenarios, collective thinking, and debate are the mainstays of the prospective approach. Exploring the future consists in working out a range of future visions, which have to be sufficiently varied to apprehend the possible evolutions of the system being studied, sufficiently contrasted to be discussed, and also pertinent, and to help in decision-making.

Such exploration is carried out by building scenarios, an essential step in the prospective approach. These were introduced in the United States in the 1960s by Herman Kahn when he was working with the RAND Corporation, then at the Hudson Institute. Kahn defines the scenario as "a hypothetical succession of evens built with a view to highlighting causal sequences and decision nodes" (Kahn and Wiener, 1967). At the same time, in France, an interdepartmental government organization known as DATAR (the Office for Regional Planning and Development) launched the first prospective exercise on the image of France in the year 2000.

Nowadays, the term scenario encompasses different conceptions as well as a variety of methods and practices: the "experts" narrative of possible futures—probable, imaginable, plausible—, the systemic analysis, the quantitative or qualitative prospective model casting into the future a set of interrelated data and describing the future reality of the territory in an abstract manner. A scenario can be the reproduction of a scenario drawn up in a similar context and considered as being of interest for the situation in question. The aim of scenarios integrated into the prospective approach is to describe the future according to different options: a trend hypothesis—business as usual—vs hypotheses intentionally contrasted, introducing from the onset, or along the way, one or several "breaks"; forecasting vs backcasting approaches; building a general model vs variations of the benchmark model. There are three main schools of scenario building (Houet, 2015; Amer et al., 2013; Bradfield et al., 2005):

- The "intuitive logics" consists in drawing up, in group, plausible narratives of causal chains without resorting to mathematical algorithms (Wack, 1985; Huss, 1988; Wright et al., 2013; Derbyshire and Wright, 2017).
- The "probabilistic modified trends" made up of two different methods. The first—the trends-impact analysis—forecasts future trends based on past trends and introduces unprecedented events likely to bring about deviations from the extrapolated trend. The second—the cross-impact analysis—adds a higher complexity level with the analysis of the interdependence of events taken by two, depending on whether they have already occurred or not, in order to correct initial probabilities given by experts (Bradfield et al., 2005; Bishop et al., 2007).
- The "prospective school" (Godet, 1986) differentiates itself from the two previous families by a more elaborate approach, mechanist, and less intuitive. The construction of scenarios relies heavily on computerized mathematical models. The number of steps in the methodological method varies depending on the authors (Durand, 1972; Godet, 1986; Schwartz, 1996; Schoemaeker, 1993; Mermet and Poux, 2002).

Whatever the method, the collective thinking process is always integrated into the prospective approach, but in variable forms, and is more or less present in the various stages of the approach. It is indispensable for identifying the key variables of the dynamic of the system being studied, for analyzing the stakeholders' strategies and, in synthesis, for bringing forward major issues for the future. It is as much useful in the subsequent stages, to move from problems to prospects, then from prospects to processes. True, the future will never match a given scenario exactly, but the latter provides a working basis, a contribution to the ongoing debate in a prospective forum (Mermet, 2005).

Since the 1990s, there have been an increasing number of studies and references on scenario methods in the scientific literature, as well as a proliferation of examples of scenarios in the public domain (Martelli, 2001). The scenario method is now widespread because it is seen as an effective dialogue tool between the various players involved in the implementation of a policy (Levêque and Urien, 2005).

1.1.3 The fields of prospective: territorial prospective and environmental prospective

1.1.3.1 Territorial prospective

The aim of territorial prospective is to work out visions, prospects, and orientations concerning the evolution of a territory and its inhabitants so as to provide information and help to take a stance and strategic options in the most complex cases (Loinger and Spohr, 2004). Planning should not be mistaken for the territorial prospective exercise. Guy Loinger points out that planning consists in putting into perspective a set of development projects, sectorial programs, policies, etc., based on predictive analyses. Whereas planning does not take into account the opinion of stakeholders and the social acceptability of the planned future, in the case of territorial prospective, "it is a question of collectively building a vision of the future with other working methods and starting from broader and more 'questioning' representations of the past and the present" (Loinger and Spohr, 2004). There are different movements of territorial prospective that are classified in three main families: "cognitive prospective," which questions the future based on assessments, situation evaluations, diagnoses, and surveys; "participative prospective" in which the future is worked out with the participation of stakeholders; "strategic prospective," which sets a set of goals to reach within a certain time limit, and a plan of action to succeed in reaching them.

In France, prospective is almost exclusively associated with public policy and planning. The "French-style" prospective conducted by DATAR from 1963 to 2014 goes together with the production of scenarios illustrating cartographically the voluntarist development promoted by the State (Delamarre, 2002). We will mention among these, the "scenario of the unacceptable"—an image of France in the years 2020 to be proscribed—, a contrario, the "network polycentrism"—scenario desired for France in 2020—, and finally, all 28 scenarios on spatial systems in France, Europe, and the world, drawn up within the framework of "Territoires 2040," DATAR's latest prospective exercise (Cordobes, 2010).

From now on, the territories of EU member countries, from city to region, handle prospective themselves, anticipate evolutions, and define their own strategy (Van Cutsem and Roëls, 2012). Indeed, an abundance of strategic prospective exercises has been observed, conducted at both the local and regional levels, and generally underpinned by scientific research work.

For example, since 2007, the ESPON program, financed by the EU Structural Funds, has been carrying out scientific studies on regional development which are intended for providing the knowledge needed for the first stage of any territorial prospective study. The European Environment Agency also carries out prospective studies. An illustration of this is the PRELUDE project, which explores plausible long-term evolutions for land use and their environmental impact (EEA, 2007). It is modeled using contrasting scenarios combining information relating to five areas of intervention and action which are determining in the trajectories of land use change: protection of the environment, solidarity and equity, governance and public intervention, optimization of agriculture, technology, and innovation. Modeling is carried out by applying a grid with cells of 18 km a side on the European area. The results of simulations are presented in cartographic form. This project illustrates the current will to combine environmental and anthropic data in a fine-scale territorial prospective.

1.1.3.2 Environmental prospective

In response to the consequences, most often harmful (pollution, destruction of environments and species, exploitation of nonrenewable resources, soil erosion, etc.) of anthropic action on natural environments, biodiversity, health, and the economy, the scientific community and territorial decision-makers have become involved in studying the future of the environment with the aim of slowing down, stopping the degradation of the ecosystems, or even improving the quality of the natural habitats and our environment. Some of the concrete actions that have been deployed internationally include UNESCO's Man And Biosphere (MAB) program, which was established by UNESCO (United Nations Educational, Scientific and Cultural Organization) in 1971 and aimed at the time to reconcile the human with its environment.

In parallel with the MAB program, Donella and Dennis Meadows, with Jorgen Randers and William W. Behrens III, published in 1972, on behalf of the Club of Rome, the book entitled "The Limits to Growth." Based on an approach based on the work of Jay W. Forrester in the early 1960s on systems dynamics, this global foresight work presented different scenarios of global evolution of natural resources and population. Although not based on a spatially explicit approach, the different results of simulations of evolution by 2100 were particularly worrying: they showed a significant decrease in natural resources and an equally significant increase in pollution and atmospheric CO_2 concentration. The consequences of the world demography were to be a major reduction of the population toward the end of the 21st century. This foresight work has had the merit of raising public awareness, politicians, and industry on the risks generated by economic development that does not take into account the limits of natural resources. Moreover, the comparison between the values simulated in 1972 and those currently observed are quite similar for variables such as the average atmospheric CO_2 content, the birth rate, the world population, and the reduction of arable land, for example. The two updates of these models published in 1992 and 2004 clarified and confirmed the evolutionary trends initially presented. This work stimulated a group of researchers who, in the following years, began researching the evolution of ecosystems, ecosystem services, and society. The accumulation of work highlighting the risks for mankind of an economy based on ever more consumption-consuming growth of exhaustible resources and increasingly polluting led the United Nations, in 1987, to define the concept of "Sustainable development" in the famous Brundtland report. The World Commission on Environment and Development was asked by the UN General Assembly to decide on the establishment of a timetable to promote a change based on sustainable development by the year 2000 and beyond. Indeed, this commission had pointed out that the ecological imbalances observed over the last decades had accentuated the disparities between human populations, accentuating poverty, health, and food problems. Because of the idea of sustainability, the concept of sustainable development then means looking ahead. But how?

In 1988 the United Nations Environment Programme of the UNO and the WMO (World Meteorological Organization) established the Intergovernmental Panel on Climate Change (IPCC). The main objectives of the IPCC are to study the evolution of climate and its impacts on ecosystems and society and to propose strategies in order to prevent risks due to global warming. The first assessment report was published in 1990 and the last and fifth in 2014. IPCC gathers more than 800 scientists that expertized more than 9000 publications about climate change evidences, its evolution, the current and potential consequences on ecosystems and society, and the options to mitigate the climate change and its expected consequences on the territories. Due to the very large literature review, very different approaches contribute to design the scenarios of climate evolution and its potential impacts on many aspects of our society. Some of these studies are spatially explicit but they do not always integrate the stakeholders demand or issues in the definition of the problematics and/or in the perspective of use of the results for the society. Most of the works analyzed by the IPCC concerning adaptation planning are based on expert opinions and some multicriteria optimization approaches that integrate climate change scenarios.

In 2001, the UNO launched an assessment of the impact of human activities on ecosystem services called "Millennium Ecosystem Assessment" (MA) that was published in 2005. The scope of this work was to establish an overview of the state of the environment, ecosystem services, and human wellbeing and to assess their evolution within 2050 by integrating scenarios of human activities development and climate change. The scenarios of the MA integrate feedbacks between social and ecological systems and they also take into account the linkages between global and local socioecological processes by using qualitative storyline and quantitative modeling mainly based on IMAGE and IMPACT models. IMAGE model (Integrated Model to Assess the Greenhouse Effect) estimates the consequences of climate changes interaction on Land Use and Land Cover and the population at the global scale. IMAGE also estimated the Greenhouse Gas (GHG) releases provoked by these changes and the consequences on biodiversity. The IMPACT Model (International Model for Policy Analysis of Agricultural Commodities and Trade) was developed to assess the consequences of global changes on food availability and poverty by integrating data about climate change, water demand and availability, crops production, economy and regulation of the agriculture market. The model simulates the operation of commodity markets and the behavior of economic stakeholders in order to estimate the supply and demand for food products. The use of these two models is applied at national and continental scales. Even if they integrated a spatial representation of the results, we cannot consider that they are spatially explicit.

Finally, the French consortium AllEnvi (Alliance nationale de recherche pour l'Environnement) that gathers 12 founders from the highest research institutions produced in decembre 2016 a report on the evaluation of 307 scenarios applied on environmental prospective (De Menthiere et al., 2016). The study shows that the scenarios can be organized into 11 huge families that indicate the different trends of social and environmental evolutions in different countries. These scenarios are applied at national, continental (Asia, Europe, Mediterranean basin, etc.), or global scales. Most of them are based on expert knowledge and are not spatially explicit.

We can conclude that, in environmental prospective, most of the researches are based on methodologies combining expert knowledge and modeling techniques of land use, integrating the evolution of climate change. Most of them are not spatially explicit and when they are, it is not clear if they take into account the issues raised by the stakeholders.

1.2 Stage 2 of prospective: taking into account geographic and sociospatial differentiations

In the last 20 years or so, scientific work carried out in environmental sciences and research work initiated on sustainable development in the field of social sciences—economy, geography, sociology, political science—led to delving into a number of subjects, including multilevel relationship, and existing differentiations across space. Such questioning had the effect of bringing changes to prospective approaches.

1.2.1 Spatial differentiations and multilevel interactions

Within the framework of studies carried out by the IPCC, the global atmospheric models used are applied on the lower levels using downscaling methods (IPCC, 2007). However, despite the steady improvement of the models used, climate information resulting from forecasts is still too approximate at the local level to meet the requirements of impact studies and adaptation measures. The climate's spatial variability has long been demonstrated by geographers (Yoshimo, 1975; Oke, 1987; Beltrando, 2010; Bonnardot et al., 2012; Quenol, 2013). Due to topography and land use, local temperature variations can be equal or higher than the temperature rises simulated by IPCC scenarios. Taking into account this multiscale spatial variability in scenarios, and broadly speaking, the functioning of local-scale geosystems has now become a challenge for the scientific community and is still to this day one of the stumbling blocks of prospective climate modeling.

Furthermore, giving further thought to sustainable development and to the necessary adaptation to global changes brings about the notion of systemic complexity of which the multilevel interactions are one of the dimensions (Allen and Starr, 1982; Wu, 1999). As stressed by the French biophysicist and philosopher Henri Atlan, such complexity lies in the fact that an observer lacks information to account for the global significance of local phenomena, taking into account the information that the observer can acquire on these two levels (Atlan, 1979). The economist, Olivier Godard, in an incisive article published in 1996 in the journal Futuribles, draws attention to the error consisting in believing that sustainable development implies adopting the same approach at the various territorial scales—global, continental, national, and local (Godard, 1996). He enjoins researchers and prospectivists "not to mechanically transfer reasonings from one scale to another," arguing that ecological constraint has an absolute meaning only at the global level; at all other levels, exchanges, substitutions, and imbalances can occur. Thus ecological constraints are relative; in that sense, sustainable development is not fractal. Likewise, he makes the case that sustainability is nonadditive. Depending on the territorial scales considered, recommendations and priorities will not be the same (Godard, 1984, 1996).

1.2.2 Challenging generic models

The challenging of "universal" models applicable to all scales, from global to local, appears in a number of studies, and more particularly as regards the environmental prospective rooted in simple causality and biophysical determinisms (Fernandez et al., 2011). Questions are raised on the normative content of the hypotheses used for defining cause-and-effect relationships, and on the physical limits of the system being studied. The critical analysis conducted on the DPSIR (Driving forces-Pressures-State-Impacts-Responses) model is a perfect example of the foregoing. This model belongs to the family of analysis tools aimed at producing useful information for decision-making, in the context of public policies. In this perspective, they are designed with a view to reducing complexity via a chain of causal links, in order to measure, with easily understandable indicators, the state of a phenomenon, and to identify and monitor the pressures affecting it. Using the DPSIR model is strongly encouraged by the European Environment Agency as well as by the Organization for Economic Cooperation and Development (OECD) and the FAO (United Nations Food and Agriculture Organization). Now, work carried out in social sciences and based on its use in water prospective in Europe gave rise to reservations on using it without prior consideration of the territory being studied. The criticisms stress that the Driving Forces are "anonymous" and "a-temporal" and that neither the stakeholders nor the trajectory of the territory on which the model is built is taken into account; the perimeter of the model is solely centered on physical or biophysical elements—for example, watershed—yet, sociosystems are part of particular geographies which transcend physical limits (Fernandez et al., 2011). Then, the authors advise to apply the DPSIR model from various angles: resources, policies, main players, with variable scales. The advantage of these variations is, on the one hand, to better grasp the complexity of spatialized problems, "what is seen as passive in a scenario can become the level of response in another"; on the other hand, to build more relevant contrasted prospective scenarios, by taking into account differentiated operating logics depending on the nature and extent of the territorial context: Europe, administrative region, portion of river basin, etc. (Fernandez et al., 2011).

These criticisms of the normative models with a universal scope echo the analysis carried out in parallel on their use in public policies, and more precisely on the EU water framework directive (Kallis and Nijkamp, 2000). Ward et al. (1997) examined critically the European Union's environmental policy. They stressed that the environmental regulation at the high European level often does not account adequately for the variety of problematic situations experienced in the Member States, for example, between urban northern areas and rural southern ones.

1.2.3 Combining generic and local prospective scenarios

It is, notably, to avoid this pitfall that new approaches of multiscale and participative prospective modeling have come up within the framework of European research programs. The PLUREL project will be taken as an example. The PLUREL project was designed between 2007 and 2010, in the context of the sixth framework program for research and technological development; its aim is to define the possible trajectories of European territorial units by 2025. Its originality lies in the combination of two interlocked series of multiscale modeling. The first consists of four contrasted scenarios, adapted from IPCC's baseline scenarios in 2000. The second applies these four scenarios at the level of the urban areas of six European metropoles varying in size, territorial organization, and spatial planning of urban development. In all six towns, people involved in local governance were requested to transcribe, in experts' words, the principles of the four prospective scenarios in the development dynamics of their respective territories, including local constraints and the urban planning legislations specific to each European state. Simulations are carried out on the Ispra Joint Research Center's Moland modeling and simulation platform. The results of the simulations reflect the combination of processes applied both at the European scale and that of the urban area being studied, and are mapped at that scale (Chery, 2010; Chery and Jarrige, 2012).

1.2.4 From spatial change to spatial prospective

Geographers went late into anticipating the future; they rather tended to direct their research toward the organization of space, observing spatial differentiations, assessing the lines of force of French space in the European context (Eckert, 1996; Brunet, 1990). In 1998 Denise Pumain even wondered whether geography would be able to invent the future (Pumain, 1998). Since then, the challenge was taken up by geographers, most of them belonging to the theoretical and quantitative geography movement. Their research work first led them to study processes generating spatial change, then to explore the issue of anticipating change. At the beginning of the year 2000, spatial change was addressed from a theoretical and conceptual point of view by the geographers of the EPEES (Espaces Post-Euclidiens et Evénements Spatiaux) group, who devised the concept of spatial event (EPEES, 2000). Their thoughts focused on the transformation process of spatial systems. The self-organization theories that formalize system dynamics from their components' multiple interactions provide a key to understanding the relationship observed between temporal rhythms and changes in space. As an example, the apparent opposition between fast and numerous spatial changes and the near-stationary organization of space becomes understandable in the light of self-organization mathematical models which describe how the dynamics of a system reconciles micro, fast, and seemingly random fluctuations, and a slow dynamics at the macro level. These researchers also looked into the context which makes that an event triggers a process which will produce a new spatial repartition with different consequences for the same event, depending on the place where it happens, and also into the "unexpected" effects occurring in some places.

Studies carried out on spatial change led geographers to take a critical look at prospective. They questioned the fact that spatial dynamics were ignored. How can people write scenarios of the future evolution of a space without knowing its spatiotemporal trajectory? (Lajoie, 2005). Prospective cannot ignore that environmental and anthropic phenomena to come will affect space in a differentiated manner. Moreover, they questioned the notions of stake and stake-laden space referred to in prospective diagnoses. Public policy apprehends the evolution of a place in terms of stakes generally linked to social or environmental risk as well as the expectations of both the society and the authorities. A stake-laden space is then seen as a lever for action. Now, stake-laden spaces are often declared such in terms of priorities rather than as a result of analyses truly devoted to detecting them. The stake-laden spaces identified by geographers can be completely different, even in contradiction with those chosen by public policies. For Voiron-Canicio (2013), a stake-laden space is defined by three precise situations: either a space—area or place—where tensions crystallize, whether they be latent or expressed by permanent or episodic conflicts; or a space with a vital interest due to some remarkable elements—populations or resources, such evaluation being assessed in terms of the prejudice that would be caused if they disappeared, and the benefit resulting from keeping them or reintroducing them; or else a space of strategic interest, not for a category of players in particular, but for the functioning of a territory or the sociosystem considered, because of the strong probability of its transformation impacting the rest of the system. The stake-laden space thus defined puts into question the notion of general interest as it is usually regarded by public policies, that is, assessed in respect of the estimated needs of individuals rather than those of the land, considering the place in a functional manner, on an area defined at an administrative level, focusing on the short term rather than on the long term (Voiron-Canicio and Olivier, 2005a,b; Voiron-Canicio and Dutozia, 2017; Voiron-Canicio, 2013; Dutozia, 2013; Liziard, 2013).

All these geographical studies are based on the same premise: space is considered as both organized and organizing. Spatial systems carry a spatial heritage which exerts constraints on projects, decisions, and a society's behavior, but which is also a potential that agents reexamine and reassess constantly, depending on their needs and aspirations, in regard to the dynamics and potentialities of other territories (Voiron-Canicio, 2006).

The spatial prospective devised by Cécile Helle and Laure Casanova falls within this framework. It extends these works while directing them toward the measurement of the sensitivity of spaces faced with change, in a perspective of aid to decision-making (Casanova, 2010; Casanova and Helle, 2012). Attention is drawn to the preparation of spatial changes and the possibility to intervene on the evolution trajectory of a space, depending on the leeway available. For these authors, the purpose of spatial prospective is to "envisage the ways in which space can change and the incidences of such change, to better prepare it and be prepared to it, which amounts to questioning the modalities of a space's future differentiation" (Casanova, 2010). The mechanisms and factors of evolution of territorial systems are revealed by a diagnosis combining the spatial, temporal, functional, and perceptive dimensions.

Spatial prospective consists in studying the possibilities for the transformation of spaces, starting from the premise that some logics of evolution are more favorable to change than others. The approach is carried out in two phases: identifying the stage of evolution of the territories and identifying the "possibles" of territorial action. The first phase studies the territory's potentiality in connection with its specific qualities, its resources—those exploited and those latent, in reserve. Qualifying this potentiality "should enable one to apprehend the capability of a territorial system to continue as it is or to become another system likely to meet the requirements of (present and future) societies" (Casanova, 2010). In the second phase, the "possibles" of territorial action ensue, on the one hand, from the territory's sensitivity to change, that is, from the differentiated behavior of spaces faced with the same event as a result both of their specific qualities and their stage of development; and on the other hand, from the territory's degree of freedom, that is, from the part of constraint linked to the existing spatial organization, which is likely to influence the territory's future evolution. It is necessary to evaluate the degree of freedom in order to assess the leeway as regards development. The operational aim of spatial prospective for future actions of development lies in detecting both the moments and environments most favorable for change (Casanova, 2010).

These considerations and research in spatial analysis have opened up a new field of exploration of the spatial future. They generated a prospective approach the specificity of which is based on the importance given to the spatial dimension, and to the attention paid to spatiotemporal factors in anticipating changes.

1.3 The emergence of geoprospective

In the mid-2000s, several research movements converged on the issue of anticipating the evolution of spatial systems, by using a number of common or similar principles, methods, and operating modes, without referring to the term geoprospective in all the cases. And when the term was used, the prefix geo covered different contents according to the areas of application.

1.3.1 The first mentions of geoprospective

As far as we know, the term "geoprospective" was first introduced by the Bureau of Geological and Mining Resources (BRGM) in France, at the beginning of the 1980s, without being defined with precision, within the scientific and technical frameworks of studying the evolution of the geophysical properties of radioactive waste storage sites in order to ensure their containment (Gadalia and Varet, 1982). The aspect linked to assessing risks, in particular those eventually leading to the loss of containment of the radioactive waste storage site's geological structure, was already there at the origin of geoprospective. It was only in the mid-1990s that the term was defined more precisely: "The geoprospective approach aims to work out plausible and coherent scenarios of this natural evolution and to assess its consequences in order to draw profitable lessons in terms of the capacity of a site to accommodate a project, and even of devising the project itself. It thus contributes to scientific objectives (demonstrating the project's feasibility), also technological and operational (project optimization) and to decision-making (selection of sites, project acceptability)" (Godefroy et al., 1994). The term was used with this meaning in the works of the BRGM throughout the years 1980 and 1990 (Courbouleix, 1983; Godefroy, 1983; Gros, 1983; Canceill et al., 1985; Afzali, 1989; Afzali et al., 1990; Garcin, 1993; Godefroy et al., 1994) and until the mid-2000s (Casanova et al., 2004), while resorting to modeling techniques and tools for simulating the physical evolution of storage sites. For this reason, at a symposium organized by the National Agency for the Management of Radioactive Waste (ANDRA), the BRGM and Paris Ecole des Mines (Godefroy et al. 1994) went back on the evolution of geoprospective methods with, notably, the move toward models integrating artificial intelligence techniques and expert advice for constructing realistic scenarios.

These studies thus laid the foundations for the research claiming to belong to geoprospective, that is, integrating "the spatial expression of phenomena the impact of which can be relativized in comparison with different project opinions..." (Godefroy et al., 1994). Taking the spatial dimension into account (in parallel with the temporal dimension) in geoprospective studies is therefore clearly assumed by the BRGM, while explaining the importance of the specific transposition of methods and tools to local circumstances. However, whereas research work concerning the storage of nuclear waste is still ongoing, the term seems to have fallen into disuse at the BRGM after 2004. In parallel, in 1998 Jean-Paul Ferrier published "Le contrat géographique ou l'habitation durable des territoires" (the geographic contract or sustainable housing in territories) in which the author introduces the term "geoprospective" in a more geographical context linked to considerations on sustainable development, without defining it precisely.

1.3.2 New geographical studies claiming to belong to geoprospective or spatialized prospective

It was only in the mid-2000s that geoprospective sparked renewed interest. Geoprospective is tied up with spatial modeling. The 2004 situation report of the CNRS' section 39 "Spaces, Territories and Societies" clarifies the scientific issues: "Existing expectations in geoprospective direct research towards the notions of unpredictability and emergence which are the main difficulties in that field. One of the stances taken by researchers consists in considering that emergences are not totally unpredictable. Analyzing and modeling spatial interaction help to grasp the complexity of dynamics, to simulate the geosystems' possible futures and to produce spatialized decision-making tools" (CNRS, 2004). The ESPACE laboratory, a joint research unit of CNRS, was a trailblazer by including geoprospective in its 2004−07 scientific program. Indeed, a line of research is devoted to working out a new kind of territorial diagnosis, more powerful than the standard diagnosis, aimed at searching, in the present, harbingers of change and emergence that could weigh heavily in the future organization of territorial systems and help in territory geoprospective.

Christine Voiron-Canicio, in her research on urban paralysis risks in times of disaster published in 2005, uses the term "geoprospective" to describe the approach combining spatial analysis, GIS, and spatial simulations, designed for anticipating the consequences of a crisis situation and help the town of Nice to become better prepared (Voiron-Canicio and Olivier, 2005a,b). In 2006, she defined the scope of geoprospective more precisely: "Its purpose, as in the case of territorial prospective, is to get to know and foresee in order to organize and decide, but its specificity is to anticipate the evolution of a territory by understanding its spatial dynamics and to spatialize, on the medium and large scale, the evolution scenarios, development recommendations and their spatial impacts. Therefore geoprospective is inconceivable without spatial modeling" (Voiron-Canicio, 2006). In parallel, Thomas Houet, in the context of a research on environmental prospective, suggested a spatially explicit simulation method of prospective scenarios aimed at detecting the influence of measures for adapting to the reform of the European Union's Common Agricultural Policy (Houet, 2006). He described this approach as spatialized prospective.

In the years that followed, new research works claiming to be geoprospective or spatialized prospective were carried out. Moreover, a social demand appeared for the scientific community to work out, test, and validate methods enabling to spatially simulate phenomena, and to quantitatively evaluate the consequences of policy choices. In 2011, researchers from four CNRS laboratories—ESPACE, EVS, GEODE, and LETG—working in that field, together with the CNRS Méthodes et Applications pour la Géomatique et l'Information Spatiale research group (GDR MAGIS), organized a seminar on the theme "Geoprospective: contribution of the spatial dimension to prospective research." That seminar brought together some 50 participants and provided the first opportunity for exchanging views and debating on the various conceptions of this new research field as well as on achievements based on the use of the spatial element in territorial and environmental prospective. As a follow-up of the seminar, a "geoprospective" group was created within the GDR MAGIS, and two surveys were devoted to geoprospective, one in an issue of the review, L'Espace géographique (EG 2012-2), and the other in the online review Cybergéo.

1.3.3 A construction at the interface of several fields of research

Geoprospective is mainly driven and disseminated within the scientific community by geographers belonging to the movement of "theoretical and quantitative geography" and spatial analysis. Spatialized modeling and simulation, which are their common denominator, take various forms and are most of the time hybridized with methods stemmed from other lines of research, such as Land Change, Land System, Companion Modeling, and, of course, the prospective research stream.

1.3.3.1 "Theoretical and quantitative geography" and spatial analysis

In the past, a number of French geographers have taken a critical look at prospective: "a highbrow and ambitious word supposed to give quality to guesswork about the future"

(Brunet et al., 1992), while stressing the necessity to scrutinize, within spatial systems, the retroactions and risks of breakage or bifurcation. These issues and, more broadly, the prospective questioning, concern theoretical and quantitative geography as well as spatial analysis. By confronting structures and dynamics, forms, and flows, and by exploring the processes of spatial change, the latter is eminently spatiotemporal. *De facto*, spatial dynamics are carriers of geoprospective; "Geoprospective seen as an attempt to integrate spatial differentiation into the prospective approach is an obvious field of application of spatial analysis" (Charre, 2003).

Since the last four decades, the theoretical and quantitative geography movement endeavors to define the processes that generate spatial dynamics. To do so, it relies on various theories—more particularly on self-organization theories—and also on modeling and simulation. Two kinds of tools are used, on the one hand, macrodynamic models based on nonlinear differential equations, and on the other hand, tools that consider space as a collection of particles—cellular automata, multiagent system—which they use for trying to formalize the change of scale. Nowadays, this field of research faces a dual challenge. On the one hand, a shift of perspective with regard to change, with the aim to explain it *ex ante* and no longer *a posteriori* only. "By allowing to achieve, not an accurate prediction, but the exploration of a diversity of possible futures, will these models help to study the change in geographical structures in a nomothetic way?" (Pumain, 1998). Furthermore, will they help in the prospective formulation, and even in decision-making. Indeed, there is a great demand for geoprospective-oriented spatial models that would be within the reach of the highest possible number of executives so as to help them simulate the spatial impacts of an envisaged measure and evaluate how development policies can reinforce or constrain territorial dynamics. The specificity of the contribution of spatial analysis to prospective lies both in these spatiotemporal questionings and in the methodological, theoretical, and applied corpus designed for attempting to find answers (Voiron-Canicio, 2006).

1.3.3.2 Land change science

In the mid-1980s, two important programs concerning the transformations of ecosystems and land systems were launched: the Land Use and Land Cover Change Project (LUCC) in 1994 and its successor the Global Land Project in 2005. Both were interdisciplinary projects within the International Geosphere-Biosphere Program (IGBP) (Lambin et al., 1999) bringing together researchers in landscape ecology, biogeography, political ecology, resource economics geographical information and remote sensing, etc., aiming to understand the land use/cover change dynamics and their relationship with global environmental change. A new line of research referred to as land change or land system science emerged from these works; its aim was to try to understand the land use/cover change dynamics by interlinking social systems and ecosystems (Verburg et al., 2015). Research was carried out in four directions: observation and monitoring of land changes, understanding of these changes as a coupled human—environment system, spatially explicit modeling of land change, and assessments of system outcomes (Gutman et al., 2004; Turner et al., 2007). The favored methodology is that of "Integrated Assessment of the land system" (Kok et al., 2004). The main theme in the LUCC community is the modeling of land use dynamics and its social and environmental impacts. Scenarios is a second recurrent theme. Both models and scenarios can be developed by using participatory approaches (Kok et al., 2004). Research work carried out in that context strongly contributed to the development of prospective modeling and spatial simulation on land cover/use changes on various scales (GLP, 2005).

1.3.3.3 Spatialized participatory approaches

These approaches are characterized by the involvement of the stakeholders at every stage of the prospective process. The Companion Modeling approach is one of the best-known methods, and also the nearest to geoprospective. This approach appeared in 1996, then was developed and formalized by a community of researchers, the ComMod Network; it uses agent-based, GIS and role-playing game models as tools to help solve complex environmental issues involving several stakeholders. It consists in producing models shared by all to represent the functioning of the system being studied, helping in dialogue, and in fine, reach a solution accepted by all stakeholders (Bousquet and Le Page, 2004).

The participatory territorial prospective works lie at the crossroads of participatory approaches and territorial prospective (Piveteau, 1995; Lardon, 2013; Lardon an Roche, 2008; Lardon et al., 2016). They use spatial representation on various media—maps, landscape block diagrams, graphic models—as a mediation tool to foster the participation of stakeholders and their involvement in the collective action. "Prospective-action" makes use of the knowledge of researchers and stakeholders, mediatized via a participatory device based on spatial representations (Lardon and Roche, 2008) to produce both scientific knowledge and knowledge for action (Lardon et al., 2016).

Nowadays, these lines of research hybridize geoprospective much more than prospective does, although the approach originated from the latter (Fig. 1.1).

There is an imbalance between assessment prospective as a decision support, which is well represented on various scales, and academic prospective, both environmental and territorial, driven by a restricted circle of scientists. The reasons are various. The prevailing scientific posture is more focused on the analysis of past dynamics than on those of the future, which are more uncertain and controversial. Such uncertainty results in divergent analyses and difficulty validating affirmations concerning the conditions of the future. Eleonora Barbieri Masini et al. (1993) wrote on that subject: "the scientific nature of prospective is its most controversial characteristic, and in fact, for numerous researchers, it is not included in the qualities of prospective." Moreover, Laurent Mermet and Xavier Poux (2002) consider that the great majority of researchers involved in research work on the environment is in near

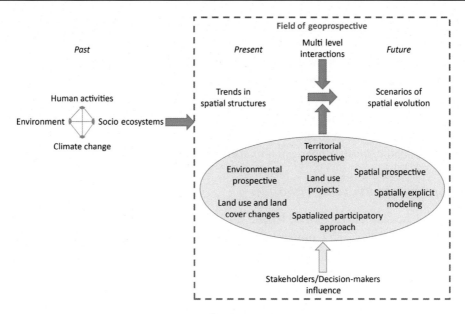

Figure 1.1
Geoprospective, at the interface of several fields of research.

total ignorance of theoretical questionings and methodological resources in works specialized in prospective.

1.4 Conclusion

Geoprospective is an emerging field of research driven by a reduced number of researchers, mostly French. Its singularity is due at the same time to its concept, its projects, and its practices. The concept is based on the central and essential place allocated to the spatial dimension in the medium-term and long-term anticipation processes of the evolution of the system being studied: ecosystem and territory.

As is the case with prospective, the target is to provide aid for action in public policy, planning, or management. However, the primary objective is of a scientific nature. The project of geoprospective is to consider, through research, the space being studied in its complexity, by trying to identify its structures and dynamics, and to understand the anthropic and environmental dynamics as well as their spatial interactions on different scales, and their impacts on the system being studied in the context of uncertainty about the future. This is a key issue, which falls within both fundamental and applied research. By associating theory and practice, the geoprospective posture tends to reduce the fundamental research/applied research dichotomy, which has become artificial in many respects in geography, but is still rooted in minds (Charre, 2003).

Geoprospective is not a school, it is a posture open to different lines of research, which encompasses various methods and practices that take part in its construction, none of them being exclusive. The formalization of the approach is built on a small number of principles and basic rules which were consensual in the founding works and were adopted *de facto* (Voiron-Canicio and Olivier, 2005a,b; Voiron-Canicio, 2006; Houet, 2006; Gourmelon et al., 2012; Houet and Gourmelon, 2014):

- Taking into account the spatial dimension of the problem is the vital lead of the approach. The spatial reasoning underpins the various stages of the process of anticipating the future.
- The prospective is carried out using spatialized representations of possible futures. The scenarios and models, whether quantitative or qualitative, aim to detect the spatial impacts of changes to come.
- The end application of the knowledge resulting from the geoprospective approach is aid to action.
- Geoprospective is a scientific approach. Spatial representations and models of the future cannot be validated by conventional methods, but nevertheless must meet the criteria of likelihood, pertinence, coherence, and usefulness.

Those researchers who choose to link their works to the field of geoprospective implicitly adhere to these rules, by inscribing their research in the wake of the founding works which they reference. Nevertheless, divergences exist over two points, the kind of modeling approach used and the inclusion of stakeholders in the course of the geoprospective approach. The choice of spatial modeling of the future varies according to the theme, the usual practices of the disciplines, and depending on the researcher's methodological background: quantitative model vs qualitative model, spatialized scenarios vs spatially explicit scenarios. The strongest discrepancies concern the place allocated to stakeholders in the development of the approach. The stakeholders' implication is strong in geoprospective approaches regarding territorial prospective, the integrated management of a resource or an area—littoral, mountain,.—, risk adaptation; it is more limited and still rather rare in the case of climate, hydrology, or biodiversity issues.

This unstabilized formalization can be disconcerting and is being debated. Should the approach be made more normative, and a School of geoprospective be created? Or should the freedom of practices be retained and the experimentation of methods combining diverse contributions be encouraged?

References

Afzali, H., 1989. Géoprospective: développement de la modélisation. Evaluation des vitesses d'altération et d'érosion dans divers contextes. BRGM, Département Stockage, 27p. (in French).

Afzali, H., Fourniguet, J., Peaudecerf, P., 1990. Géoprospective: développement de la modélisation. Évaluation des vitesses d'altération et d'érosion dans divers contextes. Direction générale. Science, recherche et développement, 37p. (in French).

Allen, T.F.H., Starr, T.B., 1982. Hierarchy: Perspectives for Ecological Complexity. University of Chicago Press, Chicago, IL.

Amer, M., Daim, T.U., Jetter, A., 2013. A review of scenario planning. Futures 46, 23−40.

Atlan, H., 1979. Entre le cristal et la fumée. Essai sur l'organisation du vivant. Seuil, Paris (in French).

Barbieri Masini, E., Bell, W., Boulding, E., 1993. La prospective et les tendances à l'unité et à l'adversité. Revue Internationale des Sciences Sociales 137, 387−461 (in French).

Beltrando, G., 2010. Les géographes-climatologues français et le changement climatique aux échelles régionales. EchoGéo 2010, 14 (in French). <http://echogeo.revues.org/11816> (accessed 17.05.19.).

Berger, G., 1957. Sciences humaines et prévision. Revue des Deux Mondes, n° 3 : 417−426, Etapes de la prospectives, sous la direction de Jean Darcet, 1967. Presses Universitaires de France, 337p. (in French).

Berger, G. (Ed.), 1967. Etapes de la prospective. PUF, Paris.

Bishop, P., Hines, A., Collins, T., 2007. The current state of scenario development: an overview of techniques. Foresight 9, 5−25.

Bonnardot, V., Carey, V., Madelin, M., Cautenet, S., Quénol, H., 2012. Using atmospheric and statistical models to understand local climate and assess spatial temperature variability at fine scale over the Stellenbosch wine district, South Africa. Int. J. Vine Wine Sci. 46 (1), 1−13.

Bousquet, F., Le Page, C., 2004. Multi-agent simulations and ecosystem management: a review. Ecol. Model. 176, 313−332.

Bradfield, R., Wright, G., Burt, G., Cairns, G., Van Der Heijden, K., 2005. The origins and evolution of scenario techniques in long range business planning. Futures 37, 795−812.

Brunet, R., 1990. Evaluation et prospective des territoires. In: Conférences sur la géographie. Ministère de la recherche et de la technologie, Paris, pp. 25−44 (in French).

Brunet, R., Ferras, R., Thery, H., 1992. Les mots de la géographie, dictionnaire critique. Reclus Montpellier. La Documentation Française, Paris (in French).

Canceill, M., Courbouleix, S., Fourniguet, J., Godefroy, P., Gros, Y., Manigault, B., Peaudecerf, P., 1985. Etude geoprospective d'un site de stockage: Simulation de l'evolution d'un site a l'aide du programme "Castor". Sciences et techniques nucléaires, Bureau de Recherches Géologiques et Minières, vol. 9, 69p. (in French).

Casanova, L., 2010. Les dynamiques du foncier à bâtir comme marqueurs du devenir des territoires de Provence intérieure, littorale et préalpine: éléments de prospective spatiale pour l'action territoriale, thèse, Avignon (in French).

Casanova, L., Helle C., 2012. Ce que les dynamiques foncières révèlent du devenir des territoires: éléments de prospective du sud-est français, L'Espace géographique 2012/2, Tome 41 (in French).

Casanova, J., Brach, M., Millot, R., Négrel, Ph., Petelet-Giraud, E., 2004. Projet PALEOHYD II. Paléohydrologie et géoprospective: modèles conceptuels et processus d'acquisition de la chimie des eaux dans les massifs granitiques. Rapport d'avancement. Rapport BRGM/RP-52880-FR, 66p. (in French).

Charre, J., 2003. Programme scientifique 2004-2007 de l'UMR ESPACE. Université d'Avignon et des Pays de Vaucluse, Avignon (in French).

Chery, J.-P., 2010. Les espaces périurbains en Europe: un grand écart entre description et prospective. In: Prospective périurbaine et autres fabriques de territoires. Datar. Territoires 2040, pp. 61−76 (in French).

Chery, J.-P., Jarrige, F., 2012. Scénarios prospectifs et modélisation des changements d'utilisation des sols: les dynamiques de périurbanisation dans la région de Montpellier à l'horizon 2025. 48ème colloque de l'Association de Science Régionale de Langue Française (ASRDLF), 2011, Schoelcher. Available from: <hal-00655815> (in French).

CNRS, 2004. Rapport de conjoncture, section 39 du Comité national. CNRS, Paris <http://www.cnrs.fr/comitenational/doc/conjoncture.htm#2004> (accessed 17.05.19.) (in French).

Cordobes, S., 2010. Les plis de territoire 2040. In: Délégation à l'Aménagement du Territoire et à l'Action Régionale, Territoires 2040. La Documentation Française, Paris (in French).

Courbouleix, S., 1983. Etude géoprospective d'un site de stockage. Climatologie: évolution du climat et glaciations. Rapport B.R.G.M., no 83 SGN 143 GEO, Orléans, France, 136p.

De Jouvenel, B., 1967. The Art of Conjecture. Basic Books, New York, NY.

De Menthiere, N., Lacroix, D., Schmitt, B., Bethinger, A., David, B., Didier, C., et al. (Eds.), 2016. Visions du futur et environnement: Les grandes familles de scénarios issues d'une analyse de prospectives internationales relatives à l'environnement. Rapport du GT Pros-pective au Conseil d'AllEnvi, vol. 1: rapport final de l'étude ScénEnvi, vol. 2: recueil des fiches prospectives (in French).

Delamarre, A., 2002. La prospective territoriale. La Documentation Française, Collection Territoire en Mouvement, Paris (in French).

Derbyshire, J., Wright, G., 2017. Augmenting the intuitive logics scenario planning method for a more comprehensive analysis of causation. Int. J. Forecast. 33, 254–266.

Didier, E., 2009. En quoi consiste l'Amérique? Les statistiques, le New Deal et la Démocratie. La Découverte, Paris (in French).

Durand, J., 1972. A new method for constructing scenarios. Futures 4, 325–330.

Dutozia, J., 2013. Espaces à enjeux et effets de réseaux dans les systèmes de risques, Thèse de doctorat de Géographie, Université Nice Sophia Antipolis, Nice (in French).

Eckert, D., 1996. Evaluation et prospective des territoires. Reclus Montpellier. La Documentation française, Collection Dynamiques du territoire, Paris (in French).

EPEES, 2000. Evénement spatial. L'Espace Géographique 29-3, 193–199 (in French).

European Environment Agency, 2007. Land-use scenarios for Europe: quantitative and quantitative analysis of a European scale. Copenhague, EEA Technical Report No. 9.

Fernandez, S., Bouleau, G., Treyer, S., 2011. Reconsidérer la prospective de l'eau en Europe dans ses dimensions politiques. Développement Durable et Territoires 2 (3), (in French).

Gadalia, A., Varet, J., 1982. Etude géoprospective d'un site de stockage. L'activité volcanique. Rapport BRGM 83 SGN 010 STO, 20p. (in French).

Garcin, M., 1993. GEOPROSPECT TD (Time Dependant): Démonstrateur d'atelier de géoprospective. BRGM, Département Géologie, 30p. (in French).

Garcin, M., Godefroy, P., Djerroud, A., Rousset, M.C., 1994. Application des techniques de l'intelligence artificielle à la géoprospective: le projet Expect. Colloque Géoprospective—, 1994. UNESCO, Paris, France, 6p. (in French).

GLP, 2005. Global Land Project. Science plan and implementation strategy. IGBP Report No. 53/IHDP Report No. 19. IGBP Secretariat, Stockholm.

Godard, O., 1984. Autonomie socio-économique et externalisation de l'environnement: la théorie néo-classique mise en perspective. Economie Appliquée XXXVII (2), 315–345 (in French).

Godard, O., 1996. Le développement durable et le devenir des villes: bonnes intentions et fausses bonnes idées. Futuribles 1996, 29–35. Available from: <hal-00624329> (in French).

Godefroy, P., 1983. Etude géoprospective d'un site de stockage. La prise en compte de l'activité sismique. Rapport BRGM 83 SGN 301 GEG, 144p. (in French).

Godefroy, P., Courbouleix, S., Fourniguet, J., Garcin, M., Gros, Y., Peaudecerf, P., 1994. Evolution des concepts, méthodes et outils de la géoprospective. Colloque Géoprospective, 18–19 avril 1994. UNESCO, Paris, pp. 89–91 (in French).

Godet, M., 1979. The Crisis in Forecasting and the Emergence of the Prospective Approach: With Case Studies in Energy and Air Transport. Pergamon Press, Oxford.

Godet, M., 1986. Introduction to la prospective: seven keys ideas and one scenario method. Futures 18, 134–157.

Godet, M., 2001. Creating Futures: Scenario Planning as a Strategic Management Tool. Economica, London.

Gourmelon F., Houet T., Voiron-Canicio C., Joliveau T., 2012. La géoprospective, apport des approches spatiales à la prospective. l'Espace géographique 2, 97–98 (in French).

Gros, Y., 1983. Etude géoprospective d'un site de stockage. Tectonique prospective: durée des phases compressives et distensives récentes, évolution du champ de contraintes dans les 100 000 ans à venir. Rapport BRGM 83 SGN 210 GEO, 67p.

Gutman, G., Janetos, A., Justice, C., Moran, E., Mustard, J., Rindfuss, R., et al., (Eds.), 2004. Land Change Science: Observing, Monitoring, and Understanding Trajectories of Change on the Earth's Surface. Kluwer Academic, New York, NY.

Houet, T., 2006. Modélisation prospective de l'occupation du sol en zone agricole intensive: evaluation par simulations dynamiques de l'impact de l'évolution des exploitations agricoles dans la France de l'Ouest. Norois 198 (1), 35–47 (in French).

Houet, T., 2015. Usages des modèles spatiaux pour la prospective. Revue Internationale de Géomatique 25 (1), 123–143 (in French).

Houet, T., Gourmelon, F., 2014. La géoprospective—Apport de la dimension spatiale aux démarches prospectives. Cybergeo: Eur. J. Geogr. (in French). <http://journals.openedition.org/cybergeo/26194> (accessed 17.05.19.).

Huss, W.R., 1988. A move toward scenario analysis. Int. J. Forecast 4, 377–388.

IPCC, 2007. IPCC. Climate Change: The Physical Science Basis. Summary for Policymakers. Contribution of the Working Group I to the Fourth Assessment of the Intergovernmental Panel on Climate Change. <https://www.ipcc.ch/report/ar4/wg1/> (accessed 17.05.19.).

Kahn, H., Wiener, A.J., 1967. The Year 2000. A Framework for Speculation on the Next Thirty-Three Years. Macmillan, New York.

Kallis, G., Nijkamp, P., 2000. Evolution of EU water policy: a critical assessment and a hopeful perspective. J. Environ. Law Policy 3, 301–335.

Kok, K., Verburg, P., Veldkamp, T., 2004. Integrated assessment of the land system: the future of land use. Land Use Policy 24, 517–520.

Lajoie, G., 2005. Modélisation et prospective territoriale. In: Guermond, Y. (Ed.), Modélisation en Géographie: Déterminismes et Complexité. Hermès, Paris, pp. 107–143. (in French).

Lambin, E.F., Baulies, X., Bockstael, N., Fischer, G., Krug, T., Leemans, R., Moran, E.F., Rindfuss, R.R., Sato, Y., Skole, D., Turner, B.L., Vogel, C., 1999. Land-use and land-cover change (LUCC): implementation strategy. In: A Core Project of the International Geosphere—Biosphere Programme and the International Human Dimensions Programme on Global Environmental Change. IGBP Report 48. IHDP Report 10. IGBP, Stockholm.

Lardon, S., 2013. Developing a territorial project. The territory game, a coordination tool for local stakeholders. FaçSADe 38, 1–4.

Lardon, S., Roche, S., 2008. Représentations spatiales dans les démarches participatives: production et usages. Revue Internationale de Géomatique 18 (4), 423–428 (in French).

Lardon, S., Marraccini, E., Filippini, R., Gennai-Schott, S., Johany, F., Rizzo, D., 2016. Prospective participative pour la zone urbaine de Pise (Italie): l'eau et l'alimentation comme enjeux de développement territorial. Cahiers de géographie du Québec 60 (170), 265–286 (in French).

Levêque, C., Urien, R., 2005. Préface. In: Mermet, L. (Ed.), Étudier des écologies futures: un chantier ouvert pour les recherches prospectives environnementales. Ecopolis 5, 13–17 (in French).

Liziard, S., 2013. Littoralisation de la façade nord-méditerranéenne: Analyse spatiale et prospective dans le contexte du changement climatique, Thèse de géographie, Université Nice Sophia Antipolis (in French).

Loinger, G., Spohr, C., 2004. Prospective et planification territoriales: Etat des lieux et propositions, Note du Centre de Prospective et de Veille Scientifique No. 19 (in French).

Martelli, A., 2001. Scenario building and scenario planning: state of the art and prospects of evolution. Futures Res. Q. *Summer* 17 (2).

Martin, B.R., 1996. Technology Foresight: capturing the benefits from science-related technologies. Res. Evaluation 6 (2), 158–168.

Martin, B.R., 2010. The origins of the concept of foresight in science and technology: an insider's perspective. Technol. Forecast. Soc. Change 77, 1438–1447.

Meadows, D., Meadows, D., Randers, J., Behrens, W., 1972. The Limits to Growth. Universe Books, New York, NY.

Mermet, L. (Ed.), 2005. Étudier des écologies futures: un chantier ouvert pour les recherches prospectives environnementales. Ecopolis 5 (in French).

Mermet, L., Poux, X., 2002. Pour une recherche prospective en environnement—repères théoriques et méthodologiques. Natures, Sciences, Sociétés 10 (3), 7–15 (in French).

Oke, T., 1987. Boundary Layer Climates, second ed. Routledge, London.

Piveteau, V., 1995. Prospective et territoire: apport d'une réflexion sur le jeu. Etudes Gestion des territoires, 15, Cemagref Editions (in French).

Pumain, D., 1998. La géographie saurait-elle inventer le futur. Revue Européenne des Sciences Sociales, Tome XXXVI 110, 36–69 (in French).

Quenol, H., 2013. Climate analysis at local scale in the context of climate change. Pollution atmosphérique, mai 2013.

Schoemaeker, P.J.H., 1993. Multiple Scenario development: its conceptual and behavioral foundation. Strategic Manag. J. 14, 193–213.

Schwartz, P., 1996. The Art of the Long View: Planning for the Future in an Uncertain World. Currency Doubleday, New York, NY.

Turner, B., Lambin, E., Reenberg, A., 2007. The emergence of land change science for global environmental change and sustainability. Proc. Natl Acad. Sci. U.S.A. 104 (52), 20666–20671.

Van Cutsem, M., Roëls, A., 2012, Cosmopolis, No. 3–4.

Verburg, P., Crossman, N., Ellis, E.C., Neihimann, A., Hostert, P., Mertz, O., et al., 2015. Land system science and sustainable development of the earth system: a global land project perspective. Anthropocene 12, 29–41.

Voiron-Canicio, C., 2006. L'espace dans la modélisation des interactions nature-société. Actes du Colloque International Interactions nature-société, analyses et modèles, La Baule, mai 2006. Available from <halshs-02133213> (in French).

Voiron-Canicio, C., 2013. Déceler les espaces à enjeux pour l'aménagement. In: Masson-Vincent, M., Dubus, N. (Eds.), Géogouvernance, utilité sociale de l'analyse spatiale. Collection Update, Editions Quae, Versailles, pp. 171–182 (in French).

Voiron-Canicio, C., Dutozia, J., 2017. Anticiper et simuler les dynamiques de changement pour diagnostiquer et améliorer la résilience d'un système territorial urbain. Risques urbains. <https://www.openscience.fr/Anticiper-et-simuler-les-dynamiques-de-changement-pour-diagnostiquer-et> (accessed 17.05.19.) (in French).

Voiron-Canicio, C., Olivier, F., 2005a. Using a GIS to anticipate the consequences of urban paralysis when a disaster strikes: application to the city of Nice (France). In: Systèmes d'information géographique et gestion des risques, ISTED, Paris-La Défense. Available from: <halshs-02098378> (in French).

Voiron-Canicio, C., Olivier, F., 2005b. Simulations et détection des espaces à enjeux dans un système urbain, en situation d'inondation. Actes du Colloque International SAGEO 2005, Avignon. Available from <halshs-02133218> (in French).

Wack, P., 1985. Scenarios: uncharted waters ahead. Harv. Bus. Rev. 63, 139–150.

Ward, N., Lowe, P., Buller, H., 1997. Implementing European water quality directives: lessons for sustainable development. In: Baker, S., Kousis, M., Richardson, D., Young, S. (Eds.), The Politics of Sustainable Development. Global Environmental Change Series. Routledge, London.

Wright, G., Bradfield, R., Cairns, G., 2013. Does the intuitive logics method—and its recent enhancements—produce effective scenarios? Technol. Forecast. Soc. Change 80, 631–642.

Wu, J., 1999. Hierarchy and scaling: extrapolating information along a scaling ladder. Can. J. Remote. Sens. 25 (4), 367–380.

Yoshimo, M., 1975. Climate in a Small Area: An Introduction to Local Meteorology. University of Tokyo Press, Tokyo.

CHAPTER 2

Anticipating the impacts of future changes on ecosystems and socioecosystems: main issues of geoprospective

Emmanuel Garbolino[1] and Jacques Baudry[2]

[1]*Climpact Data Science, Nova Sophia—Regus Nova, Sophia Antipolis Cedex, France,*
[2]*INRAE Centre Bretagne-Normandie, UMR 0980 BAGAP, Rennes Cedex, France*

Chapter Outline

The development of public environmental policies requires scientific bases to define their action modes. Predicting, assessing, as much as possible, the environmental consequences is a major issue. When these policies become more global and no longer sectoral (reserves for flora and fauna, water quality management, etc.), the problems to be addressed become more complex. This complexity is expressed through interactions between ecological processes, to which are added the interactions between actors of the territories and the diversity of their actions in these territories and their impacts on resources.

It is in this decision-making framework for the management of ecosystems and socioecosystems that the objective of the geoprospective is inscribed. In this chapter, we first present a reminder concerning the genesis of the environmental concerns, which emerged from the second half of the 20th century and which encouraged researchers and decision-makers to establish methods and tools in territorial foresight. This first part is then followed by a description of the positioning in the field of managing risks relating to the changes caused by global warming and the transformation of the territory. Following this conceptual positioning, the third part introduces a methodological geoprospective framework for adapting to climate change and changing land use, this framework being used for the development of methods, models and tools in partnership with the concerned decision-makers.

2.1 Global changes, ecosystems, and prospects

Unlike previous climatic changes that the earth has known, including since the beginning of the Holocene, the origin of the rapid climate change that we are currently experiencing is anthropogenic in nature: it is induced by the coproducts of our activities involving combustion and the fermentation of large quantities of organic substances at a global scale. It is also induced by the artificialization and transformation of natural environments, which produce imbalances in ecosystem services regulating GHGs (greenhouse gases; Woodwell et al., 1983; Houghton et al., 1985) and albedo's modifications (Otterman, 1974; Charney et al., 1975).

This process began during the industrialization of Western society and it has amplified and generalized, until causing an increase in temperatures large enough for researchers to understand, in the mid-20th century, the causal relationship between the evolution of the atmospheric content of GHGs and the development of industry and means of transport with thermal engines. The current problem is not so much the fact that the climate evolves or that its evolution is due to human activities: the main difficulty is that the kinetics of this warming is such that a part of living beings will not be able to adapt to these new conditions.

Ecological studies require, in the huge majority of cases, investigations in environments that may be more or less anthropized. As Ernst Haeckel defined in 1866, the purpose of ecology is to study the relationships between living beings and the world around them. These relationships are deduced from the influence of environmental factors on living beings. These environmental factors induce effects on the distribution of living beings on a global scale. The spatial dimension is therefore closely linked to the work of ecologists, at least in the phases of data collection and analysis of the results. The data used in ecology have greatly diversified since the birth of this discipline. They correspond to in situ observations of living beings either visually or with sensors such as images from airborne systems, sound wave sensors, and radars. Environmental data (climate, soil, air) are often used to understand the role of these parameters on the spatial and temporal distribution of the species studied.

Since the end of the 20th century, the ecologists have shown that anthropogenic action is causing the sixth mass extinction in the earth history (Leakey and Lewin, 1996), due to the habitat transformation and destruction, overexploitation of natural resources, pollution, introduction of invasive species, and climate change (Couvet and Teyssèdre-Couvet, 2010; Normander et al., 2012; Primack et al., 2012; Kolbert, 2014).

Furthermore, seen from the perspective of Safety Sciences, climate change can also be considered as a dysfunction or a gap in the assessment of the risks inherent in our activities on a global scale. This phenomenon and its anthropogenic origin were identified in the 1950s by researchers who were able to have a global observation network of climate variables such as temperatures, precipitation, and atmospheric CO_2 content (Willett, 1950; Shapley, 1955; Plass, 1956; Mitchell, 1961).

It is thanks to this change in the scale of perception of phenomena going from the local to the global that the problem of global warming has been able to emerge. In the following years, the publication of the Club of Rome report called "The limits to growth" (Meadows et al., 1972) was a major event in the scientific community (see Chapter 1: The Origins of Geoprospective) for at least two reasons: the first is the scale of the very pessimistic results on the future of humanity, natural resources, and the level of pollution (including the atmospheric rate of CO_2); the second deals with the methodology used, which is based on systems dynamics applied on a global scale.

At a time when environmental questions were starting to become subjects that civil society was taking up, the scientific community and stakeholders could have envisaged that it was scientifically and technically possible to propose scenarios for the evolution of certain parameters describing our world, without the process being spatially explicit.

The years following the publication of the Meadows report saw the development of methodologies associating prospective and spatial aspects, within certain cases the participation of stakeholders involved in the management of the environments of a given territory requiring the participation of ecologists. However, the approach was not spatially explicit since the authors remained at the level of a geographically global approach. This report, however, contributed to the emergence of the concept of sustainable development which was introduced in the late 1980s in the report of the World Commission on Environment and Development (Brundtland, 1987).

Landscape ecology has also greatly contributed to the understanding of spatial phenomena that have an impact on the habitats of living organisms and on their life cycle, these impacts can thus alter the ecosystem services from which we benefit and they cause the emergence of risk situations. The core definition of landscape ecology has been set up in 1984. "Landscape ecology considers the development and dynamics of spatial **heterogeneity**, spatial and temporal interactions and exchanges across heterogeneous landscapes influences of spatial heterogeneity on biotic and abiotic processes, and **management of spatial heterogeneity**" (Risser et al., 1983).

According to Baudry (2002), a landscape is a mosaic of various types of land use, plant cover, bare soil, and buildings. Various networks are associated with this mosaic (network of hedges, streams, roadsides, etc.). The mosaic has a composition and a structure; the various elements differ by their size and their adjacency. Each element, each composite structure has its operating rhythm. A forest changes much more slowly than a cornfield. The various species of plants and animals also have their own rhythm (annual plant/ perennial plant, variable life span of animals) and displacement capacities vary widely. This discipline thus introduced the following three key concepts allowing to study the structuring of landscapes and the environments which compose them, in particular from the point of view of population dynamics, biogeochemical flows, movements of individuals, and trophic interactions.

Heterogeneity: it is a property of landscapes. The measures are based on the diversity and spatial distribution of their components (Fahrig et al., 2011).

Fragmentation: a property of landscape components. Commonly, the term is used when a large patch of a component (i.e., a forest) is split into smaller parts because of forest clearance for agriculture or it is cut by a highway (Villard and Metzger, 2014).

Connectivity a landscape characteristic for a type of component (habitat) in interaction with the dispersal ability of a species. Habitat patches are connected when the species of interest can move from one to the other, despite the fact that the land between is not a habitat (Baudry et al., 2003; Baguette et al., 2013).

The hierarchical structure of landscapes. Each element is embedded in a mosaic, itself part of a larger mosaic and so forth. This hierarchy has several consequences. The first, and it is the *raison d'être* of landscape ecology, is that the ecological state of any landscape element is driven by both its internal features and the features (structure, species composition, etc.) of the surrounding elements. The size of the "surroundings" depends on the dispersal capacity of the species present as well as how far are the abiotic fluxes (nutrients, pollution) affecting the element are coming from. Therefore it is of utmost importance to manage at the same time both the various elements and the global landscape (O'Neill et al., 1986; Suarez Seoane and Baudry, 2002).

Another consequence of hierarchical structure appears when we analyze either the structure or the dynamics of landscapes. This is done with various metrics (heterogeneity, connectivity, density, rate of change, etc.). All metric values vary when the extent of space or time analyzed is changing. They level off at some point that is landscape specific. For instance, in landscapes with many small elements (e.g., fields of a few hectares) heterogeneity increases rapidly when the extent of analysis increases, but it levels off rapidly over a few hundreds of hectares as the whole diversity of elements and structures is captured. In landscapes with very large elements (e.g., fields of several tens of hectares), the extent at which the whole diversity is captured is much greater. The phenomenon is called "scale dependence." It is very important to consider when comparing landscapes. To say that one landscape is more heterogeneous or changes more rapidly, you must control the extent (area, duration) of analysis.

These concepts of landscape ecology have been translated into spatial indicators allowing to study the degree of fragmentation of an environment or its degree of connectivity with other landscape structures and environments. Monitoring these indicators then shows how a landscape evolves, according to the biotic and abiotic factors of the environment. Because it is expressed in a global way, the influence of climate change plays a preponderant role today on the structuring of environments and landscapes, alongside anthropic action which modifies land uses and ecosystem functions. Landscape ecology therefore provides a spatially explicit light on the influence of these global changes which have local impacts.

Since the first report of the IPCC (Intergovernmental Panel on Climate Change) showed the likely effects of global warming in 1990, other works focused on the currently observable impacts of global warming on ecosystems and the environment (melting glaciers, forest decline, rising sea levels, etc.). Among the contributions regarding the impacts of human activities on ecosystems, the "Millennium Ecosystem Assessment" published in 2005 had a significant impact on the community of researchers and decision-makers. Launched in 2001 by the UNO, this work aimed to establish a global overview of the state of the environment, ecosystem services, and human well-being and to assess their evolution within 2050 by integrating scenarios of human activities development and climate change. Even if this assessment was provided at national and continental scales, it was not really spatially explicit. But the results have shown the potential evolution of our environment toward 2050 that would know a significant decrease of ecosystem services and biodiversity.

Other public initiatives developed in the following years at the national and regional levels in many countries. Public authorities and states are more and more called upon by the scientific community and collectives of all kinds to act in favor of mitigating global warming and adapting to its consequences.

We have therefore gone from an era of relative ignorance, even denial, of the proven and probable impacts of our activities on the global climate system, to an era of awareness of the gravity of the phenomenon that pushes us today to adopt an objective and responsible approach to risk management. The contribution of the UN and the results of the IPCC are significant in raising awareness.

As a matter of fact, the latest scenarios proposed by the IPCC members show an average rise in temperatures by 2100 varying from 1.5°C, for the most optimistic scenario, to 4.8°C for the most pessimistic scenario and, if they are confirmed in the future, climate change will have, among other things, an impact on the nature and distribution of the species of ecosystems. This aspect is particularly important for our society because such modifications should bring about changes in the functioning of ecosystem services (biomass for use of food —energy—construction, etc., prevention of erosion risks, recreation in natural areas, landscaping appeal, local thermal regulation, water filtration, etc.).

The fifth report of the IPCC (2014) thus forecasts an increase in the occurrence of natural hazardous phenomena (floods, forest fires, erosion, spread of invasive species, etc.) due to the increasingly frequent occurrence of extreme events, such as drought or heat waves, but also due to the increased mortality of plant and animal taxa (Fink et al., 2004; Ciais et al., 2005; Reichstein, 2005; Fischer, 2007; Niua et al., 2014), in turn causing effects on ecosystem services.

These consequences will have economic impacts due to the increase in the destruction of property and infrastructure, the intervention missions of the emergency services and health professionals, or also due to expenditure on insurance cover for individuals and organizations.

Faced with these worrying prospects, the public authorities are now seeking to assess the economic, social, and environmental consequences of global warming on the territory in order to anticipate the means of prevention and protection to be implemented. This is evidenced by the creation of structures intervening specifically on this issue (e.g., in France, the General Commission for Sustainable Development, 2011, the National Observatory on the Effects of Climate Warming, 2011) and the publication of documents such as "the national strategy of adaptation to climate change" (ONERC, 2007), "the National Plan of adaptation to climate change" (MEDDTL, 2011), and "the European strategy of adaptation to climate change" (European Commission, 2013, 2009, 2007).

European Union proposes adaptation measures to reduce the vulnerability of forests to climate change (2010). The IPCC (2012) also published a report dedicated to decision-makers, with a view to managing the risks of disasters and extreme phenomena induced both by climate change and by poor control of urbanization and land management.

These documents stress that decision-makers need to have research results adapted to spatial scales (continent, country, region, department, intermunicipal, municipalities, localities, districts, and plots) enabling them to develop and apply adaptation strategies to climate change in line with security needs, given the evolution of the territory's vulnerability to the emergence of new risks or the reinforcement of some of them.

Because of these uncertainties about the future of the climate and its impacts, it is necessary to set up a scenario based approach in order to provide decision-makers with examples of possible futures taking into account the history of the evolution of parameters to model, but also by integrating, as far as possible, societal choices (spatial planning, protection of some areas, development of activities, etc.). This research object, linked to the potential emergence of hazardous situations on the territory, is that of geoprospective for which we propose a methodological framework taking into account global changes and benefiting from the triple disciplinary anchoring of Ecology, Geography, and Safety Sciences.

2.2 Risks, territories, and climate change

Risk research often concerns geographers who, alongside other disciplines (ecology, sociology, ergonomics, economics, chemistry, physics, process engineering, etc.), have participated in the development of "Safety Sciences," both from a conceptual point of view and from a methodological and knowledge dissemination point of view. The contribution of geographers to the study of the sciences of risks and dangers is significant insofar as, to understand the origin and the consequences of their manifestations, it is necessary to look at their relations with the territories on which they are located and for which they participate in the transformation in space and time.

This contribution also relates to the study of the ways in which human societies are organized in the face of risks, and more particularly to their prevention by the actors concerned (citizens, public authorities, and private structures).

The advent of landscape ecology, promoting research on interactions among landscape elements, gave an impetus to the study of the spread of disturbances as fire, floods, and pest outbreaks. Landscape ecology also develops ways to combat plant and animal populations' risks of extinction. The landscape ecology program also seeks to understand the drivers of landscape dynamics. Two major sources of complexity are the interactions between social and ecological systems and **cross-scale interactions** within each domain. Multiple factors drive landscape dynamics over a large range of scales from local to international and back. These cross-scale interactions occur in both the biophysical and social components of the socioecosystems. We illustrate this with the interactions between global changes and biodiversity dynamics. Changes in climate, for instance, will drive the dispersal of different species. The propagules of migrant will reach novel habitats if landscape connectivity is favorable. Locally, landscape patches with the suitable environment (microclimate, fertility...) will be colonized. This will change landscape patterns and the regional species pool (Fig. 2.1).

In the social realm, global changes lead to novel policies aiming at transforming transport systems, sources of energy, and landscape planning and management. Locally land planning and land management can offer or not suitable habitat. If not a feedback to policy makers may transform planning options and so forth (Fig. 2.2).

This led to a research question on socioecological systems. Therefore both the ecological and the human dimensions are progressively integrated, facilitating flows of information to

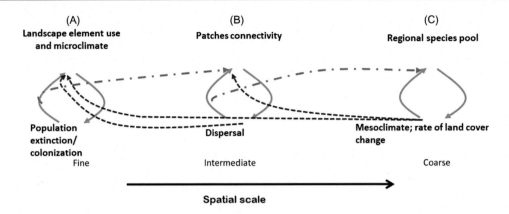

Figure 2.1

Cross-scale interactions in the ecological domain drive plant and animal population dynamics.

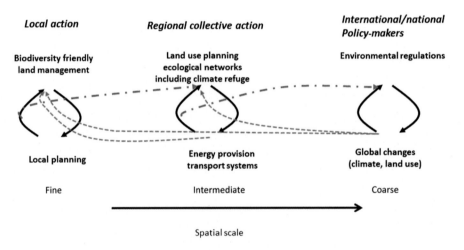

Figure 2.2

Human actions occur at very different scales to control biodiversity dynamics.

design landscapes appropriate to minimize risks. With its spatial and temporal dimension, landscape ecology is conceptually well equipped to bridge with geography and history.

The development of this research has contributed to the social and technical construction of risks (November, 2006) which is perceptible by the means implemented by decision-makers for the prevention and protection of territories. One of the simplest and widely used definitions of the term "major risk" is that of the "https://www.georisques.gouv.fr" site (site consulted in September 2019) which is under the supervision of the Ministry of the Ecological and Solidary Transition: "A potentially dangerous event named HAZARD is a

MAJOR RISK only if it applies to an area where human, economic or environmental EXPOSURES are present." This definition, among other things, introduces the geographic dimension of risks by the term "area."

It is also inspired by other definitions which have been proposed in the past, in particular that set out by Crichton in 1999 concerning the triangle of risk: "'Risk' is the probability of a loss, and this depends on three elements, hazards, vulnerability and exposure." In this definition appears the term vulnerability which will be taken up later in the definition given by the ISDR (International Strategy for Disaster Reduction) of the UN (United Nations) which defines the risk as follows (UN/ISDR, 2004): "The probability of harmful consequences, or expected losses (deaths, injuries, property, livelihoods, economic activity disrupted or environment damaged) resulting from interactions between natural or human-induced hazards and vulnerable conditions. Conventionally risk is expressed by the notation

$$\text{Risk} = \text{Hazards} \times \text{Vulnerability}.$$

Some disciplines also include the concept of exposure to refer particularly to the physical aspects of vulnerability. Beyond expressing a possibility of physical harm, it is crucial to recognize that risks are inherent or can be created or exist within social systems. It is important to consider the social contexts in which risks occur and that people therefore do not necessarily share the same perceptions of risk and their underlying causes."

In this definition, the term "vulnerability" replaces that of "exposures" proposed by the Ministry of the Environment, which represents a simplification, while not all issues are necessarily vulnerable to the same hazard. Moreover, in 2003, Nick Brooks proposed to distinguish two types of vulnerability for a given territory:

- Biophysical vulnerability: it is linked to the level of damage to the issues, whether human or material. It therefore depends on the physical impact of the hazard on the issues, in terms of both its intensity and its frequency. This vulnerability is also analogous to the "sensitivity" of the system studied in the face of the hazard. For example, the use of lethal effect thresholds according to thermal flows makes it possible to characterize the biophysical vulnerability of the population in a given territory to a fire.
- Social vulnerability: it represents the capacity of a system to cope with a dangerous event, which in this case joins the definition of resilience which is proposed below. A system is therefore more or less vulnerable and resilient if it is able, at least in part, to face adversity. Social vulnerability is therefore distinguished from biophysical vulnerability by the fact that it does not depend solely on the frequency and intensity of the hazard but that it depends on the relationships and interactions between the elements of the system, in particular from an organizational and socioeconomic point of view: for example, the use of property insurance is a factor in reducing the vulnerability

of a system because it makes it possible to at least partially compensate for the economic losses induced by a hazard.

Following a fairly extensive bibliographic review of the concept of vulnerability, Hans-Martin Füssel (2007) offers a framework for reflection to apply this concept to the problem of global climate change. According to the author, this framework must integrate the following six dimensions including the definition of the system (identify the elements of the system and their links), the hazard that may affect the system or its elements, the spatial scale of study of the system and its vulnerability, the time scale for understanding the dynamics of the system and events, the area concerned (biophysics, socioeconomic, or both), and finally the exposure or sensitivity of the system and its components at the considered hazard.

To integrate more specifically the territorial dimension in risk management, D'Ercole and Metzger (2009) propose that "territorial vulnerability refers to the idea that there are, within any territory, localizable elements likely to generate and to spread their vulnerability to the whole of a territory, provoking effects which can disturb, compromise, even interrupt its functioning and its development. In this logic, the territorial vulnerability analysis aims primarily identifying, characterizing and prioritizing the areas from which vulnerability is created and disseminated within the territory. It therefore allows defining the areas for which risk prevention actions would be highly effective, thus taking the opposite view from the usual interventions aimed at risk reduction, most often ad hoc and chosen on a contingent basis."

These authors also specify that taking an interest in the vulnerability of a territory is a fortiori taking into account the vulnerability of its main exposures, such as life support networks, also called essential infrastructure. They propose to focus on the vulnerability of major issues as a priority because they constitute the essential elements for the functioning of the territory, regardless of the type of hazard. This posture highlights social space as an object of study.

To complete this point of view, we think that it is advisable to consider at the same time the vulnerability of the exposures and the nature, the frequency, and the intensity of the hazards to understand the vulnerability of the territory. In other words, the vulnerability of the territory is inseparable from the vulnerability of major exposures or not, which are most often dependent on each other, to hazards of various origins that should be known because not all exposures are vulnerable to the same way to these hazards.

Finally, for the IPCC (2014), vulnerability represents the "The propensity or predisposition to be adversely affected. Vulnerability encompasses a variety of concepts and elements including sensitivity or susceptibility to harm and lack of capacity to cope and adapt" (Fig. 2.3).

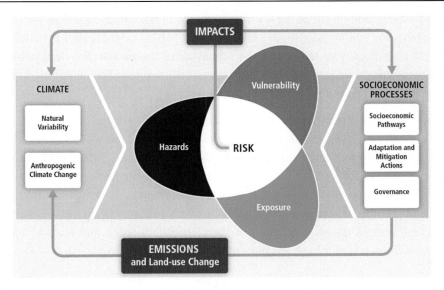

Figure 2.3

Summary for Policymakers. In: Climate Change 2014: Impacts, Adaptation, and Vulnerability. Part A: Global and Sectoral Aspects. Contribution of Working Group II to the Fifth Assessment Report of the Intergovernmental Panel on Climate Change [Field, C.B., V.R. Barros, D.J. Dokken, K.J. Mach, M.D. Mastrandrea, T.E. Bilir, M. Chatterjee, K.L. Ebi, Y.O. Estrada, R.C. Genova, B. Girma, E.S. Kissel, A.N. Levy, S. MacCracken, P.R. Mastrandrea, and L.L. White (eds.)]. Cambridge University Press, Cambridge, United Kingdom and New York, NY, USA. *"Illustration of the core concepts of the WGII AR5. Risk of climate-related impacts results from the interaction of climate-related hazards (including hazardous events and trends) with the vulnerability and exposure of human and natural systems. Changes in both the climate system (left) and socioeconomic processes including adaptation and mitigation (right) are drivers of hazards, exposure, and vulnerability."* (IPCC, 2014).

The definition of risk proposed by the IPCC completely integrates that proposed by Crichton (1999), but it supplements it by adding the biophysical and socioeconomic dimensions of the risks related to climate change. It shows the role of relationships and interactions between socioeconomic processes (from the point of view of GHG emissions and land use change) and the climate system (which undergoes anthropogenic and natural forcings) in the manifestation of risk situations.

It may be noted that in this proposal of risk model presented by the IPCC, the resilience of the systems is not formally mentioned as such, but it is replaced by the use of adaptation and mitigation measures making it possible to face with the consequences of global warming, such as the reduction of GHG emissions and the preservation of natural environments. The contribution of the work of ecologists is preponderant here insofar as it makes it possible to assess the old, current, and future consequences of climate changes on the structure and functioning of ecosystems, these changes having effects on human societies.

Often used as an antonym of vulnerability, resilience is also debated within the community of geographers and ecologists (Reghezza-Zitt et al., 2012; Dauphiné and Provitolo, 2013). In 1973, Holling introduced resilience as "a measure of the persistence of systems and of their ability to absorb change and disturbance and still maintain the same relationships between populations or state variables."

This definition comes directly from the observation of ecological processes and it brings us back to the issue of the conservation of ecosystem properties and the processes that govern them, the latter having a spatial dimension. In 2007, Rebotier introduced territorial resilience as "the capacity of a socio-spatial system to recover from a disturbance and to reduce the impacts expected during a subsequent disturbance, in particular through learning and integrating return experience in the characteristics of the system." In this case, territorial resilience is always a property of the elements of space, but it also concerns human organizations which have to deal with dangerous phenomena.

We will retain here the following definition of the IPCC (2014) which seems to us more complete and better adapted to the context of global change: "the capacity of social, economic, and environmental systems to cope with a hazardous event or trend or disturbance, responding or reorganizing in ways that maintain their essential function, identity, and structure, while also maintaining the capacity for adaptation, learning, and transformation."

This definition has the advantage of highlighting the conservation of the systems structure and functions that are impacted by climate change, in particular through reorganization processes. It also mentions adaptation as an element allowing the resilience of the system. Already in 1992, Dovers and Handmer introduced adaptation as a component that increases the level of resilience of a system. On this subject, the IPCC (2014) defines adaptation as being "the process of adjustment to actual or expected climate and its effects. In human systems, adaptation seeks to moderate or avoid harm or exploit beneficial opportunities. In some natural systems, human intervention may facilitate adjustment to expected climate and its effects."

Following these references, we propose the following conceptual framework to introduce the concepts of risk, vulnerability, and resilience of systems subject to climate change. The risk thus results from the interaction between:

- One or more dangerous phenomena, called hazards, characterized by their probability of occurrence, their intensity, their kinetics of triggering and evolution, and their spatial extent.
- One or more exposures (human lives, goods, infrastructure, ecosystems, etc.) characterized by their vulnerability, their resilience, and their spatial influence.

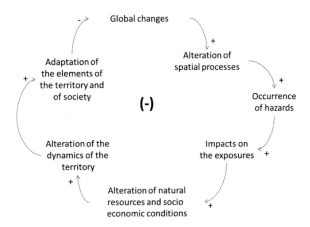

Figure 2.4

Simplified model of the dynamics of the impact of climate change on the territory and the response of society.

This interaction can result in the modification of the physical structure (state) and the functional properties of the exposures, as well as the change in the organization of the relationships between the exposures that make up the territory and the spatial processes, which can have the consequence to increase the vulnerability of the territory taken as a whole. The alteration of spatial processes can thus lead to the emergence of risk situations (Fig. 2.4).

The resilience of the exposures also belongs to their capacity, both intrinsic and organizational, to recover a state, functional capacities and to reestablish interactions allowing them to ensure their integrity, or even to generate the emergence of a new balance. This is the case for the evolutionary processes of living beings which sometimes manifest themselves as a result of dangerous, even catastrophic events, which will select within the same population for a given specie, individuals with a genetic heritage and/or behaviors and interactions with their environment allowing them to resist conditions normally harmful to other individuals. In this context, the resilience of the territories in the frame of climate change will be the capacity to adapt not only to the elements of the territory, but also the society and its response in terms of policies to decrease GHGs emissions and conserve ecosystem services to reduce impacts, intensity, and kinetics of global warming.

This model (Fig. 2.4) is represented using a causal graph: positive interactions (+) vary in the same direction and therefore have an amplifying role, while negative interactions (−) vary in the opposite direction from the original variable thus conferring on the relationship an attenuating role.

This model shows the major role of territorial adaptation, which may or may not depend on anthropogenic action, in the adaptation of our society. This variable introduces negative, therefore stabilizing, feedback to regulate the overall functioning of this system.

Given the complexity of the studied system due to the uncertainty which weighs on the evolution of its behavior and the interactions and phenomena which may still escape our knowledge, it is essential to already adopt a geoprospective approach to assess the impact of decisions on the future of the system and to more finely assess the performance of solutions for adapting the territory. The objective of applying such an approach is to provide a dynamic framework for risk analysis and management depending on scenarios based on a reproducible approach, adapted to the different spatial scales concerned, controllable, and adjustable over time.

2.3 Proposal of a geoprospective methodology framework for adaptation to climate change

In the context of assessing the potential impacts of climate change on nature and its services from which our society benefits, the geoprospective is part of a risk management approach that must be taken. To some extent, this is a return to the roots of the geoprospective's security objective. But it is enriched by the study and modeling of the spatial dynamics of the elements of the territory (natural and anthropic phenomena) to assess the consequences of planning decisions in terms of exposure to risks and impacts on the environment, these impacts may again have consequences for our society due to reciprocal human—nature interactions.

We propose the following methodological framework (Fig. 2.5) which aims to structure research and development works for decision support dedicated to stakeholders in land management, to support climate change mitigation strategies and adaptation to these impacts. The different stages of this methodological framework are presented below.

2.3.1 Coconstruction with the stakeholders of the problematic of the envisaged solution

Because of the operational nature required for mitigation and adaptation to climate change, this methodological framework is based on a research-intervention approach proposed by Herbert Simon (1977) which is part of a process of coconstruction of the problem and solutions envisaged with stakeholders conventionally comprising four complementary phases:

- A phase of **diagnosis of the problem** for which it is necessary to identify the needs of the stakeholders. This phase includes the development of a first model of

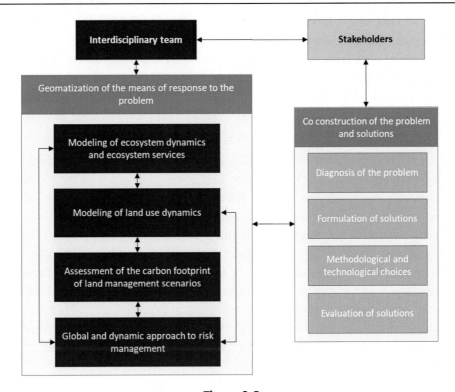

Figure 2.5

Methodological framework of the geoprospective approach applied to adapting to climate change and reducing its impacts.

representation of the studied system with regard to the concerned problematic. Different modeling methods can be used to develop this first model which aims to communicate with stakeholders and researchers. Then, at the end of this consultation, a first model is developed, the latter then being brought to evolve during the following consultations and the results of the work. Starting from the definition of risk introduced previously which highlights the process of interactions between dangerous phenomena and exposures, we propose to integrate the dynamics of systems (Forrester, 1961) as support for the production of a reference model of the problematic based on the methodology of the causal graph. This model thus aims to highlight the relationships between the variables of the territory, whether or not they have a spatial representation.

• A phase of **designing and formulating possible ways of solving a problem**, providing to assess the possibilities of answering questions asked by stakeholders. This phase most often requires a multiscalar spatiotemporal and interdisciplinary approach taking into account the complexity of the studied system and the phenomena that occur there, to propose a methodological framework allowing to approach the problem.

- A phase of **choosing a mode of action** that consists in selecting the methodological and technological means to develop the solution. This phase highlights the knowledge and skills of the experts constituting the interdisciplinary team formed for decision support.
- An **evaluation phase of the chosen solution** that can lead to a reelaboration of the previously designed model, particularly when it comes to scenarios for which the planning options are not completely fixed. This phase consists of the development of the models and their presentation to stakeholders and decision-makers.

This framework implies that several learning loops be employed to progressively build a better integration of the stakes. There may be first some adaptations of the models to include more variables needed to improve the choice of possible pathways to solutions. Then, it may occur that when choosing a mode of action, the chosen pathways are difficult to implement, this requires reframing the problem to produce novel pathways. For instance, the area under scrutiny is too small to include all the major source of GHG. A possible third learning loop is to expand the causes of global warming or the range of risks. In the system under study, the management objectives are so diverse that it faces a wicked problem with many social interactions and antagonistic expectations and biophysical processes hierarchical structured and highly nonlinear. Threshold effects are expected at all scales, legacies are unknown and intractable processes foreseen, as the release of methane from permafrost.

2.3.2 The geomatization of the means of response to the problem

The geomatization process is based on the following steps:

2.3.2.1 Modeling of ecosystem dynamics and ecosystem services

This modeling is necessary for understanding the dynamics of ecosystems and their services with regard to the evolution of climate forcings assessed by the IPCC. It is the basis of what can be called an Ecosystem (Based) Approach introduced in the Convention on Biological Diversity, which aims to promote the sustainable development of human activities by preserving natural resources. This modeling is based on the following stages which are applied to plant ecosystems and which could be generalized, in our approach, by the integration of animal populations:

- *Calibration between biotic variables (plants) and environmental variables (climate and soil types)*: this step is particularly relevant because it will subsequently influence the results of the modeling. It makes it possible to produce bioindicators of the climate and soil types. To characterize the statistical relationship between the distribution of plants and climatic variables, it is necessary to adapt the calibration to a relevant scale so as to take into account in the best possible way the climatic range in which a specie can grow. For example, if we are interested in microendemic species colonizing a small area such as a watershed, climate calibration will require gathering botanical observations and meteorological measurements of the study area. On the other hand, if the studied plants are species whose distribution extends beyond the watershed, it will be necessary to widen the spatial influence of botanical observations and climate measurements. As part of the work proposed by Garbolino et al. (2007, 2012), the climatic calibration of 4000 species was carried out on the scale of France so as to take into account as much as possible the different climatic situations that taxa can encounter. In addition, the calibration must be based on a probabilistic approach taking into account the intermittence of plant observations in the climate range, which is not usually the case with geometric methods such as PCA (component analysis main) and the correlations which strictly apply only to continuous data whereas observations in the natural environment are generally discontinuous. Finally, it is also necessary to calibrate the species according to their different levels of abundance because this information provides details as to the colonization capacity of the environments by taxa. Thus, during the modeling of the distribution of plants, the abundance gives indications on the possible evolution of ecosystems structure and functions.

- *Modeling of vegetation potential according to the current climate*: the objective of this stage is to determine a baseline for which the potential distribution of vegetation is estimated according to the parameters of the current climate and the types of soil (Garbolino, 2014). This baseline is then compared with models of the probable distribution of plants according to the IPCC scenarios in order to identify the part attributable to climate change in future changes in plant distribution. This modeling makes it possible to complete observation plots or distribution maps of plant formations or species. It is based on the estimation of the probability of the presence of calibrated plants in the range of climatic variables throughout the territory. To model the new species distribution, it is necessary to incorporate landscape connectivity. It is the possibility for the plant propagules to move across the landscape from a source (current distribution) to the newly available sites. Several landscape features can impede this movement: the climate itself in the mountainous areas, crossing a pass may necessitate colonizing forest upslope. Then, of course, plants move in different ways. Vegetative reproduction implies a slow movement of plants may be too slow to keep pace with the spatial movement of climate. Wind dispersal is the most efficient, as seed can be carried hundreds of kilometers away. Dispersal by animals is quite common; plant dispersal depends on animal movements. Some seeds are dispersed by ants (*Primula*), others by birds (fleshy fruits), which can fly long distances. People also disperse plants by boat, plane. An important pathway for invasive species, as they have efficient dispersal mechanisms, climate change may enhance their colonization ability. Allergies, competition with "native" species are prime risks associated to such processes.

- *Modeling of vegetation potential according to climate change scenarios*: it is established on the basis of climate calibration of plants and outputs from IPCC models. It provides maps of vegetation potential by identifying the probably suitable areas for the taxa according to the expected evolution of the climate. Thus it makes it possible to identify the likely dynamics of plants up to 2100 (Garbolino, 2014). In this case, the probable absence of a taxon can be interpreted as a possible death of the latter or as a potentially stressful situation making it more vulnerable to climatic hazards, diseases, and parasites.

- *Prospective analysis of the evolution of the functioning of ecosystems*: it is based on the analysis of the consequences of vegetation dynamics on the functioning of ecosystems (replacement of taxa, mortality or stress of some of them, etc.). It can also be enriched by the use of vegetation growth models or by studying the evolution over time of vegetation state indicators such as net primary productivity (NPP). This analysis makes it possible to identify, and even quantify, the repercussions of these changes on the services provided by ecosystems, such as trees growth and its use for wood supply; CO_2 sequestration by ecosystems, etc. Changes in biotic interactions are major thread for ecological processes within the climate change context. Discrepancies between pollinators and plants to pollinate already occur. Hatching of birds and food availability is another problem. This can lead to species extinction. The biogeochemical cycles are also affected as mild winters foster the transformation of organic matter into its chemical components.

- *Search for analogies with paleoenvironments*: studies of paleoenvironments based, in particular, on palynological analysis that of plant and animal macroremains or fossils provide indications on the impact of the climate, but also of anthropic action on plant landscapes and environments. These studies make it possible, for example, to estimate the possible progression of plants according to climatic forcings (Huntley, 1995), thus helping to validate forecasting models. Furthermore, modeling the distribution of plant species during climate evolution of the Holocene can also be useful for comparing distributions with models using the outputs from the IPCC scenarios (Salzmann et al., 2009).

2.3.2.2 Modeling land use dynamics

Trends in land use change are observed globally for crops, urban areas, grasslands, and forests. These trends have been highlighted, in particular, thanks to the long time series of earth observation satellites on a global scale, but also to old archives (maps, cadasters, registers, etc.). These transformations are most often heterogeneous: they do not have the same intensity or the same trajectories depending on the regions of the globe. This observation requires the application of integrated modeling of biophysical and socioeconomic factors and their trends on land use according to the spatiotemporal scales of perception of decision-makers. Taking all the factors into account is difficult because of the interactions between them and the existence of feedback phenomena, some of which still elude us and which induce uncertainties.

Changes in land uses in rural areas may take different aspects, depending on the changes in climate, warm periods, and dry/wet seasons. The usual patterns will be modified. Novel crops and novel pest will come in. The population dynamics is a major risk for food safety, climate variability, as well as it will make difficult for the farmers to adapt their cropping systems.

For urban expansion, warming is a problem and will require novel forms of urbanisms. But, the overriding variable may be the lack of water leading to collapse of cities or the creation of expansive infrastructures to transfer water from "sources" to cities. "War for water" is the consequent risk. The necessity to plant trees to stop irrigate crops will, in turn, lead to new land use changes.

Adapting to climate change and reducing its intensity require developing strategies that most often have a spatial dimension, for example, forest management, limiting the artificialization of soils, using public transport, or the development of energy systems with low GHG emissions (photovoltaic fields, wind farms, biomass-based energy systems, etc.).

The impact on biodiversity may be reduced by protecting refuge zone. These zones at the local level have a special microclimate that is cooler than the regional climate or warms up at a slower pace.

The purpose of modeling land use dynamics is to provide a formalized knowledge base for its management using scenarios where expert intervention is necessary. Given the interactions between the natural environment and the anthropic environment, this modeling requires the integration, within the same modeling framework, of concepts, methods, and tools from various disciplines such as geography, economics, sociology, ecology, and engineering sciences. According to Verburg et al. (2004), the modeling of land use dynamics must integrate six main components: (i) the level of analysis, multiscale dynamics (Veldkamp et al., 2001) (nesting of scales/cross scales),[1] (ii) the drivers of change linked to land use, be they economic, social, ecological, environmental, historical, and cultural, (iii) spatial interaction and the neighborhood effect due to spatial autocorrelation—with respect to the environmental factor gradients and the distribution of land use classes,[2] (iv) the temporal dynamics for which the changes are generally nonlinear and the thresholds play an important role, which requires taking into account (v) the retroactive effects, (vi) the level

---

[1] A scale can be defined as the spatial, temporal, quantitative or analytical dimension used to measure and study objects and processes (Gibson et al., 2000).
[2] It can be taken into account by the use of cellular automata for which constraints can be defined according to different spatial scales (White et al., 1997; White and Engelen, 2000).

of integration of complexity based on the interactions of the subsystems constituting the system studied.

The use of a first form of modeling based on the causal graph method can be used as a first step because of the presence of feedback loops between the elements of the territory and the phenomena observed (Fusco, 2012; Hinojos Mendoza, 2014). Tools for modeling and simulating spatial dynamics from artificial intelligence and operational research can then be applied in geoprospective (Voiron, 2006; Voiron-Canicio and Olivier, 2005; Voiron-Canicio, 2012; Emsellem et al., 2012; Gourmelon et al., 2012; Houet and Gourmelon, 2014). Let us cite, for example, the use of multiagent systems (Parker et al., 2001, 2002) and cellular automata for the simulation of spatial dynamics of land use (White and Engelen, 1997; Dubos-Paillard et al., 2003; Langlois, 2010), or even probabilistic approaches based on Bayesian networks allowing the integration of uncertainties in the creation of management scenarios (Fusco, 2012). Such models can be coupled together according to the integration levels of the complexity of the studied system.

These methods and tools are able to take into account the spatial distribution and the forms of the main drivers of transformation of land use and landscapes such as the physical elements of the territory (slopes, altitudes, types of soil and rocks, slope exposure, etc.), biotic elements (vegetation formations and species, etc.), and socioeconomic elements (buildings, infrastructures, activity zones, demographic trends, movements, etc.). The involvement of stakeholders in the establishment of development scenarios is necessary for the construction and evaluation of realistic scenarios. Such an approach simulating the dynamics of land use thus makes it possible to structure exchanges with stakeholders and to evaluate the scenarios on repeatable, modifiable, and quantifiable bases adapted to the scales of perception of decision-makers.

2.3.2.3 Assessment of the carbon footprint of land management scenarios

The scenarios for planning and developing economic activities (agricultural, industrial, services, etc.) proposed must be part of an approach to reduce GHG emissions and/or adapt to climate change. These scenarios therefore have a specific carbon footprint due to the energy required to carry them out. In a geoprospective approach intended to reduce the impacts of climate change, it is necessary to assess the carbon footprint of each of these scenarios to compare them with a methodology based on a systemic analysis, using the causal graph developed previously, the building blocks of the scenarios and those that will be influenced by them.

Several methods exist for carrying out such an assessment, such as the life cycle assessment (LCA), which can be adapted to the study of the carbon footprint of a good or consumer service. This method, proposed in the early 1970s, underwent various methodological developments to take into account the specificities of certain products, their natural and sociotechnical environment (Guinée et al., 2011). More recently, methods have attempted to take into account the dynamic aspects of product life cycle emissions with those arising from the global change scenarios developed by the IPCC (Levasseur et al., 2010).

Methods more specifically adapted to the assessment of the carbon footprint have been developed in recent years on a national scale, such as the Bilan Carbone of the ADEME (Environment and Energy Management Agency), or internationally with the GHG Protocol proposed by the World Resources Institute and the World Business Council for Sustainable Development.

All of these evaluation methods use reference databases on emission factors, the majority of which are open access. They give the main GHG emission factors for the various processes, tools, means of transport, etc., identified in the development scenarios. These methods also make it possible to assess GHG emissions according to land use practices in order to carry out a truly comprehensive assessment of releases between different scenarios.

2.3.2.4 Global and dynamic approach to risk management

The geoprospective approach has a security objective insofar as it makes it possible to establish land use and use scenarios that may be subject to risk analysis. These scenarios are based on a systemic approach integrating the temporal dimension to simulate the dynamics of the elements of the territory and the phenomena encountered. Because of this evolving nature of the system studied, we propose a global and dynamic approach to risk analysis (Fig. 2.6) unlike conventional approaches to risk assessment (Andrews and Moss, 2002). This approach is global since it is concerned with all the risks that can be identified in the system and which can significantly contribute to the transformation of the territory during the manifestation of dangerous phenomena. It is also dynamic because it integrates the dynamics of the system, through simulation methods (multi-agent system (MAS), Systems Dynamics, discrete events, etc.), with the aim of identifying risk situations and evaluating the performance of risk control measures (Garbolino et al., 2009, 2010, 2016; Leveson, 2004, 2012). This approach can also be based on work on the development of a risk ontology in order to enrich the modeling of studied systems (Provitolo et al., 2014).

The previous stages of the proposed geoprospective approach lead to the production of spatial dynamic scenarios, based on models of vegetation dynamics and land use. Their carbon footprints are evaluated to optimize them in view of the objectives of mitigating the impacts of climate change. This risk control process is based on the following stages:

- **The identification of risk situations**: it assesses the emergence of risk situations arising from the interactions between hazards and the exposures existing in the territory. It integrates the dynamic aspects by comparing the scenarios at different time steps. It seeks for the potential causes of these risk situations with the aim of subsequently defining means that can, if necessary, act on the origin of the risk situations.

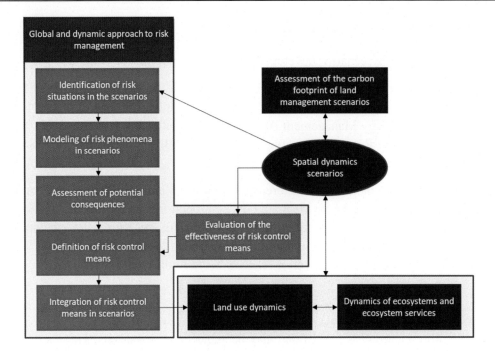

Figure 2.6
Global and dynamic risk management approach to promote an adaptation and mitigation strategy for the consequences of climate change on the territory.

- **Modeling of hazardous phenomena**: some changes in the territory can lead to bringing potentially dangerous elements into contact. For example, the potential increase in woody mortality can lead to the production of highly flammable fuel in the Mediterranean area. Forest fire trigger and propagation models can thus be used to assess the territories' future exposure to this risk. The disappearance of part of the plant cover can, for its part in a sloping area, lead to increasing soil erosion and favoring runoff which can cause flooding. These phenomena can also be modeled to understand their extent, intensity, and potential consequences on inhabited areas. Domino effects also including the exposure of industries and infrastructures and the manifestation of technological hazards must also be modeled within the framework of a global and systemic approach.
- **Assessment of potential consequences**: models and simulations of natural and technological hazards provide maps showing the extent of their effects on the territory. Coupled with dynamic exposures distribution maps, it is possible to produce maps of potential exposure to the risks that will arise in the future. The assessment of human, economic, and environmental consequences requires the integration of additional databases to qualify the vulnerability of the exposures facing these hazards.

- The **definition of risk control means**: these means are grouped into three categories which can sometimes relate to the same means. These are the means of prevention, protection, and intervention. They are based on physical barriers and organizational lines of defense. In the context of climate change, it is necessary to define means making it possible to mitigate the effects of climate change (actions on the hazard) and to adapt to the consequences of the latter (actions on the exposures). The impact of the implementation of these means in terms of risks must also be assessed. These means may concern land use strategies allowing conservation of the natural CO_2 sequestration functionalities, sustainable forest management measures making it possible to promote their development, budgetary anticipation of the intervention costs of airborne means of combating fires in altitude areas where this threat should spread in the future, the revision of regulatory systems for the prevention of already existing hazards (natural and technological risk prevention plans; crisis management plans) considering future developments in the territory and the risks, as well as information, awareness-raising, and training actions for the populations on the risks induced by climate change in their territory and the means of prevention and protection available.

- **Integration of risk control means in the scenarios**: it aims to identify the consequences on the territory of risk control measures. These measures can be implemented at different times during the construction of the new land use scenarios. They can have consequences on the development of buildings, on the conservation or not of natural spaces, on the dynamics of economic activities, etc. During all of the risk control stages, new variables or new control parameters can be added to the spatial dynamics scenarios due to the emergence of new phenomena appearing over time.

- **Evaluating the effectiveness of risk control methods**: the new spatial dynamics scenarios established and optimized in terms of their GHG emissions are then evaluated from the point of view of the exposure to dangerous phenomena. This assessment is carried out by comparing the new scenarios with the previous ones in order to identify changes in terms of risk control. The process "definition of risk control means— integration of these means into the scenarios—evaluation of the effectiveness of these risk control means" can be iterative in order to obtain several optimal risk control scenarios. It is essential that stakeholders are involved in the development and choice of scenarios throughout this process in order to validate realistic scenarios that can guide choices for land management.

- **The reduction of risks generated by climate change: a wicked problem**: The problem we are dealing with has processes at multiple spatial and time scales, nonlinear responses to drivers and the consequences affect differently people. A wicked problem has no "good solution" because when a solution to part of the problem is found, this may has strong implications in changing the structure of the system and potentially generate worst consequences. Therefore, on the one hand, scientist, policy makers, and decision-makers can generate alternative scenarios; on the other hand, choices must be explained and discussed with the various stakeholders. Stakeholders have different objectives, values, knowledge, and power. The final choices must be evaluated when implemented, to start again with the learning loops describes above.

In order to help decision-makers to formalize and evaluate the impact of land management choices in the face of the likely impacts of climate change on natural and anthropogenic environments, the methods and tools described above can be integrated into a geomatization process allowing to define a spatial decision support system (SDSS). These computer systems appeared in the mid-1970s and were called SDSS in the mid-1980s (Armstrong et al., 1986). In 1991, Densham explained that the SDSS was "explicitly designed to provide the user with a decision-making environment that enables the analysis of geographical information to be carried out in a flexible manner."

Chevallier (1993, 1994) subsequently gave a more complete definition: "A SDSS aims to support a complex activity, that is, a process of developing and evaluating scenarios and variants, to identify the best actions based on 'a situation, objectives and criteria. [] So, in addition to the functionality of a GIS, a SDSS must propose the possibility of integrating mechanisms allowing:

- identify, describe and manipulate a set of data with decision-making scope such as actions, scenarios, evaluations;
- jointly exploit original data and various forms of representation;
- to link spatial reference databases and specialized application software to develop simulations and perform analyzes specific to the problem;
- have the means to evaluate, compare scenarios, and then choose using the techniques of multi-criteria analysis."

The development of a SDSS is therefore part of a geomatization process which must be based on taking organizational and human aspects into account so as to take into account the objectives, sometimes conflicting, of stakeholders and to assess the impact of SDSS in the organizations targeted. The participatory aspect provides to identify the real needs of the users. Such an approach must make it possible to avoid certain pitfalls observed during the creation of this kind of tool, namely, a partial or ill-suited response to the problems of users which ultimately result in underuse, or even abandonment of the tool.

The development of such a geomatic infrastructure (Chevallier and Caron, 2002) aims to consider the process of geomatization of the solutions provided to decision-makers "as a factor of organizational evolution." This holistic approach underpins the development of the geomatic infrastructure "as a component of an organization considered as a whole" and which takes into account operational needs and constraints. The development of online and data sharing solutions such as Web-based and service-based SDSS makes it possible to envisage collaborative spaces to facilitate, structure, and maintain exchanges between interdisciplinary teams and stakeholders (Jankowski et al., 1997, 2006; Sikder and Gangopadhyay, 2002; Sugumaran and DeGroote, 2011; Jelokhani-Niaraki and Malczewski, 2012).

Beyond the design of the SDSS intended for stakeholders, it should be added that the transition to a geomatization project must be enriched by the contribution of landscape ecology because of the integrative function of this discipline. As a matter of fact, landscapes have a strong social component; they are socioecological systems. These systems have their own processes and are managed by a social (or sociotechnical, or sociopolitical) system, in short by humans who make decisions with intentional or unintended effects on ecological processes. This led to the development of public policies to design and manage landscape for environmental purposes. Landscape ecology is the backbone of the Green and Blue Infrastructure strategy as it provides evidences for policy design and implementation. Many outputs of the landscape ecology research are incorporated in evidence-based policies (EBPs). However, is an EBP the most appropriate manner to foster planning and actions? Several researchers address this question (Head, 2016). We can list some major shortcomings of EBP: (1) it requires an experimental design with replicates: not possible with landscapes, (2) usually results are at fine spatial scales, and (3) the results do not incorporate the situations of action. The later point may explain why though evidences of climate changes and its causes are mounting, policies to combat the problem are not efficient (Newman and Head, 2017).

In this decision-making process, the last step linked to the geomatization process relates to learning to take biodiversity into account in the territories.

A "development project" is a collective project on a territory aimed, here, at managing biodiversity by modifying elements of the landscape, their arrangement, or their management.

The challenge of learning loops is to allow self-assessment of the pathways to sustainable management of biodiversity. Indeed, there can be no innovation without continuous reflection, otherwise individuals or collectives remain on their values and a priori from the start. In fact, humans tend to seek and select information that confirms their point of view. At the same time, these developments must be assessed.

The learning device has three loops.

> **The first loop** is an improvement of the actions decided at the start, the first step being an inventory of biodiversity, an acquisition of knowledge. Vast ambition requires adjustments, in terms of methods and means. It is therefore an adaptation loop to the concrete situation.

Table 2.1: The triple learning loop for taking biodiversity into account in the regions.

	Context	Baseline frame	Action	Exit
Apprenticeship scheme	**3** Transformation ←	**2** Reframing ←	1 Adjusting routines	
Aim	Make development and management proposals within the framework of territorial policies	Establish relationships between the distribution of species and human activities	Inventory of biodiversity	
Baseline	**Model** scenarios of biodiversity dynamics linked to management options	**Understand** the role of landscape and practices in the dynamics of biodiversity	**Map** species	
Action plan	**Integrate** biodiversity into public action choices	**Manage** biodiversity	**Protect** biodiversity	
Report to biodiversity	**Biodiversity is a management object** for itself and for the ecosystem services it provides	The dynamics of biodiversity are linked to human activities and the structure of the landscape	Biodiversity is an autonomous category	
Stakeholders	All users of the territory and the services in charge of regional planning and their external support	All users of the territory	Citizen science approach by trying to attract more people to make observations	
Personal engagement	What is my commitment in public life with regard to biodiversity?	What relationships between my personal practices and biodiversity?	I like to observe biodiversity	

From right to left, we have successive loops that enrich the process.

The second loop is to reframe the action, it can be to narrow priorities, define new sampling plans, for example, moving from "remarkable spaces" to the diversity of landscape situations.

The third loop concerns the redefinition of the context. For example, moving from a context of knowledge acquisition to a context of using knowledge for action (planning, management plan). New players are coming into play: design office doing modeling, urban planners, and of course elected officials. They were able to participate in the acquisition of knowledge as citizens, but there they become decision-makers.

This table is proposed for discussion on

1. Reasoning on the need for learning;
2. The lines to build the reasoning;
3. The content of the boxes.

Citizen science and learning publications grow like Jordan et al. (2016). This seems essential for the development of practices in areas of biodiversity (Table 2.1).

In practice, the collective or collectives who work to define the action plan(s) must navigate between various fields ranging from biodiversity observation, with the diversity of the methods defined above, the design of plans of action, and, always, the acquisition of new knowledge which forces to revisit the observations, the development of the action plan.

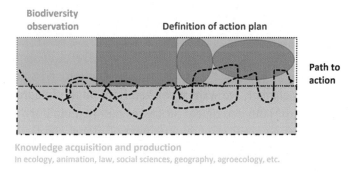

Figure 2.7
The path to action: navigate between various fields and implement learning loops. There is no time arrow that would indicate a cumulative linear path.

Table 2.2: Example of questions and activities dedicated to knowledge, observation, and action plan steps.

Knowledge	Observations	Action plan
C1: What is a habitat for a species?	O1: Observe the diversity of vegetation formations on the territory	P1: What are the legal obligations in terms of biodiversity? For who?
C2: What is habitat fragmentation? What are the related problems?	O2: Map vegetation formations	
C3: How to define ecological continuities?	O3: Observe the spatial distribution of some species	
C4: What is citizen science?		

This navigation is chaotic in appearance (Fig. 2.7) because the process cannot be linear with, simply predefined steps which would simply follow one another. It is therefore essential to record the progress of the work, the step back with the acquisition of knowledge, and the step back which consists of revisiting the data and the drafts of the action plan. Consultation and the construction of scenarios will explain this navigation.

We can define a certain number of elements belonging to the three fields of Fig. 2.1 in Table 2.2.

At this stage, the main challenge in practice is how to navigate?

In our geoprospective approach, those questions help researchers and ecological engineers to identify the most relevant information in order to implement the methods and tools for assessing the different planning scenarios. The integration of both approaches, that is, ecosystem-based and participatory ones, is a key step in order to help the stakeholders to appropriate themselves the main issues and solutions.

2.4 Conclusion

Born from Earth Sciences and Safety Sciences in the early 1980s, the geoprospective was transposed by geographers to provide a framework for reflection and action in sustainable land use planning. In this chapter, we wanted to shed light on the integration, in the geoprospective approach, of both ecological processes and participatory approach, in order to provide a framework for its application in a context of global changes carrying risks for the environment and society (decrease of biological resources, emergence of diseases, increase of wildfires etc.) . Given the systems concerned, deemed complex by the interactions between the elements of space, the definition of territorial planning, and the uncertainties about their evolution, we propose to apply a global and dynamic approach in order to adapt our strategies and technics to make our territories and activities resilient. The current challenge is now related to the awareness of decision-makers and stakeholders about the importance of the application of such an approach, as soon as possible due to the impacts that current (and future) human activities (will) provoke on our socioecosystems.

References

Andrews, J.D., Moss, T.R., 2002. Reliability and Risk Assessment. John Wiley and Sons Ltd, p. 540.

Armstrong, M.P., Densham, P.J., Rushton, G., 1986. Architecture for a microcomputer based spatial decision support system. In: Second International Symposium on Spatial Data Handling, Seattle, WA.

Baguette, M., Blanchet, S., Legrand, D., Stevens, V.M., Turlure, C., 2013. Individual dispersal, landscape connectivity and ecological networks. Biol. Rev. 88 (2), 310–326.

Baudry, J., 2002. L'écologie du paysage, une discipline récente. Séminaire national du projet PROSPEA. PRojet pour le Service Public d'Enseignement Agricole (2002-01-29-2002-01-30) Paris (FRA). In: Loi d'orientation agricole, Savoirs et Qualifications professionnelles. Analyse des enjeux et perspectives pour l'évolution des savoirs. Chantier PROSPEA. Educagri Editions, Dijon (FRA).

Baudry, J., Burel, F., Aviron, S., Martin, M., Ouin, A., Pain, G., et al., 2003. Temporal variability of connectivity in agricultural landscapes: do farming activities help? Landsc. Ecol. 18 (3), 303—314.

Brooks, N., 2003. Vulnerability, risk and adaptation: a conceptual framework. Tyndall Centre for Climate Change Research Working, Paper 38, 16p.

Brundtland, G.H., 1987. Our Common Future. Report of the World Commission on Environment and Development. United Nations, 318p.

Charney, J., Stone, P.H., Quirk, W.J., 1975. Drought in the Sahara: a biogeophysical feedback mechanism. Science 187, 434—435.

Chevallier, J.J., Caron, C., 2002. Développement d'infrastructures géomatiques: déterminisme technologique ou approche holistique. In: Symposium on Geospatial Theory, Processing and Application, Ottawa.

Chevallier, J.J., 1993. Systèmes d'aide à la décision à référence spatiale (SADRS): méthode de conception et de développement. Actes de congres GIS/SIG'93, Ottawa, 23—25 mars 1993.

Chevallier, J.J., 1994. Système d'aide à la décision à référence spatiale. MoSIT 2, 11—15.

Ciais, Ph, Reichstein, M., Viovy, N., Granier, A., Ogeé, J., Allard, V., et al., 2005. Europe-wide reduction in primary productivity caused by the heat and drought in 2003. Nature 437 (7058), 529—533.

Couvet, D., Teyssèdre-Couvet, A., 2010. Écologie et biodiversité: Des populations aux socioécosystemes. Editions Belin, France, 288p.

Crichton, D., 1999. The risk triangle. In: Ingleton, J. (Ed.), Natural Disaster Management. Tudor Rose, London, pp. 102—103.

D'Ercole, R., Metzger, P., 2009. La vulnérabilité territoriale: une nouvelle approche des risques en milieu urbain. Cybergeo Eur. J. Geogr. Dossiers Vulnérabilités urbaines du sud, document 447. <http://www.cybergeo.eu/index22022.html>.

Dauphiné, A., Provitolo, D., 2013. Risques et Catastrophes—Observer, Spatialiser, Comprendre, Gérer. In: Colin, A. (Ed.) (2ème édition), Collection U, 412p.

Densham, P.J., 1991. Spatial decision support systems. In: Maguire, J., Goodchild, M.S., Rhind, D.W. (Eds.), Geographical Information Systems: Principles and Applications. Longman Publishing Group, pp. 403—412.

Dovers, S.R., Handmer, J.W., 1992. Uncertainty, sustainability and change. Glob. Environ. Change 2 (4), 262—276.

Dubos-Paillard, E., Guermond, Y., Langlois, P., 2003. Analyse de l'évolution urbaine par automate cellulaire: le modèle SpaCelle. L'Espace Géogr. 32 (4), 357—378.

Emsellem, K., Liziard, S., Scarella, F., 2012. La géoprospective: l'émergence d'un nouveau champ de recherche? L'Espace Géogr. 41, 154—168.

European Commission, 2007. Adapting to climate change in Europe—options for EU action. COM 354 final, 32p.

European Commission, 2009. Adapting to climate change: towards a European framework for action, COM 147 final, 17p.

European Commission, 2013. An EU strategy on adaptation to climate change. 11p.

Fahrig, L., Baudry, J., Brotons, L., Burel, F.G., Crist, T.O., Fuller, R.J., et al., 2011. Functional landscape heterogeneity and animal biodiversity in agricultural landscapes. Ecol. Lett. 14 (2), 101—112.

Fink, A.H., Brücher, T., Krüger, A., Leckebusch, G.C., Pinto, J.G., Ulbrich, U., 2004. The 2003 European summer heatwaves and drought? Synoptic diagnosis and impacts. Weather 59 (8), 209—216.

Fischer, E.M., 2007. The Role of Land—Atmosphere Interactions for European Summer Heat Waves: Past, Present and Future. ETH Zurich, 167p.

Forrester, J.W., 1961. Industrial Dynamics. MIT Press, p. 464.

Fusco, G., 2012. Démarche géo-prospective et modélisation causale probabiliste, *Cybergeo : Eur. J. Geogr.* [En ligne], Systèmes, Modélisation, Géostatistiques, document 613, mis en ligne le 20 juillet 2012. URL: http://journals.openedition.org/cybergeo/25423; https://doi.org/10.4000/cybergeo.25423

Füssel, H.M., 2007. Vulnerability: a generally applicable conceptual framework for climate change research. Glob. Environ. Change 17, 155−167.

Garbolino, E., 2014. Les Bio-indicateurs du Climat: Principes et Caractérisation. Presses des MINES, Développement durable, 129p.

Garbolino, E., De Ruffray, P., Brisse, H., Grandjouan, G., 2007. Relationships between plants and climate in France: calibration of 1874 bio-indicators. CR Biol. 330, 159−170.

Garbolino, E., Chery, J.P., Guarnieri, F., 2009. Dynamic systems modelling to improve risk analysis in the context of SEVESO industries. Chem. Eng. Trans. 17, 373−378.

Garbolino, E., Chery, J.P., Guarnieri, F., 2010. Modélisation dynamique des systèmes industriels à risques. In: Collection "Sciences du Risque et du Danger", Notes de synthèse et de recherche. Lavoisier, 112p.

Garbolino, E., De Ruffray, P., Brisse, H., Grandjouan, G., 2012. The phytosociological database SOPHY as the basis of plant socio-ecology and phytoclimatology in France. Biodivers. Ecol. 4, 177−184.

Gibson, C.C., Ostrom, E., Anh, T.K., 2000. The concept of scale and the human dimensions of global change: a survey. Ecol. Econ. 32, 217−239.

Gourmelon, F., Houet, T., Voiron-Canicio, C., Joliveau, T., 2012. La géoprospective, apport des approches spatiales à la prospective. L'Espace Géogr. 41, 97−98.

Guinée, J.B., Heijungs, R., Huppes, G., Zamagni, A., Masoni, P., Buonamici, R., et al., 2011. Life cycle assessment: past, present, and future. Environ. Sci. Technol. 45 (1), 90−96.

Haeckel E., 1866. Generelle Morphologie der Organismen. Allgemeine Morphologie der organischen Formen-Wissenschaft, mechanisch begriindet durch die von Charles Darwin reformirte Descendenz-Theorie. 2 vols. G. Reimer, Berlin, 462p.

Head, B.W., 2016. Toward more "evidence-informed" policy making? Public. Adm. Rev. 76 (3), 472−484.

Hinojos Mendoza, G., 2014. Identification des risques de perte de biodiversité face aux pressions anthropiques et au changement climatique à l'horizon 2100: Application de la conservation dynamique au territoire des Alpes-Maritimes. Thèse de l'Ecole Nationale Supérieure des Mines de Paris, Spécialité « Science et Génie des Activités à Risques », 321p.

Holling, C.S., 1973. Resilience and stability of ecological systems. Annu. Rev. Ecol. Syst. 4, 1−23.

Houet, T., Gourmelon, F., et Françoise Gourmelon, La géoprospective − Apport de la dimension spatiale aux démarches prospectives, Cybergeo : Euro. J. Geo. [En ligne], Systèmes, Modélisation, Géostatistiques, document 667, mis en ligne le 08 février 2014. URL: http://journals.openedition.org/cybergeo/26194; https://doi.org/10.4000/cybergeo.26194

Houghton, R.A., Boone, R.D., Malillo, J.M., Palm, C.A., Woodwell, G.M., Myers, N., et al., 1985. Net flux of CO_2 from tropical forests in 1980. Nature 316, 617−620.

Huntley, B., 1995. How vegetation responds to climate change: evidence from palaeovegetation studies. In: Pernetta, J., Leemans, R., Elder, D., Humphrey, S. (Eds.), Impact of Climate Change on Ecosystem and Species: Environmental Context. UICN Publication, p. 98.

IPCC, 2012. Managing the Risks of Extreme Events and Disasters to Advance Climate Change Adaptation. Cambridge University Press, 1131p.

IPCC, 2014. Climate Change 2014: Impacts, Adaptation, and Vulnerability. Part A: Global and Sectoral Aspects. Cambridge University Press, 693p.

Jankowski, P., Nyerges, T.L., Smith, A., Moore, T.J., Horvath, E., 1997. Spatial group choice: a SDSS tool for collaborative spatial decision-making. Int. J. Geogr. Inf. Sci. 11 (6), 577−602.

Jankowski, P., Nyerges, T., Robischon, S., Ramsey, K., Tuthill, D., 2006. Design considerations and evaluation of a collaborative, spatio-temporal decision support system. Trans. GIS 10 (3), 335−354.

Jelokhani-Niaraki, M., Malczewski, J., A Web 3.0-driven Collaborative Multicriteria Spatial Decision Support System », Cybergeo : Eur. J. Geo. [En ligne], Systèmes, Modélisation, Géostatistiques, document 620, mis en ligne le 15 septembre 2012. URL: http://journals.openedition.org/cybergeo/25514; https://doi.org/10.4000/cybergeo.25514

Jordan, R., Gray, S., Sorensen, A., Newman, G., Mellor, D., Newman, G., et al., 2016. Studying citizen science through adaptive management and learning feedbacks as mechanisms for improving conservation. Conserv. Biol. 30 (3), 487–495.

Kolbert, E., 2014. The Sixth Extinction: An Unnatural History. Henry Holt and Co., New York, NY, p. 336.

Langlois, P., 2010. Simulation des Systèmes Complexes en Géographie: Fondements Théoriques et Applications. Lavoisier, Paris, 332p.

Leakey, R., Lewin, R., 1996. The Sixth Extinction: Patterns of Life and the Future of Humankind. Anchor Books, New York, NY, p. 288.

Levasseur, A., Lesage, P., Margni, M., Deschenes, L., Samson, R., 2010. Considering time in LCA: dynamic LCA and its application to global warming impact assessments. Environ. Sci. Technol. 44 (8), 3169–3174.

Leveson, N., 2004. A new accident model for engineering safer systems. Saf. Sci. 42 (4), 237–270.

Meadows, D.H., Meadows, D.L., Randers, J., Behrens, W.W., 1972. The Limits to Growth. A Report for the Club of Romes Project on the Predicament of Mankind. Universe Books, New York, NY, p. 205.

MEDDTL, 2011. Plan national d'adaptation au changement climatique. Ministère de l'Écologie, du Développement durable, des Transports et du Logement, 187p.

Mitchell Jr., J.M., 1961. Recent secular changes of global temperature. Ann. N.Y. Acad. Sci. 95 (1), 235–250.

Newman, J., Head, B.W., 2017. Wicked tendencies in policy problems: rethinking the distinction between social and technical problems. Policy Soc. 36 (3), 414–429.

Niua, S., Luob, Y., Li, D., Caod, S., Xiab, J., Li, J., et al., 2014. Plant growth and mortality under climatic extremes: an overview. Environ. Exp. Bot. 8, 13–19.

Normander, B., Levin, G., Auvinen, A.P., Bratli, H., Stabbetorp, O., Hedblom, M., et al., 2012. Indicator framework for measuring quantity and quality of biodiversity—exemplified in the Nordic countries. Ecol. Indic. 13, 104–116.

November, V., 2006. Le risque comme objet géographique. Cah.Géogr. Québec 50 (141), 289–296.

O'Neill, R.V., de Angelis, D.L., Waide, J.B., Allen, T.F.H., 1986. A Hierarchical Concept of Ecosystems. Princeton University Press, Princeton, NJ.

ONERC, 2007. Stratégie nationale d'adaptation au changement climatique. La documentation Française, 95p.

Otterman, J., 1974. Baring high albedo soil by overgrazing: a hypothesized desertification mechanism. Sciences 186, 531–533.

Parker, D.C., Berger, T., Manson, S.M., 2002. Agent-Based Models of Land-Use/Land-Cover Change. LUCC Report Series No. 6, 124p.

Parker, D.C., Manson, S.M., Janssen, M.A., Hoffman, M., Deadman, P., 2001. Multi-agent systems for the simulation of land-use and land-cover change: a review. CIPEC working paper CW-01-05; Forthcoming. Ann. Assoc. Am. Geogr. 93 (2), 35.

Plass, G.N., 1956. Effect of carbon dioxide variations on climate. Am. J. Phys. 24, 376–387.

Primack, R.B., Sarrazin, F., Lecompte, J., 2012. Biologie de la Conservation. Dunod, Paris, p. 384.

Provitolo, D., Dubos-Paillard, E., Müller, J.P., 2014. Une ontologie conceptuelle du domaine des risques et catastrophes. In: Phan D (Ed.), Ontologies et modélisation par SMA en SHS,. Lavoisier, Hermes, 558p.

Rebotier, J., 2007. What role for institutions in resilience?: an interpretation through the case of Caracas. In: Conference Building Resilience of territories IRD-UCV. Valparaiso, Chili, (in French).

Reichstein, M., 2005. Severe impact of the 2003 European heat wave on ecosystems. Potsdam Institute for Climate Impact Research. <https://www.pik-potsdam.de/news/press-releases/archive/2005/severe-impact-of-the-2003-european-heat-wave-on-ecosystems>.

Risser, P.G., Karr, J.R., Forman, R.T.T., 1983. Landscape Ecology Directions and Approaches. The Illinois Natural History Survey, Champaign, IL, 16p.

Rufat, S., Djament-Tran, G., Le Blanc et Serge Lhomme, A., 2012. What Resilience Is Not: Uses and Abuses, *Cybergeo : Eur. J. Geogr.* [En ligne], Environnement, Nature, Paysage, document 621, mis en ligne le 18 octobre 2012. URL: http://journals.openedition.org/cybergeo/25554; https://doi.org/10.4000/cybergeo.25554

Salzmann, U., Haywood, A.M., Lunt, D.J., 2009. The past is a guide to the future? Comparing Middle Pliocene vegetation with predicted biome distributions for the twenty-first century. Philos. Trans. R. Soc. 367, 189–204.

Shapley, H., 1955. Climatic change. evidence, causes and effects. Sci. Soc. 19 (1), 88–89.

Sikder, I., Gangopadhyay, A., 2002. Design and implementation of a web-based collaborative spatial decision support system: organizational and managerial implications. Inf. Resour. Manag. J. 15 (4), 33–47.

Simon, H., 1977. The New Science of Management Decision. Prentice Hall, Upper Saddle River, NJ, p. 175.

Suarez Seoane, S., Baudry, J., 2002. Scale dependence of spatial patterns and cartography on the detection of landscape change. Relationships with species' perception. Ecography 25, 499–511.

Sugumaran, R., DeGroote, J., 2011. Spatial Decision Support Systems: Principles and Practices. CRC Press, Taylor & Francis Group, p. 469.

UN/ISDR, 2004. Living with Risk: A Global Review of Disaster Reduction Initiatives, II. United Nations Publication, Annexes, 127p.

Veldkamp, A., Verburg, P.H., Kok, K., de Koning, G.H.J., Priess, J., Bergsma, A.R., 2001. The need for scale sensitive approaches in spatially explicit land use change modeling. Environ. Model. Assess. 6, 111–121.

Verburg, P.H., Schot, P.P., Dijst, M.J., Veldkamp, A., 2004. Land use change modelling: current practice and research priorities. GeoJournal 61, 309–324.

Villard, M.A., Metzger, J.P., 2014. Beyond the fragmentation debate: a conceptual model to predict when habitat configuration really matters. J. Appl. Ecol. 51 (2), 309–318.

Voiron, C., 2006. L'espace dans la modélisation des interactions nature-société. Actes du colloque international « interactions Nature-Société, analyse et modèles ». UMR 6554 LETG La Baule, 6p.

Voiron-Canicio, C., Olivier, F., 2005. Anticiper à l'aide d'un SIG, les conséquences de la paralysie urbaine en temps de catastrophe: application à la ville de Nice. In: Systèmes d'Information géographique et gestion des risques. ISTED, Paris, pp. 55–58.

Voiron-Canicio, C., 2012. L'anticipation du changement en prospective et des changements spatiaux en géoprospective. L'Espace Géogr. 41, 99–110.

White, R., Engelen, G., 1997. Cellular automata as the basis of integrated dynamic regional modelling. Environ. Plan. B 24, 235–246.

White, R., Engelen, G., 2000. High-resolution integrated modelling of the spatial dynamics of urban and regional systems. Comput. Environ. Urban. Syst. 24, 383–400.

White, R., Engelen, G., Uijee, I., 1997. The use of constrained cellular automata for high-resolution modelling of urban land-use dynamics. Environ. Plan. B 24, 323–343.

Willett, H.C., 1950. Temperature trends of the past century. In: Centenary Proceedings. Royal Meteorological Society, London, England, pp. 195–206.

Woodwell, G.M., Hobbie, J.E., Houghton, R.A., Melillo, J.M., Moore, B., Peterson, B.J., et al., 1983. Global deforestation: contribution to atmospheric carbon dioxide. Science 222, 1081–1086.

CHAPTER 3

Knowledge challenges of the geoprospective approach applied to territorial resilience

Christine Voiron-Canicio[1] and Giovanni Fusco[2]

[1]University Côte d'Azur, Laboratory ESPACE, Nice, France, [2]Université Côte d'Azur, CNRS, UMR ESPACE, Nice, France

Chapter Outline

3.1 Introduction

The future evolution of territorial systems, whether global or local, is a key issue for human societies, which these address with a variety of aims and from a variety of angles. This results in approaches that are specific to the chosen viewpoint. As we have seen, the

DOI: https://doi.org/10.1016/B978-0-12-818215-4.00003-1
© 2021 Elsevier Inc. All rights reserved.

specificity of spatial prospective is to handle the issue while putting emphasis on the spatial point of view. The primary questioning underlying the geoprospective approach concerns the way in which territories will behave when faced with the disruptions of a multiple nature that will occur in a not too distant future. How will they impact the territory, and in return, what will the latter's possible responses be depending on the changes?

In practice, this questioning breaks down into a multiplicity of scientific questions, and also practical ones, notably operational. Looking for answers to these questions is then done according to the methods determined by the goal to reach, but also by the theoretical and epistemological foundations on which the research is based (Mermet and Poux, 2002). Among these, the toolkits of system dynamics and of complexity theories, and the sustainable development paradigm, guide the reasoning process, and *de facto*, the lines of research, toward notions such as upholding on the long term, a system's adaptive capacity to changes in its environment to ensure its survival, and the forms of regulation used for achieving this.

These notions dovetail with resilience. Applied to territories, the latter can be understood in three possible manners: as a state, as a process, and as a property. The most frequent approach consists in detecting whether all or part of a territory has managed to hold on after a shock, and after diagnosing the state of resilience, in attempting to report on the process that generated it. The posture is *a posteriori*. Less widespread is the viewpoint considering that resilience existed prior to the shock and was revealed by the latter. Resilience is then considered as an intrinsic quality of the system, and research focuses on the resilience potential preexisting the shock. Here the posture is ex ante. Therefore the relation to time is different, and this has an influence on the way to conduct a geoprospective approach. In the first case, the reactions observed in the past as a response to a type of shock will be used to infer the territory's future behavior, whereas in the second case, the future behavior will be assessed in accordance with the territory's adaptive potential. This second option is the most in line with the geoprospective principles and is the one which we will mainly develop.

The focus will be on the systemic interpretation framework and the questionings it infers on territorial resilience. How to move from the theoretical frame of systemic resilience to the operational prospective framework? How does geoprospective contribute to it? The potentialities of geoprospective will be examined on two theme registers: anticipating territorial resilience to risk and the contribution of geoprospective to the knowledge of urban resilience.

3.2 Territorial resilience through the systemic prism

3.2.1 Systemic resilience and complexity

The notion of resilience in system dynamics is different from the traditional sense issued from physics: the time of return to equilibrium for an object after a shock. The conceptual base is the systemic resilience property. Indeed, the first quality of an open system is its adaptability to changes in its environment to ensure its survival when affected by a disruption. Thus ecologic resilience, as defined in Holling's seminal work (1973), is the ability of a system to absorb a disruption and to integrate it into its functioning while making changes but without altering its identity, its structure, and its essential functions. This reorganization, which is specific to living systems, is a system of permanent reorganization (Atlan, 1979).

From complex systems theory (Morin, 1974, 1990, 1994; Haken, 1977; Prigogine and Stengers, 1979), we will retain four fundamental notions which are as many keys to enter territorial systems and understand their behavior: complexity, adaptiveness, self-organization, and creativity.

The complexity and adaptiveness of systems are interrelated. The more complex behaviors are, the more they will show flexibility to adapt to the environment. They will be able to modify themselves in response to external changes—exceptional events, tensions—, and will also be able to alter the immediate environment, to adjust it, adapt it to their needs (Morin, 1974). Systemic complexity increases, on the one hand, with the growing number and increased diversity of elements, and, on the other hand, with the less and less determinist character of interrelations (Levin, 2002). Self-organization, which characterizes complex systems, makes them evolve in an unpredictable way toward various forms of organization. Their dynamics combine phases of stability and moments of instability during which changes, of course, can occur. These dynamics resulting either from an external disruption or from the amplification of small internal fluctuations play a role at the micro level.

Self-organization carries creativity within itself. It notably generates new forms and macrostructures which are not directly predictable from the knowledge of the states of microscopic elements (Pumain, 2003). This morphogenetic ability is not limited to genetic mutation, but also influences the system's behavior. The development of heuristic competences makes the system able to envisage several possible evolution strategies, to create conditions of choice. Thus the morphogenetic emergence is coupled with the emergence of degrees of freedom.

Furthermore, systems work at different levels of space, time, and social organization. Interscale interactions determine the system's dynamics whatever the scale considered. "The adaptive renewal cycles nested across scales have been termed panarchy" (Gunderson and Holling, 2002). Strong interscale relations are the very sign of the self-organization of complex systems and differentiate them from most systems, whether simple or complicated, which are only organized by their designers' external action (generally machines and technologies).

3.2.2 What lessons for the resilience of territories?

A territory seen through the systemic prism is an open sociospatial system with a high degree of complexity. Its complexity is due at the same time to its multicomposite structure—socioeconomic, environmental, spatial, and institutional components—, to the high number of these components, to overlapping scales and organization levels, as well as to interrelations of various natures with the environment, notably through information and via the remote influence of exogenous variables.

Territories are sociospatial systems and can be distinguished from ecosystems by two specificities. On the one hand, change is driven by invention-innovation, which does not come from the natural environment but is consubstantial to sociotechnical systems (Holling and Sanderson, 1996). Permanent invention-innovation enables territorial systems to develop new organizational arrangements much faster than ecosystems do, that is, over historical times. On the other hand, the momentum of change is impelled by cognition. Representations, information, memory, and learning have an influence on decision-making, with more or less awareness, and introduce anticipation and strategy in the way systems operate. The perception that stakeholders have of their territory affects its evolution. The self-fulfilling prophecies which partly guide social evolution are an extreme case of the foregoing (Pumain, 2003).

A territorial system, like an ecologic system, evolves according to a cyclic transformation process formalized by Holling as the "adaptive renewal cycles"—growth, collapse, reorganization, renewal, and reestablishment. The change of a territorial system will be dependent on the way it reacts to disruptions of various origins and natures. These come either from the external environment—natural hazard, technological disruption, political or economic crisis—or from internal phenomena generated by the way the territorial system works—population aging, specialization in an activity, land artificialization. These slow or sudden disruptions continuously affecting the system are the reason for its variability. Either the system adapts by integrating the disruption into its functioning by means of adjustments or it resists and the existing structure may be reinforced in order to counter the disruption's negative effects. Adaptation reflects a degree of flexibility in the system's functioning whereas resistance strategies rather show some rigidification.

The spatial systems, in constant evolution, are very seldom in equilibrium, and for a short period. This has two fundamental implications. On the one hand, a system characterized by high variability will have more chances than a system with high stability to put up with a disruption without collapsing, because it will be more flexible and more able to incorporate it into its functioning (Aschan-Leygonie, 2000). On the other hand, this variability induces the possibility of, and therefore, various trajectories which are possible responses to a disruption as well as organizational solutions such as the appearance or disappearance of territorial subsystems.

The theoretical framework of adaptive systems and also the research applied to sociospatial systems bring out two categories of resilience criteria, one linked to the territorial system—spatiotemporal variability of its characteristics, level of robustness and adaptivity—, the other linked to the disruption—nature, extent, sudden appearance vs gradual. Nevertheless, the same research works warn against generalizing these criteria and against a deterministic vision of their effect. Geographic, historical, and organizational territorial contexts in which these criteria occur play a major role and explain the different impacts on resilience that can be observed from a territory to another. Moreover, the ways in which the system reacts to a disruption, the bifurcations and emerging structures which they generate make the evolution of a territorial system unforeseeable, and *de facto*, unpredictable. From then on, how to anticipate its future? What should be the targets of a prospective analysis?

The systemic reasoning leads to new questionings on how territories function and on their spatiotemporal trajectory, which guides the prospective approach toward research questions where resilience and future evolution of territorial systems are closely interrelated.

3.3 New questionings

3.3.1 Going deeper into the relation between territorial sustainability and resilience

The concept of territorial resilience is related to the paradigm of sustainable development, of which it is one of the dimensions. Despite recommendations issued at world summits in the years 1990s, the beginning of the years 2000s has undermined the fundamental principles of transversality and adaptation intrinsic to sustainable development. Of the three parts of sustainable development, the environment tends nowadays to prevail in sectorial public policies, so much so that sustainable development is often confused with the environment (Voiron-Canicio, 2015). Another confusion adds to it, this time with the economy of resources. The environment is often reduced to the minimalist vision of a municipality's environmental performance (Rumpala, 2003). However, a return to the original conception of sustainable development took place recently. As an example, at the Conference on Housing and Sustainable Urban Development (UN Habitat III) held in Quito in 2016, the new urban agenda clearly interrelated resilience and sustainability, by defining the ideal sustainable city as more inclusive, safer, more resilient, and fairer. To this resilient structure is associated concerted development as the harmonization variable of sustainable development (Da Cunha and Thomas, 2017).

Furthermore, at the beginning of the years 2000s, research on resilience gave new impetus to sustainability science, an emerging field of research "that seeks to understand the fundamental character of interactions between nature and society and to encourage those interactions along more sustainable trajectories" (Kates et al., 2001; Clark and Dickson, 2003; Independent Group of Scientists, 2019). In this new context, the fight against climate change is seen as an opportunity to implement sustainable development, the aim of which is not only to slow down climate changes but also to allow the survival of ecosystems and human societies (Magnan, 2009). Improving the resilience of a territory appears as a means of increasing the chances to come close to the objectives of sustainable development (Folke et al., 2002; Da Cunha and Thomas, 2017; Scherrer, 2017). "Sustainability is viewed as a process, rather than an end-product, a dynamic process that requires adaptive capacity in resilient social−ecological systems to deal with change" (Berkes et al., 2003). From then on, the hybridization process that takes place between resilience and sustainable development has to be analyzed. As far as public action is concerned, does resilience help to operationalize the utopian vision of sustainable development? Is it the path to follow to move toward sustainable territorial development, and how?

3.3.2 Reexamining the link between resilience, adaptation, and adaptivity

The question of a territory's resilience has first been studied in the context of natural or technological risk, and mainly from the technical-functional viewpoint—performance of protective structures, reducing the vulnerability of technical networks, constructing new resilient buildings. The current research questions relate to the evaluation of stakeholders' ability to imagine and anticipate disruptions likely to affect a territory. The challenge is twofold; on the one hand, examining a territory's learning capacity based on feedbacks on past events and more globally on collective memory, and on the other hand, working out evaluation methods of adaptive capacity.

Faced with a disruption, the response of the territorial system's stakeholders can be to resist, give up, or adapt, and the adaptation of a territorial system to an exceptional event generally combines all three behaviors, which occur either simultaneously or successively. A territorial system's resilience at a given moment results from that complex combinatory. However, interrelations between adaptation and resilience need to be defined better. In fact, the two notions are often confused. What differences can be identified between the mechanisms of resilience and those of adaptation? If adaptation and resilience are complementary, how to express it in strategies and actions? More globally, what does adapting mean for a territory? What to adapt to, when and where? (Magnan, 2010).

Adaptation is usually understood according to three perspectives, as a process, a state, or a strategy, and most of the time, separately. Nowadays, research on their articulation is needed to define trajectories of adaptation so as to help work out resilience strategies (Fig. 3.1).

Furthermore, it is indispensable to acquire further knowledge about the determinants of a territory's adaptive capacity, and about adaptation modes—reactive vs anticipative—in order to better take into account the complexity of the adaptive process, and thus avoid to

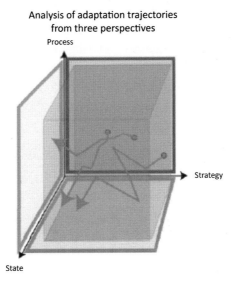

Figure 3.1

Understanding adaptation trajectories by interconnecting three perspectives: state, process, and strategy.

reduce the adaptive capacity to a strictly economic and technological vision, or to link, in a determinist manner, level of development and capacity to adapt to climate change (Smit et al., 1999; Smit and Wandel, 2006; Vincent, 2007; Adger et al., 2009). Likewise, the factors affecting the adaptive capacity do not act individually, but interact by generating synergies or, on the contrary, negative retroactions. Alexandre Magnan suggests handling these various issues by coupling them in a research framework aimed at assessing a territory's adaptive capacity to climate change (Magnan, 2009). He identifies four main lines of research to deepen the understanding of adaptive capacity: factors influencing adaptive capacity, the spatial and temporal scales that are relevant for adaptation, the relationship between vulnerability and adaptive capacity, and the link between adaptation and sustainable development. More precisely, he invites researchers to identify the thresholds beyond which the positive effects of a given adaptive factor—territorial cohesion, network structuring, economic diversification, etc.—fade out, and can even be reversed.

Today there is some consensus on the structural, organizational, sociocultural, and technical conditions that contribute to reinforcing a territory's resilience, and enable it to face an exceptional situation, whatever its origin (UNISDR, 2012; Laganier, 2013; Giacometti and Teräs, 2019). Territories are characterized by their different capacity to bring together these favorable conditions in order to improve resilience: capabilities of resistance, of collective response, of self-organization, of learning, and of adapting. Among these capacities, the

OECD singles out three major ones: the absorptive capacity—the ability to resist the negative impact of shocks; the adaptive capacity—the ability to adapt to new conditions; and the transformative capacity—the ability to change fundamental structures (OECD, 2014). Ongoing research endeavors to detect these capacities in various geographic contexts. However, the focus is mainly on political, socioeconomic, and cultural factors. But the spatial dimension of resilience is still relatively unexplored.

3.3.3 Going deeper into the role of the spatial dimension in resilience

3.3.3.1 Space and resilience of systems

Space is essential in any resilient system. It is not just a matter of identifying the portion of space occupied by various elements or of understanding how a more or less long distance can have an effect on certain phenomena, but of seeing how a system's spatial structure, that is, the more or less stable organization of its elements and functioning in space, offers many possibilities or constraints to the system's possible adaptations and transformations faced with a changing environment.

The theory of the general resilience of ecosystems proposed by Holling finds an essential complement in the concept of spatial resilience. For Cumming (2011) it underlines the role of the spatial dimension in ecosystems. Spatial resilience concerns the way in which the spatial variation of key variables, both inside and outside the system, influences (and is influenced by) the system's overall resilience through the multiple spatiotemporal scales. Trans-scale properties, just like the factors of form of ecosystems (connections by corridors, extension/density/fragmentation of habitats, etc.), come back at the center of the mechanisms through which an ecosystem is capable of showing resilience, beyond the sole diversity of the species that are present.

It is clear that analyzing the resilience of a spatial system is closely linked to the specificities of the system being studied. Beyond a few principle analogies, the spatiality of an ecosystem is not the same as that of a city, a wine-growing region, or a transnational industrial system. The theory of complex systems can thus provide a common framework for reflection, but it would not be easily operational to evaluate the resilience of specific spatial systems. In the last section of this chapter, we will expound a spatial resilience theory specific to evaluating the resilience of the city's forms and how this theory could contribute to informing an urban geoprospective. A similar effort will be required in every area of the territorial analysis, to identify every time what are the elements of the system's spatiality which are most important in terms of resilience and how recognizing more or less resilient spatial structures in the present system and in the different scenarios can help to choose a shared benchmark scenario. Nonetheless, in what follows we are going to present some key principles of the role of spatial organization in systems change, regardless of the field of study.

3.3.3.2 Interactions of spatial structures and change

The role of spatial interactions on the modes of change and, consequently, on the resilience of territorial systems should be identified better. Interactions are of two kinds: one horizontal, between places and elements located at variable distances; the other vertical, concerning the links between society, space, and the environment. The two interactions combine in the way the system confronts a disruption. Until now, research focused mainly on vertical interactions between the human society and ecosystems (Aschan-Leygonie, 2000). What is still needed is a deeper understanding of interactions between spatial structure and change and, by extension, of resilience.

In this perspective, stimulating avenues of thought are provided by the works of environmental scientists and those of physicists. Environmental scientists working on spatial patterns have since long observed the role of the components' spatial configuration in the resilience processes of species (Scheffer and Carpenter, 2003; Scheffer et al., 2012; Rietkerk et al., 2004). For their part, researches carried out by physicists on the robustness of networks have highlighted the importance of connexity in the vulnerability of scale-free networks and in the transmission of domino effects (Barabasi and Reka, 1999; Barabasi, 2002). Studies of the behavior of complex systems converge on the crucial role played by two structural characteristics, heterogeneousness and connectivity, in the response of a system to a change in its environment. For example, it has been observed that components' heterogeneousness combined with a low level of connectivity tends to produce a gradual change. Conversely, spatial homogeneousness and a strongly connected network tend to foster resistance to change up to a certain threshold beyond which all components swing synchronously to another state. However, in the face of the change which continuously impacts the spatial structure of territorial systems, interactions between the structure of the built-up area, for example, and its function—its uses—take various forms. The spatial structure can remain unchanged whereas the function has changed, or the spatial structure can be modified even though the use by stakeholders has not changed.

Determining the role of spatial structures and dynamics in a territory's resilience process is a research goal which fully concerns geoprospective. Because of the diversity of its methods, the latter enables researchers to question in many different ways the relationship between spatial change and the behavior of territories vis-à-vis the pressures exerted on them.

3.4 Spatial change and resilience seen through the prism of geoprospective

One of the entries into the issue of resilience consists in examining the mechanisms of spatial change from various angles: the territory's propensity to change, anticipating future changes, and its adaptive potential.

To do so, two main avenues open to researchers: to analyze the territorial system's trajectory and to anticipate its response to future tensions or crises, focusing on the system, not the risk. These approaches come under spatial prospective.

Researching the spatiotemporal trajectory and adaptive potential enables researchers to consider the evolutive character of adaptation. Indeed, the global and local context is continuously changing, the territory's adaptive capacity evolves as a result of previous choices—political decisions, action undertaken vs inaction. Therefore the adaptive capacity at a moment T is no guarantor of the capacity at the moment T + 1. Then, it must be constantly reconsidered in the light of simultaneous changes of the territorial system and its environment, which makes the task much more complex. So, rather than researching the real ability to adapt, it seems more relevant to assess this evolutionary process through the spatiotemporal trajectories of response to the forces of change and to adaptation to change in relation to a dynamic vision of the territorial functioning (Magnan, 2009; Voiron-Canicio, 2012).

3.4.1 Spatiotemporal trajectory and propensity to change

Although spatial change is not predictable, the territory's propensity to change can be examined in the light of the past trajectory. The aim is not to extrapolate the trend observed in the past but to identify the process of spatial change, to characterize the behavior of the spatial components under the pressure forces—demography, land speculation, climate constraint, etc.—exerted over time. The important here is not the nature of responses, but the form of reactivity of each spatial component in the overall spatiotemporal trajectory. The dynamics of change over a period is assessed qualitatively, using the spatial change index which synthesizes the behavior of the various components in the timeline: indices of similarity, turnover, diversity, intensity, quality, and speed. For example, the turnover index records how many changes occurred between adjacent pairs of years (Swetnam, 2007; Casanova, 2010). The spatiotemporal trajectory can be characterized even more accurately, as shown in the research carried out by Sophie Liziard (2013) on the prospective of the Latin Arc's littoralization. This author has worked out indicators that inform on the preferred direction that population movements could take in the future. Other indicators inform on the instability of the past trajectory, on changes of direction or temporary standstills, the stage of the densification process in each subspace—early, intermediate, advanced—and how fast the population grows. All the information provided by these indicators can be combined in order to identify stake-laden spaces: saturated spaces, those near saturation, and potential spaces for densification transfer (Liziard, 2013).

These spatial prospective analyses provide a knowledge on the spatial process of change. The qualitative information drawn from the past trajectory in no way prejudges the future trajectory but makes it possible to assess the spatial dimension of the territory's responsiveness. It helps to identify the potentials of spatial change, not for predictive purposes but as an aid for working out scenarios for the future.

3.4.2 Anticipating the early-warning signs of change

A disruption external to the system—rising sea levels, for example—or an endogenous phenomenon such as the deterioration of electrical infrastructures, creeps in progressively, slowly and continuously, until a disaster or a crisis occurs suddenly—submergence of a coastal area, power blackout—, which makes people aware of the connection between the exceptional event and its root causes.

Bearing in mind that the dynamics of change operate both with gradual signs and more sudden ones, it is useful to look for the early signs of sudden changes. This new field of research aims to determine the generic early-warning signals of a sudden change, and even of a critical point affecting ecosystems, also concerns sociosystems. The ongoing research aims notably at discovering the pertinent indicators of an approaching sudden change (Scheffer et al., 2009). A number of signs deserve attention: changes in spatial configuration, a recovery rate after an exceptional event tending to decrease with the passing time, a phenomenon described as "critical slowing-down," an increase in the system's variability, a sign that the system explores a wider variety of states, the temporal monitoring of a variable characteristic of the ecosystem—biomass, vegetation cover, etc.— that shows an increase in variance, a strong autocorrelation, and temporal asymmetry are signs considered as temporal indicators of an approaching sudden change, and even of a critical point. Recently, studies have shown that the equivalents of these indicators—spatial variance, spatial autocorrelation, spatial asymmetry-, —are characteristic of a spatial system evolving toward sudden change (Dakos et al., 2009).

3.4.3 Anticipating the occurrence of tensions through anticipative monitoring

Another challenge consists in anticipating risk-bearing future tensions. It is no longer a matter of analyzing the past trajectory, but of detecting, in the present, weak signals of the future, via anticipative monitoring. The concept of anticipative monitoring implies to project oneself into a future which is not only probable in the light of the recent past, but equally plausible, or even merely possible. There is a dual objective: detecting situations likely to create tensions in the functioning of all or part of the territory, which could lead to new vulnerabilities, and assess the overall resilience capacity of the territorial system. This second objective is not easy to reach. However, exploring the future can be envisaged. The research posture requires to look for antagonistic forces in preparation or likely to occur and create tensions and for new risks. For example, the emergence of a technology elsewhere is a risk for a region of traditional industry, but is also an opportunity if the territorial system is reactive, and capable of transforming the shock into a creative disturbance by provoking knowledge spillover and diversification. The posture consists in broadening the field of vision in order to perceive the global–local interactions and, more precisely in

- Paying attention to concomitant phenomena yet that play on different temporalities, to delays before the effects of public policies and local developments are felt. For example, a local policy in favor of carbon-free mobility—incentives to buy electric cars, opening of new electric tram lines, etc.—requires that the regional power supply be secured beforehand, either by completing power infrastructures guaranteeing a stable electricity supply or by rolling out charging stations in smart-charging mode enabling the electric vehicle's battery to give energy back to the grid at peak demand.
- Detecting weak signals. A weak signal is a furtive information, seemingly innocuous, fragmentary, with no apparent utility, equivocal, uncertain, etc. (Lesca, and Lesca, 2011). Picking it up necessitates to be on the alert but without a definite objective or theme, and receptive to information coming from various sources. Once it has been picked up, the weak signal must quickly be put together with other pieces of information for cross-checking. The interpretation is carried out collectively by comparing various opinions. The weak signal is a key element of geoprospective. It constitutes not only a point of vigilance, but also a basis for scenario building.
- Looking for opposing forces in the making or likely to occur and cause tensions, for risks and the resulting new vulnerabilities, as much human as material and organizational. An emerging technology elsewhere can be a risk, but is also an opportunity, a creative disturbance by provoking knowledge spillover and diversification.
- Detecting differentiated dynamics in space, the places conducive to technological innovations, those creating collective innovations, and *a contrario*, spaces losing momentum (Voiron-Canicio, 2013).

3.4.4 Detecting a territory's adaptive capacity

A territory's adaptive capacity is the cornerstone of the resilience process. Geoprospective provides a range of tools to help detect, within a region or a city, elements that contribute to forging that capacity. The expression "adaptive capacity" encompasses the abilities of the various local actors to anticipate the forces of change, detect their impact on the territory, and act to prepare the latter to face it. Such anticipation, as we just saw, is achieved through anticipative monitoring. It is also achieved through researching the exogenous risk-bearing trends but also opportunities for the territory being studied and through analyzing the endogenous stressors, that is, the long-term trends that weaken the potential of a region and deepen the vulnerability of its actors (OECD, 2014). This capacity rests greatly on human competences, and more precisely on awareness, readiness to change, and the

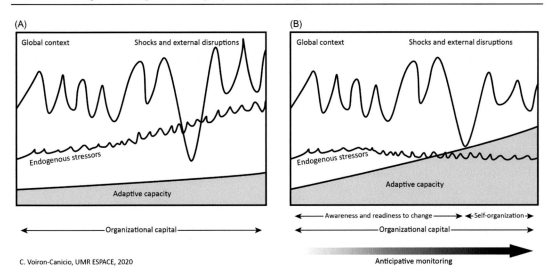

Figure 3.2

Relationship between adaptive capacity to shocks/stressors and organizational capital: (A) weak organizational capital and (B) strengthened organizational capital.

organizational capital. We saw in Section 2 that adaptive capacity does not exist in the absolute that the factors influencing it do not act in isolation but in a systemic way. This capacity is peculiar to each region and specific to its physical and human context (Fig. 3.2).

Therefore the first task is to detect, in the regional system, the various elements that scientists and planning experts agree to consider as favorable to adaptation, or on the contrary, unfavorable. Studies on the resilience of biological systems highlight diversity, connectivity, redundancy, and multiscale interconnections as criteria contributing to adaptability and mutability. All these structural features can be detected easily within a region by spatial analysis. Processing will focus on the organization of space, the structure of internal and external relations, the configuration of networks as well as the degree of functional diversification—activities, businesses, employment—, and also spatial—presence of subregions and intensity of their interactions. However, for all that, high values on these criteria do not lead to a high adaptive capacity. Furthermore, the role of these criteria in the adaptive capacity's construction process remains uncertain, and even controversial at times—the diversity of stakeholders in a quick response to a crisis, for example. On the other hand, there is more certainty on the detrimental effect of a number of elements, such as run-down communication infrastructures, relative isolation, and a low-education population. These weaknesses and constraints are undoubtedly brakes on recovery after a crisis. Over-dependence on a sector, aging population, demographic decline, skills shortages, unreliable transportation, changing climate conditions, etc., are slow burns that reinforce other stressors in a series of negative feedback loops. Here, reasoning using the via negativa (Taleb, 2012) by focusing the attention on endogenous stressors which constitute real handicaps proves more pertinent. This analysis should be complemented by looking for information on the behavior of local actors. Regarding the population, it will be useful to gather qualitative information coming from surveys, for example, on the attachment to the region, the trust in local institutions, and in the absence of surveys, to scrutinize the rate of youth emigration, notably in a crisis period. Regarding stakeholders, information on the existence of collaborative culture among regional actors and of inclusive governance practices will be favorable signs for self-organization, which is necessary to absorb change by reorganizing work and partnerships and revaluating strategies (Giacometti and Teräs, 2019; Adger et al., 2005).

Processing this spatialized information using spatial analysis tools and mapping it will result in a detailed knowledge, combining new quantitative and qualitative information that no standard database can provide on the regional system. Then, the second task consists in gathering and organizing this multiform knowledge within a knowledge base synthesizing the elements that are essential to understanding the forces specific to the region being studied. These forces, both established and latent, are potential engines for adapting to still unknown future events. Geoprospective modeling, and more precisely building scenarios of possible futures, draws on this knowledge base.

3.4.5 Anticipating the functioning of a territory on the long term using simulation

Geoprospective simulation is a valuable aid for exploring the future and evaluating the reactivity of a territory faced with pressures to come. The following example illustrates its importance. The purpose of the model is to examine what could the impacts of CC be in the functioning of an urbanized territory, which otherwise is evolving according to its own socioeconomic dynamics. Usually, the chain reactions that occur on the short term but also the lasting repercussions on the system's structure and dynamics are not well defined. An example of geoprospective modeling applied to this problem is the SERENICIM prototype. The aim of this program is to help anticipate the impacts of climate change on the functioning of a Mediterranean urban area (Voiron-Canicio et al., 2009). The SERENICIM model has been worked out by using the Nice agglomeration as a test. This region is a particularly complex urban system because of interacting environmental and anthropic dynamics, and the conflicts of use between property, tourism, peri-urban agriculture, and the industrial-commercial activity, all of which unfold on an ever scarcer space. In its current version, the systemic model simulates the functioning of the territory by creating interactions between the progressive changes—quarterly lag until 2050—of the local climate's physical parameters—temperatures, rainfalls, seasonal variability, extreme events—and the urban anthroposystem of which the structure is made up of six components: the resident population, numbers of tourists, land use, water resources, water consumption depending on the type of habitat, and the type of agriculture. The energy component is currently being introduced into the system. The territory's dynamics is modelized using a system dynamics model—stock-flux model—where interactions between the system's components are expressed by circular causal relations and either positive or negative feedback loops, which give a clear picture of the complexity of interactions otherwise difficult to perceive. The resilience of the system is assessed by analyzing the behavior of the territorial system in a context of progressive climate change, interspersed with crisis episodes due to climate paroxysms (torrential rain, marine incursion, drought, heatwave). The future risk of water shortage is yet ill-defined in urban territories, notably in the Mediterranean region. Measures taken by the authorities are of the crisis management type, by means of water restrictions bylaws, based on alert thresholds. The purpose of the simulations is to assess the impacts of such measures in the case of water shortage risk. Drought spells have been introduced randomly in the course of the simulation period. The model includes alert thresholds concerning water resources, which trigger various kinds of water consumption restrictions for domestic and farming use. Then simulations make it possible to assess the effectiveness of measures adopted in times of crisis, and their long-term effects, and more precisely, to observe whether agricultural activities remain viable with the measures adopted. It then appeared that frequent water consumption restrictions had, on the medium term, significant impacts on peri-urban irrigated agriculture, already weakened by urban pressure. The simulation exercise of drought periods combined with public action revealed that water restrictions—an emergency response to an existing or imminent water shortage—is not the best strategy for the overall resilience of the urban area on the long term. With the idea of preserving urban agriculture, this geoprospective exercise impels the territory's decision-makers and managers to establish, as from now, a water governance policy with all stakeholders, and more particularly with farmers, to think about a transition toward new farming methods that would be viable both on the short and long terms.

3.5 Resilience of forms and urban geoprospective

In this section, we are going to show another area in which the issue of resilience becomes central in urban geoprospective. To remain very close to the challenges of planning, the question will be mainly about the resilience of the city's physical forms, on which urbanism usually intervenes. Therefore we will talk mainly about the city's morphological resilience (Fusco, 2018), a viewpoint which should clearly be complemented by the resilience of human capital, of the system of governance and, more broadly, of the city's social fabric. At the same time, the physical forms make the city's spatiality a central issue, thus going beyond the mere framework of a strategic prospective that is projected on a territory but does not identify the constraints and potentials of its spatial organization.

3.5.1 From the controllable city to the complex city

Spurred on by metropolization processes and sociotechnical change, the city is undergoing accelerated transformation (Ascher, 1995; Newman and Kenworthy, 1998; Wiel, 1999; Bourdin, 2014). The city is being built differently, integrates new urban objects, is breaking up and spreading, new centralities are emerging, techno-eco-neighborhoods are produced on models that are becoming globalized, and at the same time, entire neighborhoods and sectors are collapsing, productive and commercial fabrics are declining, and urban landscapes are disappearing. These transformations raise new challenges of sustainability: to the usual issues of urban pollutions, consumption of resources and economic and social development add more existential challenges such as the places' identity, the dialectic between heritage and globalization, solidarities between the metropolitan components, and the risk of accelerated obsolescence of the new forms of urbanization faced with the constant turnover of sociotechnical functioning.

Both the geographic theories and the practice of planning and urbanism are challenged by the new functioning of urban and metropolitan areas and the associated physical forms. The need of new geographic knowledge of the city is emerging to understand its most recent transformations and provide guidelines for envisaging its future.

Scientific paradigms are also changing. Traditionally, geographers and planners were looking for "strong" theories to explain urban phenomena and foresee their unfolding, both in time and in space. On these bases, rational comprehensive planning (Mc Loughlin, 1969; Faludi, 1986; Taylor, 1998; Portugali, 2000) focused on land use and resource allocation. Its forecasting and optimization procedures left no room for uncertainties. Its functionalist vision of the urban space was also indifferent to form issues: the important thing was to regulate the quantity of urban functions and their location, giving free rein to the designers of urban forms (developers, architects, public operators), by crossing building standards, traffic requirements, and the trends of the time. Garden cities, modern cities, and postmodern complexes followed one another in time and juxtaposed in space. Their urbanistic performance has rarely been assessed a posteriori. Furthermore, no urban prospective exercise conducted before the turn of sustainable development in the years 1990s broaches the issue of the forms of urbanization. The common practice regarding urban prospective was using demographic, transport, and land-use models, to justify urban development plans, regulatory zoning plans, and local housing programs.

Echoing what we have seen in Section 1, the framework in which the scientific community nowadays repositions the knowledge of cities and their physical form is that of complexity science. The city is a complex system characterized by a dialectic between self-organization tendencies and attempts of control, which are always partial, by plans and policies. But the knowledge that we can have of such a type of system is always incomplete, approximative, and uncertain. The uncertainty of knowledge and the nonpredictability of future states are one of the major challenges to move from the "controllable city" to the "complex city."

3.5.2 Implications for planning and the need of a geoprospective approach

The imperfect knowledge of the complex city (Portugali, 2000; Marshall, 2012) shakes up the traditional approaches to planning and urbanism. In economics, Hayek (1967, 1978) had already made the case that, for a complex social system, we cannot obtain "detailed" knowledge (explanatory or predictive), but only knowledge "in principle," enabling us to sketch the system's typical behaviors qualitatively. As far as urban planning is concerned, Moroni (2015) and also Alfasi and Portugali (2007) suggest to give up producing plans based on forecasting and let urban self-organization emerge within a system of simple rules. In fact, knowledge is already injected into any rule system. As an example, Talen (2012) shows that rules are not neutral and that the making of the city could give a much larger part to urban codes as compared with set plans, mass layout, or zoning plans. Using codes and a few key principles on the urban forms being sought would be much more likely to foster urban self-organization, because it would allow more flexibility to the making of the city. But what role should we give to projects and plans in such context? Two other questions, even more existential, arise when thinking about the future of the city, justifying possible plans and projects, but also any decision concerning a set of codes to impose to urban stakeholders. First, can we base a vision of the city's future on weak and nonpredictive theories? Next, is there a theory of the complex urban form and how could it inform this vision of the city's future?

According to Blecic and Cecchini (2016), in the context of planning and urbanism, the answer to the first question is clearly positive. The vision of the future is even necessary. By admitting that the future is not predetermined by a strong theory, but that different futures remain possible, it becomes essential to mobilize the energies of urban stakeholders to give a meaning and legitimacy to the collective action and, at the same time, create the best conditions to attain the future which has been consensually identified as desirable. The tool enabling to make the desired future explicit is the project, and the approach to achieve it is, for these authors, a prospective approach, and more precisely, a geoprospective one. The absence of strong theories undermines also the technocratic approaches to the project, where scientific knowledge was formerly imposed to stakeholders in a top-down manner, and reinforces the importance of the shared project. The project will probably be more flexible and will take the form of strategic geoprospective, so as to coordinate the actions of individual stakeholders in the urban self-organization.

Dupuy (1992, 2012) provides theoretical bases to the necessity of both project and prospective: the characteristic of sociotechnical systems is the possible retroaction by the future (see also Section 1.2). If the stakeholders coordinate themselves in view of pursuing a shared project, the said project, which does not yet exist as a state (future in any case) of the system, is capable to interact with the current state and the dynamics at work to modify the course of events (without any guarantee that the system's future state will be that recommended by the project, of course).

However, the absence of strong theories and the fact that technocratic approaches are left out do not mean that the new scientific knowledge should not play a part in the working out of this shared vision. Indeed, scientific research in geography and urbanism, in connection with advances in complexity science, has produced in recent years an increasingly clearer vision of the role of form and spatial organization in the complex city. Traditionally, the city's strategic prospective paid little attention to the spatial dimension's potentialities and constraints and, in particular, to urban forms. Nowadays, the shared project can be produced within an urban geoprospective approach rooted in the specificities of the complex city, of the spatial logics of its physical forms, with their constraints and potentials, and be informed about the uncertainties of our knowledge of urban futures (Fusco, 2018). The aim of urban geoprospective is to explore the complex city's possible futures, in order to lay the foundations of an effort of collective organization in an irreducibly uncertain context. The contribution of the new scientific insights could then precisely be as follows. Scientists are neither in a position to provide a foresight of the urban future, nor to identify the universe of possibilities with certainty. However, they can help identify the forces and weaknesses of the existing city and of the outlines of the proposed projects (the shared vision).

3.5.3 Resilience as a response to unknowable futures

In this context of profound uncertainty on urban futures, a consensus seems to be emerging today on the opportunity to achieve highly resilient and so-called "antifragile" urban forms, capable of adapting and changing to benefit from the sociotechnical innovation constantly produced by the city (Holling and Sanderson, 1996; Pumain, 2012; see Section 1.1). If the concept of resilience is derived from already mentioned seminal works in ecology, that of antifragility, first proposed to the general public by Taleb (2012), was soon used in urbanism and planning (Blecic and Cecchini, 2016). From our point of view, antifragility can be seen as a highly adaptive and transformative form of resilience. In any case, regarding the resilience of urban forms when faced with sociotechnical change, the posture toward resilience will essentially be ex ante (Section 1.1): it is a matter of identifying the resilience potential of the city's forms, observable or projected, guided by a theoretical approach to morphologic resilience.

A first theory of the resilience of urban forms is based on the configuration of street networks (Hillier and Hanson, 1984; Hillier, 1996; Marshall, 2012), of which the best-known protocols are those of space syntax. This approach appears directly in a geocomputation context and specifies to urban spaces the protocols for analyzing complex networks (Freeman, 1979; Barabasi, 2002). The configurational approach shows that the founding principle of urban organization is not distance in itself, whatever its definition, but the system of the whole of reciprocal distances, that is, the configuration. The configurational properties are emerging and strongly structure the potential of urban spaces. Hillier (1996) identifies the configuration of street networks as a major interface of the city's socioeconomic functioning. A concept key of configurational analysis, in relation with resilience, is that of synergy. The latter is an interscale relationship, defined as being the degree of dependence of local configurational properties compared with the same properties evaluated globally within the urban space. If, for example, the two spatial levels are highly correlated, then local changes in the networks will have a limited impact on the configurational properties of the city as a whole and of each element of the network.

On these bases, various researches helped to better target what could be specific to urban resilience in a configurational approach. Nevertheless, these researches are carried out in a context of resilience to debilitating disruptions, but do not address the dimension of sociotechnical change, which is unavoidable in an urban geoprospective approach. The debilitating disruption is flood in Esposito and Di Pinto (2015), an earthquake in Cutini (2013), an erupting volcano in Cutini and Di Pinto (2015), and the collapse of bridges and groups of targeted network cuts in Abshirini and Koch (2017). In relation to the persistence of the network's connectivity alone, groups of configurational properties are evaluated before and after the disruption. Thus Cutini (2013) and Cutini and Di Pinto (2015) suggest to follow three configurational properties: Hillier's synergy, the average connectivity, and a relative betweenness index, called frequency index (the ratio between the maximal choice index in the configuration and the theoretical maximal number of minimal paths in a same-size configuration). Configurational resilience would increase correlatively to the increase of synergy and average connectivity and to the decrease of the frequency index (because, in this case, the configuration is less dependent on a single element with very high betweenness). Koch and Miranda Carranza (2013) suggest two other configurational properties for measuring resilience, these properties being finally used by Abshirini and Koch (2017) on a city scale: similarity and sameness. If, after a disruption, the configuration of a city's street network retains the same spatial extent of the foreground network (similarity) and the same constitution of this network in terms of its components (sameness), then the configuration is considered as resilient to the disruption.

3.5.4 Morphological resilience to urban change

The challenge for urban geoprospective is not to be limited to the resilience to catastrophic events, but to integrate a broader vision of the urban form's resilience to any sociotechnical change, whether slow or sudden. To do so, the urban morphological system as a whole must be taken into account: the street network (like in the configurational approach), but also the built fabric, the plot pattern, the topography, and, obviously, land functions and use.

A complete theory of the resilient and antifragile urban form faced with sociotechnical change remains to be built, but its foundations have been laid by new theories which include the approach of the urban form's structural complexity [like, for example, in Hillier (1996, 2012) and Salingaros (2005)] and can even draw on more traditional knowledge of urban morphology (Conzen, 1960; Caniggia and Maffei, 1979). Such a theory should also enable to identify, in contrast, the most fragile urban forms. Namely those incapable to absorb new functioning to adapt to new societal requirements by gradual changes. There is a high risk that fragile forms, even when optimal in the context that produced them, become dysfunctional, and thus obsolete, are unable to self-regenerate and necessitate strong external interventions, often in the form of complete demolition-reconstruction. The most famous example is that of the modernist social housing neighborhood of Pruitt-Igoe in St. Louis, MI, completely demolished in the mid-1970s only 20 years after being built. Unfortunately, Pruitt-Igoe was only the first of many modern neighborhoods to be condemned to early demolition. Recognizing the fragility of present or planned forms is of major interest for urbanism and urban planning.

Here, the main challenges for researchers are at the same time the convergence of relatively different theoretical approaches to urban complexity and, above all, the connection with the empirical analysis of observable urban forms and of the fragility or resilience potentials that can be deducted from analyzing their properties and their transformations. Moving from general to specific must also characterize the connection between the theory of resilient urban form and the urban geoprospective approach, which must be carried out on a given city in a given historical context, in which urban forms are also perceived and assessed by stakeholders through extremely powerful cultural filters (Duarte, 2010).

The reflection on the resilient urban form also meets that on the reversible city, an extremely vast subject explored by a seminal symposium in Cerisy (Scherrer and Vanier 2013). Urban planning has traditionally been poles apart from the idea of reversibility. It is precisely a change of social paradigm, the recognition of the uncertainties of the future, which raises the question of an organization of the physical city which could go back on decisions taken in a previous time. In fact, as underlined by Panerai (2013), complete reversibility never exists for the physical city, but neither does the absence of any reversibility. We think that the concept of resilient forms does tally with what Panerai identifies as being the urban form's dynamic tension between permanence and change, a tension which must be evaluated over time. In referring to the principles of urban morphology, Panerai points out how the city's morphologic infrastructure, its street and plot pattern, enables to envisage the city as being based on a powerful structure that sees short periods of history succeed one another. In its broad lines, the pattern is invariable, the built fabric alone is temporary, variable, evolving with the new requirements of populations and activities. That is, at least, the lesson learnt from urban forms which have been seen surviving while adapting in a highly flexible manner over the centuries.

These considerations resonate with Mehaffy and Salingaros's more comprehensive reflections (2013). By developing an analogy with the ecosystems' structure and functioning, these authors lay down founding hypotheses for a theory of the resilient urban form. Any resilient urban form should be characterized by interconnected route networks and by a redundancy of route types, resonating with the strong interconnection in the ecosystems' trophic chains. However, not every pattern would be a resilience catalyst: tree-like networks and very large blocks would be particularly fragilizing for the urban form, contrary to organic closely knit networks (like in medieval cities) or orthogonal (like in the grid patterns of bastides or Barcelona and Turin urban expansions). The resilient form should also propose great diversity and redundancy of activities, in regard to building types, objectives, functions, and populations. Taking the opposing view of functionalist urbanism, Mehaffy and Salingaros see in that great diversity, considered at the city's various scales, a potential of resilience to future changes, unknowable at the time when the urban plan was drawn up. Here fragility would be linked to the specialization of the functional zoning which produces business districts devoted solely to office towers or private housing areas with residential monofunctionality. Accepting redundancy and diversity also means giving up optimization (of the form in relation to the function, of the means in relation to the results, etc.). The form is not optimized for a particular functioning, it is redundant and its "program" is open, which gives it a certain amount of adaptability in relation to the diversity of possible functioning.

Resilient urban forms are also those having a wide range of scales in their structure, from the macroform of an entire metropolitan area, to the assembling of plots, streets, and buildings, down to the details of urban design. The finest scales are even the most important, because they provide possibilities of autonomous and quick responses to urban change. We can thus understand the great fragility of major urban projects designed in their entirety, or of slab-above-the-street urbanism: any change through small additions and progressive transformations is a problem and it often becomes necessary to reinitialize the whole of the urban form by destruction-reconstruction, an admission of failure of a nonresilient urbanism. Even in resilient forms, urban change often entails a pivotal scale of response, but trans-scale relations articulate the response at all scales. This suggests a relatively autonomous level of response by individual stakeholders at the finest scale (that of each plot and building), while changing the morphological infrastructure or building large projects will necessitate greater coordination.

Finally, resilient urban forms evolve while retaining the essential of their prior structural and informational content: only the innovation which is strictly necessary to a structure which stood the test of time is introduced (they are structure-preserving). This also brings us back to Panerai's observation (2013): to be successful, any change must rest on permanence.

In the end, as stressed by Blecic and Cecchini (2016) the urbanism of a resilient and antifragile city is an urbanism for complexity: the structural complexity is seen as a resource for the future of the urban space.

Feliciotti et al. (2016), paying considerable attention to the forms of the built environment, of plots and streets traditionally studied by urban morphology, propose an operational expression of these general principles of urban morphologic resilience. By using proxies, they show how the dimensions of connexity, diversity, modularity, efficiency, and redundancy can serve as a basis for assessing the resilience potential of real urban forms. It should be remarked that the principle of efficiency is assessed very differently from an optimization of forms in relation to a program: the form is efficient insofar as it is the result of self-organization, typically characterized by power law distributions (Salingaros 2005) and strong interscale relations. In a further publication (Feliciotti et al., 2017), the same authors show how this protocol of analysis can apply at a neighborhood scale, taking as examples the various urban forms taken in the course of the 20th century by the Gorbals neighborhood in Glasgow. Thus Victorian Gorbals' closely knit urban fabric of tenements and terraced houses was able to adapt during more than a century of sociotechnical transformations, showing great resilience capacity. The new modernist Gorbals, erected in the fifties, included very large blocks, a low level of modularity and connectivity, and high functional specialization. These fragilizing factors led to an early obsolescence and costly demolition by the eighties. The Gorbals neighborhood resulting from the transformations of the years 2000 has rectified the main faults of modernist forms. Its future is not yet written but its potential of resilience can already be evaluated as being intermediate between those of the two previous urban forms.

In conclusion, new knowledge on the complex city's form can feed the geoprospective reflection on the city of the future. A few basic indications can be proposed.

First, due to the absence of strong theories to predetermine the urban future, it is not possible to propose a new form of ideal city that would be optimal for a given program. The program of the city of the future must remain open, and a plurality of forms is possible.

A rigid urban plan does not seem to befit that background uncertainty on urban futures. For a number of authors, any idea of plan should even be abandoned and urbanism should be limited to laying down a set of rules that would guide stakeholders' self-organization. Now, whether it is a flexible, reversible, and strategic plan or a set of rules, the geoprospective approach necessitates a shared vision (in order to justify the rules/plans and coordinate individual actions).

This shared vision, to be fully geoprospective, must give an important role to the reflection on the forms of the city of the future, and not only to its functions. Forms bind the city on the long term, especially those of the infrastructure of streets and plots. Complexity theory shows that some forms seem to be more resilient than others, that is, capable of a high level of adaptivity and transformability (antifragility).

In any case, any geoprospective approach should start from the forms of the existing city, understand their potential of resistance or fragility, and understand to what extent they could face various scenarios of urban innovation. The forms identified as more resilient will have a more important role to play in a city which is bracing itself for uncertain futures. Plans and rules will have to facilitate the emergence of such forms and proscribe the production of "fragilizing" forms.

3.6 Conclusion

Understanding how territories will transform, adapt to pressures to come, and move toward sustainability is a major scientific challenge. The issue is complex, commensurate with the complexity of territorial systems of which the dynamics is unforeseeable and unpredictable. The territory is the marker of the dynamics of change, and the most accurate at our disposal. It undergoes, absorbs, and generates various evolutions and disturbances, and assimilates them in a differentiated manner in space. The resulting spatial change is the visible expression of this process. Geoprospective offers various avenues to read the mechanisms of change and help to anticipate future behaviors. Analyzing the territory's spatiotemporal trajectory informs, to various degrees, on its propensity to change and on its modes of responsiveness. Its resilience capacity is a multifaceted and place-based process. Multiple components interact, some relating to the socioeconomic structure and governance, others to the organization of space and spatial configuration. Detecting the adaptive capacities of the spaces that make up the territory requires to have some depth of field. A variety of spatial analysis tools can be used for doing so; to evaluate the resilience and fragility potential of the existing urban forms; to collect information on the behavior of stakeholders in a crisis context and detect the ferments of self-organization; to match quantitative and qualitative data so as to produce synthetical indicators; to map the resulting information; and to organize that spatialized and multivaried knowledge in a database. That spatialized knowledge base, specific to the territory being studied, is the foundation of prospective spatial modeling. Crossed with conjectures resulting from anticipative monitoring, it provides guidelines for choosing scenarios for the future.

This chapter has illustrated the contribution of geoprospective to the analysis of the relationship between resilience and spatial change. The following chapter presents the range of methods and tools available to researchers.

References

Abshirini, A., Koch, D., 2017. Resilience, space syntax and spatial interfaces: the case of river cities. ITU A|Z 14 (1), 25–41.

Adger, W.N., Hughes, T.P., Folke, C., Carpenter, S.R., Rockström, J., 2005. Social-ecological resilience to coastal disasters. Science 309 (5737), 1036–1039.

Adger, W.N., Dessai, S., Goulden, M., Hulme, M., Lorenzoni, I., Nelson, D.R., et al., 2009. Are there social limits to adaptation to climate change? Clim. Change 93, 335–354.

Alfasi, N., Portugali, J., 2007. Planning rules for a self-planned city. Plan. Theory 6 (2), 164–482.

Aschan-Leygonie, C., 2000. Vers une analyse de la résilience des systèmes spatiaux. L'Espace Géogr. (1), 64–77 (in French).

Ascher, F., 1995. Métapolis ou l'avenir des Villes. Odile Jacobs, Paris (in French).

Atlan, H., 1979. Entre le cristal et la fumée. Essai sur l'organisation du vivant. Seuil, Paris (in French).

Barabasi, A.L., 2002. Linked: How Everything Is Connected to Everything Else and What It Means. Perseus, Cambridge, MA.

Barabasi, A., Reka, A., 1999. Emergence of scaling in random networks. Science 286 (5439), 509–512.

Berkes, F., Colding, J., Folke, C. (Eds.), 2003. Navigating Social-Ecological Systems Building Resilience for Complexity and Change. Cambridge University Press.

Blecic, I., Cecchini, A., 2016. Verso una pianificazione antifragile. Come pensare al futuro senza prevederlo. Franco Angeli, Milano (in Italian).

Bourdin, A., 2014. Métapolis revisitée. Editions de l'Aube, La Tour d'Aigues (in French).

Caniggia, G., Maffei, G., 1979. Lettura dell'edilizia di base. Alinea, Firenze (in Italian).

Casanova, L., 2010. Les dynamiques du foncier à bâtir comme marqueurs du devenir des territoires de Provence intérieure, littorale et préalpine. Éléments de prospective spatiale pour l'action territoriale. Université d'Avignon, thèse de doctorat de géographie (in French).

Clark, W.C., Dickson, N.M., 2003. Sustainability science: the emerging research program. PNAS 100 (14), 8059–8061.

Conzen, M.R.G., 1960. Alnwick, Northumberland: a study in town-plan analysis. Institute of British Geographers Publication 27, George Philip, London.

Cumming, G.S., 2011. Spatial resilience: integrating landscape ecology, resilience, and sustainability. Landsc. Ecol. 26 (7), 899–909.

Cutini, V., 2013. The city when it trembles. Earthquake destructions, post-earthquake reconstruction and grid configuration. In: Kim, Y.O., Park, H.T., Seo, K.W. (Eds.), Proceedings of the Ninth International Space Syntax. Sejong University, Seoul.

Cutini, V., Di Pinto, V., 2015. On the slopes of Vesuvius: configuration as a thread between hazard and opportunity. In: Proceedings of the 10th International Space Syntax Symposium. UCL, London.

Da Cunha, A., Thomas, I., 2017. Introduction » in La ville résiliente, comment la construire?, Collectif, Isabelle Thomas, Antonio Da Cunha, Collection PUM, Montréal (in French).

Dakos, V., Nes, E.H., Donangelo, R., Fort, H., Sheffer, M., 2009. Spatial correlation as leading indicator of catastrophic shifts. Theor. Ecol. 3, 163–174.

Duarte, P. (Ed.), 2010. Les démolitions dans les projets de renouvellement urbain, représentations, légitimités et traductions. L'Harmattan, Paris (in French).

Dupuy, J.P., 1992. Introduction aux sciences sociales. Logique des phénomènes collectifs. Ellipses, Paris (in French).

Dupuy, J.P., 2012. L'Avenir de l'économie: Sortir de l'écomystification. Flammarion, Paris (in French).

Esposito, A., Di Pinto, V., 2015. Calm after the storm: the configurational approach to manage flood risk in river cities. In: Proceedings of the 10th International Space Syntax Symposium. UCL, London.

Faludi, A., 1986. Critical Rationalism and Planning Methodology. Pion, London.

Feliciotti, A., Romice, O., Porta, S., 2016. Design for change: five proxies for resilience in the urban form. Open House Int. 41 (4), 23–30.

Feliciotti, A., Romice, O., Porta, S., 2017. Urban regeneration, masterplans and resilience: the case of Gorbals, Glasgow. Urban Morphol 21 (1), 61–79.

Freeman, L., 1979. Centrality in social networks: conceptual clarification. Social Netw 1, 215–239.

Folke, C., Carpenter, S., Elmqvist, T., Gunderson, L., Holling, C.S., Walker, B., et al., 2002. Resilience and sustainable development: Building adaptative capacity in a world of transformations. In: Scientific Background Paper on Resilience for the process of the World Summit on Sustainable Development on behalf of the Environmental Advisory Council to the Swedish Government.

Fusco, G., 2018. Ville, complexité, incertitude. Enjeux de connaissance pour le géographe et l'urbaniste. Habilitation à Diriger des Recherches en Géographie. Université Côte d'Azur, halshs: <tel-01968002> (in French).

Giacometti, A., Teräs, J., 2019. Regional economic and social resilience. Nordregio Report 2019, 2.

Gunderson, L.H., Holling, C.S. (Eds.), 2002. Panarchy: Understanding Transformations in Human and Ecological Systems. Island Press.

Haken, H., 1977. Synergetics, An Introduction. Springer, Berlin.

Hayek, F., 1967. Studies in Philosophy, Politics and Economics. Routledge, London.

Hayek, F., 1978. New Studies in Philosophy, Politics, Economics and the History of Ideas. Routledge, London.

Hillier, B., 1996. Space Is the Machine. Cambridge University Press, Cambridge.

Hillier, B., 2012. The genetic code for cities: is it simpler than we think? In: Portugali, J., et al., (Eds.), Complexity Theories of Cities Have Come of Age. Springer, Berlin, pp. 129–152.

Hillier, B., Hanson, J., 1984. The Social Logic of Space. Cambridge University Press, Cambridge.

Holling, C.S., Sanderson, S., 1996. Dynamics of (dis) harmony in ecological and social systems. In: Hanna, S., Folke, C., Mäler, K.G. (Eds.), Rights to nature: ecological, economic, cultural, and political principles of institutions for the environment. Island Press, Washington, DC, pp. 57–86.

Independent Group of Scientists, 2019. The Future is Now—Science for Achieving Sustainable Development, Global Sustainable Development Report 2019. United Nations, New York.

Kates, R.W., Clark, W.C., Corell, R., Hall, J.M., Jaeger, C.C., Lowe, I., et al., 2001. Science 292 (5517), 641–642.

Koch, D., Miranda Carranza, P., 2013. Syntactic resilience. In: Proceedings of Ninth International Space Syntax Symposium. Sejong University Press, Seoul.

Laganier, R., 2013. Améliorer les conditions de la résilience urbaine dans un monde pluriel: des défis et une stratégie sous contrainte. Ann. Mines Responsab. Environ. 72 (4), 65–71 (in French).

Lesca, H.H., Lesca, N., 2011. Les signaux faibles et la veille anticipative pour les décideurs. Collection Business, Lavoisier (in French).

Levin, S., 2002. Complex adaptive systems: exploring the known, the unknown and the unknowable. Bull. Am. Math. Soc. 40 (1), 3–19.

Liziard, S., 2013. Littoralisation de la façade nord-méditerranéenne: Analyse spatiale et prospective dans le contexte du changement climatique, Thèse de géographie, Université de Nice, Sophia Antipolis (in French).

Magnan, A., 2009. Proposition d'une trame de recherche pour appréhender la capacité d'adaptation au changement climatique. Vertigo 9 (3). Available from: http://vertigo.revues.org/9189 (in French).

Magnan, A., 2010. Questions de recherche autour de l'adaptation au changement climatique. Nat. Sci. Soc. 18, 329–333 (in French).

Marshall, S., 2012. Planning, design and the complexity of cities. In: Portugali, J., et al., (Eds.), Complexity Theories of Cities Have Come of Age. Springer, Berlin, pp. 191–205.

Mc Loughlin, B., 1969. Urban and Regional Planning: A Systems Approach. Faber & Faber, London.

Mehaffy, M., Salingaros, N., 2013. Towards Resilient Architectures. I: Biology Lessons. Metropolismag.com.

Mermet, L., Poux, X., 2002. Pour une recherche prospective en environnement—repères théoriques et méthodologiques. Nat. Sci. Soc. 10 (3), 7—15 (in French).

Morin, E., 1974. La complexité. Colloque sur les interrelations entre la biologie, les sciences sociales, et la société. Mars, Paris (in French).

Morin, E., 1990. Introduction à la pensée complexe. Seuil, Paris (in French).

Morin, E., 1994. La complexité humaine. Champs Flammarion, Paris (in French).

Moroni, S., 2015. Complexity ant the inherent limits of explanation and prediction: urban codes for self-organizing cities. Plan. Theory 14 (3), 248—267.

Newman, P., Kenworthy, J., 1998. Sustainability and Cities: Overcoming Automobile Dependence. Island Press, Washington, DC.

OECD, 2014. Guidelines for Resilience Systems Analysis. OECD Publishing.

Panerai, P., 2013. Eloge de la trame. In: Scherrer, F., Vanier, M. (Eds.), Villes, Territoires, Réversibilités. Collection Colloque de Cérisy. Hermann, Paris (in French).

Portugali, J., 2000. Self-Organisation and the City. Springer, Berlin.

Prigogine, I., Stengers, I., 1979. La nouvelle Alliance. Gallimard, Paris(in French).

Pumain, D., 2003. Une approche de la complexité en géographie. Géocarrefour 78/1, 25—31 (in French).

Pumain, D., 2012. Urban systems dynamics, urban growth and scaling laws: the question of ergodicity. In: Portugali, H.M., Stolk, E., Tan, E. (Eds.), Complexity Theories of Cities Have Come of Age. Springer, Berlin, pp. 91—104.

Rietkerk, M., Dekker, S.C., de Ruiter, P.C., van de Koppel, J., 2004. Self-organized patchiness and catastrophic shifts in ecosystems. Science 305.

Rumpala Y., 2003. Régulation publique et environnement. Questions écologiques, réponses économiques. L'Harmattan, Paris (In French).

Salingaros, N., 2005. Principles of Urban Structure. Delft University of Technology, Techne Press, Delft.

Scheffer, M., Carpenter, S.R., 2003. Catastrophic regime shifts in ecosystems: linking theory to observation. Trends Ecol. Evol. 18 (19).

Scheffer, M., Bascompte, J., Brock, W., Brovkin, V., Carpenter, S.R., Dakos, V., et al., 2009. Early-warning signals for critical transitions. Nature 461, 53—59.

Scheffer, M., Carpenter, S.R., Lenton, T.M., Bascompte, J., Brock, W., Dakos, V., et al., 2012. Anticipating critical transitions. Science 338.

Scherrer F., 2017. Avant-propos in La ville résiliente, comment la construire?. Collectif, Isabelle Thomas, Antonio Da Cunha, Collection PUM, Montréal (in French).

Scherrer F., Vanier M., 2013. Villes, Territoires, Réversibilités. Collection Colloque de Cérisy. Hermann, Paris (in French).

Smit, B., Wandel, J., 2006. Adaptation, adaptive capacity and vulnerability. Global Environ. Change 16, 282—292.

Smit, B., Burton, I., Klein, R.J.T., Street, R., 1999. The science of adaptation: a framework for assessment. Mitig. Adapt. Strat. Global Change 4, 199—213.

Swetnam, R.D., 2007. Rural land use in England and Wales between 1930 and 1998: mapping trajectories of change with a high resolution spatio-temporal dataset. Landsc. Urban Plan. 81 (2007), 91—103.

Taleb, N.N., 2012. Antifragile. Things that gain from disorder. Random House, New York, NY (in French).

Talen, E., 2012. City Rules: How Urban Regulations Affect Urban Form. Island Press, Washington.

Taylor, N., 1998. Urban Planning Theory Since 1945. Sage Publications, London.

UNISDR, 2012. Rendre les villes plus résilientes - Manuel à l'usage des dirigeants des gouvernements locaux. Une contribution à la Campagne mondiale 2010-2015 Pour des villes résilientes—Ma ville se prépare!, Genève (in French).

Vincent, K., 2007. Uncertainty in adaptive capacity and the importance of scale. Global Environ. Change 17 (1), 12—24.

Voiron-Canicio, C., 2012. Forecasting change in prospective and spatial change in geoprospective. Espace Geographique, Berlin.

Voiron-Canicio C., 2013. Déceler les espaces à enjeux pour l'aménagement. In: Masson-Vincent, M., Dubus, N. (Eds.), Géogouvernance, utilité sociale de l'analyse spatiale, Collection Update. Editions Quae, pp. 171−182 (in French).

Voiron-Canicio, C., 2015. Une ville résiliente? Quid de l'innovation dans la marche vers la durabilité urbaine?. In: Hajek, I., Hamman, P. (dir.), La gouvernance de la ville durable; entre déclin et réinventions. PUF, Rennes, pp. 267−277 (in French).

Voiron-Canicio, C., Dubus, N., Loubier, J.-C., Liziard, S., 2009. Evaluer les Impacts du Changement Climatique sur le Fonctionnement d'une Aire Urbaine Littorale: Outils d'Aide à la Réflexion et d'Aide à la Décision Existants. In: 5th Urban Research Symposium Cities and Climate Change: Responding to an UrgentAgenda, Marseille, France, June 2009. halshs-00470199 (in French).

Wiel, M., 1999. La transition urbaine ou le passage de la ville pédestre à la ville motorisée. Mardaga, Liège (in French).

CHAPTER 4

Methods and tools in geoprospective

Christine Voiron-Canicio[1], Emmanuel Garbolino[2], Giovanni Fusco[1] and Jean-Christophe Loubier[3]

[1]University Côte d'Azur, CNRS, Laboratory ESPACE, Nice, France, [2]Climpact Data Science, Nova Sophia—Regus Nova, Sophia Antipolis Cedex, France, [3]University of Applied Sciences and Arts Western Switzerland, Sierre, Switzerland

Chapter Outline

Ecosystem and Territorial Resilience.
DOI: https://doi.org/10.1016/B978-0-12-818215-4.00004-3
© 2021 Elsevier Inc. All rights reserved.

4.1 Introduction

Assessing the future dynamics of natural systems and of socioecosystems, appraising their resilience capacity, anticipating spatial change to come: the spectrum of issues in geoprospective is broad. The prospective purpose, combined with sustainability, underlying behind the search for sustainable paths, brings geoprospective close to sustainability science. These two scientific currents have in common to differentiate themselves from the usual scientific approaches. Faced with the dynamics' complexity and nonlinearity, the unpredictability of processes a fortiori on the long term, and the uncertainty of the spatial future, researchers are driven to work out protocols that combine various kinds of tools and models and to conceive semiquantitative models using "qualitative" data based on stakeholders' viewpoints.

This chapter does not intend to give a full review of all the methods and tools usable in geoprospective, but to expound the abundance of approaches conceived in that context. A number of methods, such as modeling, are not specific to geoprospective. The perspective adopted is to shine the light on what their use in geoprospective entails: the specific constraints and the new questions raised, relating to the weight of past evolutions, to the unforeseen (changes, of course, and ruptures), to uncertainty. Therefore attention will be drawn to adaptations of existing models, new operating modes, and validation procedures. Furthermore, other approaches have been specifically conceived for geoprospective. Their characteristics will be presented together with their interest and the methodological development in progress for a number of them. In addition to the models of the land use and cover change (LUCC) type and companion modeling, this chapter gives much importance to the following new approaches which are hitherto hardly used in geoprospective and for some of them, under development: scenarios integrating various territorial scales, modeling of the decision-making process coupled with prospective spatial modeling, territorial geoprospective based on causal probabilistic models, graphic modeling and prospective choremes, immersive 3D simulation in landscapes of the future. Attention will more particularly be focused on the way in which space will be integrated into the processes and taken into account in the representation of results.

4.2 Types of scenarios of territorial futures

Like in prospective, scenario building plays a central role in geoprospective.

4.2.1 Qualitative, quantitative, and hybrid scenarios

Scenarios can be classified into three main families. Qualitative scenarios, narratives supported at times by diagrams, representing a group's views of the future; quantitative scenarios, resulting from analyses and models, and a combination of both: the SAS (Story and Simulation) approach (Alcamo, 2008).

The growing use of narrative-type scenarios can be explained by the fact that they allow to combine major trends and weak signals, to introduce ruptures and contingent events, and to integrate, within the same framework, heterogeneous knowledge, on different spatial and temporal scales. They express in a narrative form complex processes that models have difficulty expressing, and by doing so, they help to understand complexity (Schwartz, 1998; Mermet, 2005). However, the knowhow remains empirical. The construction process of the assumptions adopted is not always transparent, and even more rarely formalized.

Quantitative scenarios rest on scientific analyses and models aiming to assess the future state of a phenomenon, or of all or part of a system. The dynamics' determining factors and causal relationships are detected by retrospective analyses, then extrapolated for the future, most of the time from deductive patterns of the "all else being equal" type. The end-purpose is usually to predict a probable path in relation to one or several trends. Most of the environmental models rely heavily on these extrapolations (Costanza et al., 1990). However, bearing in mind that relations between variables are likely to evolve in future, other constructions of scenarios compare the extrapolations of relations based on retrospective modeling with sets of hypotheses on the possible transformations of the structures of the system being studied. Hybrid scenarios are increasingly used.

4.2.2 Scenario-building methods

The choice of the type of scenario will depend on the purpose of the prospective exercise, of its substance, and of the process (Rothman, 2008; Alcamo, 2008). De facto, there is a multiplicity of conceptions of scenarios, as well as a wide variety of construction modes and frameworks of use. The "intuitive logics" scenarios are defined by Derbyshire and Wright (2017) as "a plausibility-based approach that enables participants, usually within a workshop setting, to create narratives that describe unfolding chains of causation, which resolve themselves into sets of distinct future outcomes." They are commonly used as decision aid tools in planning and action-research (Bradfield et al., 2005; Schwartz, 1993; Wright and Goodwin, 2009).

The expression "method of scenarios" characterizes a synthetic approach that simulates, in a plausible and coherent manner, a sequence of elements leading a system to a future situation, and presenting an overall picture of the latter (Julien et al., 1975). There are two types of scenarios. Exploratory scenarios are built starting from the current situation. In addition to the so-called trend scenario—"business as usual"—contrasted scenarios are variants based on sets of assumptions opposed term by term. The second type is the normative scenario, constructed from a hypothetical state of the system in future, either desired or feared. If the future is improbable, the scenario is called "rupture scenario."

Two methodologies are commonly used in geoprospective. Alcamo (2008) has identified the steps common to a great number of exercises of environmental scenarios. The SAS approach unfolds in 10 steps (Alcamo, 2008). Step 1: a scenario team (coordinators), a scenario panel (experts and stakeholders), and a modeling team are established. Steps 2 and 3: goals and outlines of scenarios are drawn by the scenario team, and the scenario panel constructs a first draft of storylines. Step 4: the scenario team quantifies the driving forces of scenarios. Step 5: the modeling team quantifies the indicators of the scenarios. Step 6: storylines are revised based on the results of steps 4 and 5. Step 7: iterations of steps 4, 5, and 6 as necessary. Step 8: the draft scenarios are reviewed by experts and stakeholders. Step 9: scenarios are revised. Step 10: the final scenarios are published.

The method of scenarios worked out by Michel Godet for strategic prospective is one of the most formalized approaches (1986). It is in three steps. The first consists in constructing the baseline used for identifying the system's key variables, their interrelations, as well as past trends and the role of stakeholders by means of a quantified and detailed analysis. The second step sweeps over the range of possibilities by assessing, via expert's methods, the probabilities of occurrence of events. The third step is the actual working out of the scenario, describing the progression leading from the present to the final pictures retained. Evaluating the prospective scenarios is carried out by checking the four fundamentals of the construction of a scenario: pertinence, coherence, likelihood, and transparency.

4.2.3 Territorialized prospective scenarios combining several territorial and organizational scales

Scenario-driven approaches of territorial prospective are also varied. On the other hand, unlike the global prospective scenarios, or those concerning continents or countries, scenarios concerning smaller spaces (a region, a catchment, rural land) are rarer (Van Asselt et al., 1998). We present hereunder two applications of the method of scenarios concerning different regional levels, one at the scale of the European Union, and the other at the scale of a small geographic region in the South of France.

4.2.3.1 Scenarios on the territorial future of Europe (ESPON Project 3.2, 2006)

The ESPON (European Spatial Planning Observation Network) project addresses the territory of the 27 EU Member States as well as Norway and Switzerland. To start with, ESPON Network teams have looked for the most significant driving forces for the territory's evolution by constructing thematic scenarios (demography, economy, energy, transport, governance, enlargement, climate change, rural development, and sociocultural evolutions), and combining quantitative models with more qualitative visions. The following stage consisted in building integrated scenarios based on the orientations that are matter of debate within the European Union, and even of controversy within the society: for example, more economic, social and territorial cohesion, or increased extra-European competitiveness. Four scenarios were defined by the research team in collaboration with the ESPON Monitoring Committee: one baseline (trend) scenario, two prospective scenarios—cohesion-oriented scenario and competitiveness-oriented scenario. In view of the fact that the aim of territorial prospective is to help determine a strategy, both scenarios have been integrated into a proactive scenario defined as "the ideal." The scenarios' resilience has been tested by means of "four simple and short wild cards using high impact events and exploring their impact on the territory" (ESPON 3.2 final report). In the end, the distinctive features of each prospective scenario have been described and compared to those of the trend scenario. The description is of the narrative type with little or no references to regional dynamics. To compensate for this shortcoming, a mapping of the socioeconomic and environmental dynamics and of the urban organization accompanies each scenario.

4.2.3.2 Scenarios on the future of a small region—integrating multilevel dimensions

This second type of scenario building is presented taking as an example a prospective research conducted on a small geographic region, the Camargue, located in the Rhône river delta, comprising wetlands, natural areas, and agricultural land. This small region—1500 km^2 and 115,000 inhabitants—is an original socioecosystem, most of it protected—a regional natural park, a biosphere reserve, and a national reserve have been created there. In the years 2000, an interdisciplinary research bringing together researchers in human sciences and natural sciences was devoted to constructing conjectures on the long-term social and natural dynamics of this fragile environment. The scenario was built in two phases (Mermet and Poux, 2002). The first was a reflection based on the Camargue's spatiotemporal path providing information on the major changes occurred along the centuries in the use and organization of this territory. The second phase consisted in testing out several approaches of scenarios aimed at integrating different territorial scales. To start with, scenarios were constructed at the scale of the Camargue territory (meso level) according to Godet's method of scenarios: trend scenario, and contrasted and rupture exploratory scenarios resting on very different assumptions on futures, yet coherent and plausible. These assumptions were subsequently expressed in quantified evolutions on the state of the ecosystem, on the population, activities, etc., as well as in maps. These figures and maps were used as discussion supports on the socioecosystem's dynamics and on the stakes involved in its management. Another phase of scenarios consisted in integrating into the prospective reflection, the macro territorial levels in which the Camargue is embedded: regional and national context, national level, and European level. Based on the prospective scenarios drawn up in these three contexts, the major orientations in regard to demography, habitat, transport, and development infrastructures were compared with the scenarios produced at the meso scale. Finally, a prospective exercise at the micro scale was carried out in order to perceive how Camargue's stakeholders would react to the orientations of the evolution scenarios worked out at the two other scales. Interviews were carried out with large estate owners who volunteered. They helped to determine the reaction and adaptation times depending on the projected evolutions and to reveal strategies specific to each estate. The innovative element of this approach is the will to combine all three levels of organization—macro, meso, and micro—in building scenarios on the Camargue's possible futures.

Coupling the scenario with the model depends on the desired goals in the unfolding of the prospective approach. Either the prospective scenario feeds the model and modeling is carried out after writing the scenario; modeling is of the inquiry-driven scenario analysis type, and meets various objectives, mainly that of testing the coherence of the scenario, the plausibility of the relations between the scenario's variables. Or the scenario is the outcome of the simulation. This is the "classic" approach to modeling land use changes.

However, whether in working out scenarios or in the modeling phase, uncertainty is omnipresent and takes various forms.

4.3 The modeling approach in geoprospective

4.3.1 Uncertain knowledge in territorial geoprospective and its formalizations

Territorial knowledge injected into a geoprospective approach can but be uncertain: multiple observation levels of a complex system, blurring in the observation of phenomena, impossibility to precisely determine the relations between them, unpredictability of any future evolution. There is a relation between the complexity of territorial systems and that of the cognitive approaches used to understand them. The complexity of territories presents us with situations of knowledge in which we are neither certain of an event, nor of its absence. Nevertheless, we have a differentiated evaluation of the two possibilities, because we hold them as differently "plausible." Here, geoprospective can take advantage of a dialogue with artificial intelligence, which has traditionally been used for formalizing human thinking in a context of uncertainty.

Using probabilities has been the starting point, both for describing complex systems and for artificial intelligence. The renewal of the classic frequential framework by the Bayesian approach has widened the application of probabilities to uncertain knowledge situations (Jaynes, 2003). It has subsequently been the base of machine learning algorithms from the years 1980 and 1990 onwards. As regards the territory, Harvey (1969) and Withers (2002) have shown the epistemological bases and the practical interest of a Bayesian geography. Bayesianism (Drouet, 2016) assumes that the beliefs of any subject in relation to an event are graded evaluations which take the form of probabilities, non-negative quantities between 0 and 1, which can be attributed to unconnected elementary phenomena within a universe of possibilities, and the sum of which equals 1. Only the phenomena held as certain ($p = 1$) or impossible ($p = 0$) are free from uncertainty. Bayesianism considerably widens the applications (and interpretations) of probabilities: the latter not only provide a model of the asymptotic behavior of the frequency of structurally stable phenomena (frequential probabilities) but also a model for degrees of belief of rational agents in a context of uncertainty. Bayesianism also provides a rational rule for updating beliefs when new information is obtained: Bayesian conditioning, one of the key concepts of which is "likelihood": a graded evaluation of the plausibility of different hypotheses in the presence of an ascertained empirical data, corresponding to the probability that this data can be produced if each hypothesis was true. By seeking maximal likelihood, the Bayesian approach can also guide scientists in the abductive research of the model most likely to explain observations, making explicit his hypotheses and how to review his beliefs. This is particularly useful in small data situations, when the scientist has different theoretical models and a very limited number of well-formalized examples to arbitrate between them (a very frequent case in territorial analysis). Fusco and Tettamanzi (2017) apply these questions to evaluating urban densification policies and develop a heuristics for assigning likelihoods that models the scientist's reaction to the lack of plausibility of the models at his disposal.

Nevertheless, the Bayesian probabilistic approach imposes certain constraints: first, the necessity to know ex ante the universe of possible outcomes and the capacity to allocate an additive evaluation of the beliefs on these unconnected events; next, the necessity to be able to determine the likelihood of any possible pertinent event in connection with the one being studied. The last limitation concerns the quantification of the probability of an event E by the value p. We use $1 - p$, at the same time for expressing our belief in E and our uncertainty on E. This does not allow us to take a position of epistemic withdrawal where we could neither believe in E nor in $\neg$E.

Thus Shafer (1981) proposes not to consider Bayesian probabilities as a standard of rational reasoning, but as a possible model among others, that is particularly pertinent for certain situations of knowledge, but becomes unworkable for others, which leads to turn toward other models. The most general model to represent uncertainty in artificial intelligence is the imprecise probability theory (Walley, 1991), where each value of probability is replaced by an interval of minimal and maximal probability, but which is not very easy to handle. Evidence theory (Shafer, 1976) and possibility theory (Dubois and Prade, 1988) constitute more workable alternatives. Instead of giving one measure to characterize an event, each of them gives two, between 0 and 1, which will be in a duality relationship. The first allows to quantify the subject's epistemic involvement in the truth of the event, the second represents the event's plausibility, that is, its normal and nonsurprising character: "belief" and "plausibility" in the evidence theory and "necessity" and "possibility" in the possibility theory.

To conclude, geoprospective is constantly faced with uncertain knowledge. Nevertheless, in its models and approaches, it can resort to a range of formalisms enabling the subjects (the scientist, the territorial stakeholders) to have a rigorous reasoning in a context of uncertainty.

4.3.2 The specificities of geoprospective modeling

The spatial uncertainty of future change is in the forefront of uncertainties in modeling. Numerous studies that have shown different LUCC models can give different land cover projections even for the same storyline assumptions (Busch, 2006; De Chazal and Rounsevell, 2009; Hervé, 2018). The locations of future changes which differ from a family of models to another, and also between scenarios of a same model, are perplexing for local players in wait of certain knowledge. How credible are diverging results? How to take advantage of spatial uncertainty? The maps of spatial uncertainty are an answer among others. The superimposition, pixel by pixel, of the results of the multiprojections, enables to map the probabilities of a phenomenon occurring in each point (Verburg et al., 2010). These maps differ from the suitable maps which show the areas concerned by a type of change for a single scenario (Camacho Olmedo et al., 2013), whereas in the case of spatial uncertainty maps, this information comes from a number of scenarios. The frequency of changes can be quantified and expressed in probabilities so as to detect the areas with a high probability of change—hot spots—as well as stake-laden spaces (Houet et al., 2014).

As was repeatedly stressed, the place given to the past in assessing future changes is a major issue in prospective that has an impact on the choice of the modeling and also its validation. In this regard, the modeling approaches can be classified into two categories: path-dependent versus non path-dependent modeling. The path-dependent models (Brown et al., 2005; Houet et al., 2016a) are dependent on past paths (Markovian models). Therefore they are adapted to trend scenarios and to short-term simulations. Models resulting from narrative scenarios—SAS approach—and which include contrasted evolutions, rupture phenomena, and modifications of the explanatory factors of change, are de facto non path-dependent. The validation of this type of model is trickier. Thomas Houet proposes to compare the usual sensitivity tests with new proceedings based on integration tests. In the case of fictitious scenarios simulation, the idea is to introduce ruptures in the dynamics of change, in terms of quantity, process, or location, in order to verify that the model is capable to manage the modifications introduced, and give results that respect the principles of coherence and likelihood (Houet, 2015).

Another problem arises from the calibration of complex models. Unlike the models simulating the dynamics of a single phenomenon, such as the evolution of the built fabric or a type of land use, integrated complex models, that take into account the dynamics generated by the combined interactions of several phenomena, do not lend themselves to the same calibration and validation protocols. In their quality of complex systems endowed with the self-organization property, socioecosystems, like territorial systems, undergo evolutions described as quasipredictable. "Because these systems are, in a sense, creative, models of them must reflect the strategic indeterminacy of this creativity, and methodological protocols must support it" (White et al., 2015). Given that bifurcations exist, the prime interest of these models is to discover, through multiple runs, the possibilities of bifurcation inherent to the system. However, the usual calibration methods in relation to the data observed lead to blocking the possibilities for various behaviors and so far unknown spatial structures from emerging, yet such information is valuable for anticipating possible futures and making decisions in the present. In the perspective of the traditional validation model, a model capable to produce different types of behavior of a territorial system is deemed "complete but inaccurate" (Marks, 2007). Validation protocols appropriated to the models of complex systems still remain to be devised. However, the validation of a model of complex system is already achieved through repeated applications, and rests on multiple tests on a variety of similar cases by using multiple criteria in validation tests (White et al., 2015).

4.3.3 A wide range of modeling methods

The choice of the methods used in geoprospective depends on four elements: (1) the end-purpose of geoprospective and those it is meant for, (2) the scale chosen for the prospective, (3) the anticipated phenomenon (a variable vs a set of variables), and (4) the phase of the geoprospective approach. Since the latter chains several steps, different methods are de facto coupled.

- Within the framework of the SAS approach, there are a number of methods for linking narratives and quantitative scenarios. To convert qualitative information into quantifications: cross-impact analysis (Helmer, 1981); semiquantitative methods to link narratives and models (Kok et al., 2014; Houet et al., 2016b).
- The methods used for processing imprecise or uncertain knowledge can mobilize ontological approaches based on a Unified Modeling Language (UML), Fuzzy Sets (Kok, 2009; Jetter and Kok, 2014), knowledge-based systems or expert systems (Pearl, 1988; Sajja and Akerkar, 2010), or even Bayesian networks (BNs) (Jensen, 1996; Withers, 2002).
- The most frequent simulation methods use cellular automata (CA), multiagents systems (Bousquet and Le Page, 2004), logistic regressions, neural networks, artificial intelligence, and Markov chains.
- Spatially explicit modeling most frequently combines CA and geographical information system (GIS) (Feng et al., 2007; Wagner, 1997; White and Engelen, 1997), CA and remote sensing (Wang, 2001). The prospective spatial models using image processing methods and mathematical morphology—spatio-morphological methods (Voiron-Canicio, 2009, 2006)—are rarer.

- Participative modeling generally uses role plays (Le Page et al., 2011; Amalric et al., 2017; Becu et al., 2016), constructions of graphic models in workshops, but also—although more rarely—simulations of 3D possible futures in interactive mode (Loubier, 2013; Loubier et al., 2017).

Markovian models are the most used for the dynamic simulation of land use changes, such as Idrisi's Land Change Modeler, SLEUTH (Clarke et al., 1997). The Markovian process assumes that the state of a system at the t2 time can be predicted based on the evolution of its states at t0 and T1. Now, these path-dependent probabilistic models have limitations for spatializing prospective scenarios with contrasted evolutions. Therefore adaptations have been implemented by researchers belonging to the geoprospective current. SLEUTH* is an external module added to SLEUTH in order to simulate urban expansion scenarios with possible trend breaking and integrate options ensuing from stakeholders' viewpoints (Houet et al., 2016a). Furthermore, a novel approach of the Markov chains, the inverted Markovian approach, has been proposed in the context of a prospective modeling of landscape transformations (Loubier, 2013). Starting from the cartographic representation of one of the territory's possible futures, the Markov chain leading to that future has been extracted. The interest of this method is to easily "create" possible futures. Indeed, it is easier for nonspecialists to draw maps of the future than to act on transition probabilities. Moreover, comparing the probabilities of the chains resulting from these different maps helps to understand which are the interland use transfers and above all to measure the intensity of such transfers via the value of the transition probabilities.

The following sections expound a number of methods in greater detail. Other methods will be used in applications presented in later chapters.

4.4 Modeling spatially explicit prospective based on the land change approach

The study of changes in land use makes it possible to establish evolutionary trajectories to help planners to anticipate the possible futures of territorial evolution. Since the mid-1970s, this approach has seen increased interest in relating changes in land use to the modification of the albedo (Otterman, 1974) and the atmospheric content of greenhouse gases (Woodwell et al., 1983), these two phenomena participating in modifying the state of the atmosphere. It is in this context that in the early 1990s Turner and his collaborators (Turner et al., 1990, 1993) established the principles of a LUCC research program, this program having been formalized in 1995 (Turner et al., 1995) and implemented within the International Geosphere-Biosphere Program in 1999 (Lambin et al., 1999). In this context, many works have given rise to applications of artificial intelligence to simulate territorial dynamics.

The modeling of LUCC aims to provide a formalized knowledge base which may or may not be applied for the management of the territory by scenarios where the intervention of experts is often necessary. According to Veldkamp et al. (2001), it must take into account six main components:

- The level of analysis: it involves interacting the knowledge of experts from the social sciences, ecology, and geography by integrating micro- and macro-economic organizational models with environmental models, themselves structured around of GIS. Among the approaches, we can quote those of multiagent simulations and micro- and macro-economic models. A large part of the land use models is based on micro-economic aspects where managers seek to maximize the utility and the expected returns from a given space. This kind of approach is only applicable on small areas. Macroeconomic theories are used to formalize models applied to larger areas where they attempt to analyze the spatial directions of land use.
- Multiscale dynamics (nesting of scales/crossed scales): a scale can be defined as the spatial, temporal, quantitative, or analytical dimension used to measure and study objects and processes (Gibson et al., 2000). A scale range must be defined for each of the land use change and land cover processes. The spatialization of individual behaviors or groups of actors is quite difficult to achieve, while that of environmental factors and indicators is often easier. It is also advisable to seek the hierarchy of the factors intervening on the LUCC according to these scales. Two approaches are mainly used to quantify the multiscale relationships between changes in land use and factors: the first is based on the representation of data in several grids of different sizes. At each spatial resolution, the relationships between LUCC and factors of change are then statistically determined. The second approach uses multiscale statistics mainly applied in the social sciences. This approach is generally based on CA for which constraints can be defined according to different spatial scales (White and Engelen, 1997, 2000).
- Drivers of change: they concern all aspects linked to land use such as economic, social, ecological, environmental, historical, and even cultural. The selection of variables is essential and can be tricky when working on variable perception scales. It also depends on the degree of simplification of the model. The quantification of the relationships between LUCC and the drivers of change can be envisaged according to three approaches: either by trying to establish the relationships directly on the processes involved using physical laws, or by using empirical methods, or by using the knowledge of experts, the latter being implemented in CA and multiagent systems (MAS).
- Spatial interaction and neighborhood effect: the parameters of land use changes often show spatial autocorrelation with respect to environmental factor gradients and the distribution of land use classes. This spatial autocorrelation also derives from interactions between the types of land use themselves. Since the neighborhood

represents a spatial dimension that must imperatively be integrated in the models, this aspect of the modeling is generally represented from CA.

- Temporal dynamics: the changes are generally nonlinear and the threshold effects of the variables and phenomena observed on the territory play an important role. Changes in land use are often dependent on random events making their modeling difficult. Some models do not take into account temporal dynamics and are based solely on an extrapolation of trends in land use changes. Feedback effects make it even more difficult to assess the phenomena over time.

- The level of integration: in land use changes, many parameters interact with each other in a time dynamic where feedback mechanisms can be added. The integration of this complexity in the models is often envisaged according to two methods: by the coupling of systems or by objects analyzed and modeled separately. It is then assumed that the interactions and feedbacks between the elements are negligible. The second approach is based on a more global view by trying to consider the interactions and feedbacks of the subsystems. In this approach, we consider that the variables are endogenous to the system and interact with each other.

The data used in the integrated models have often a heterogeneous nature (socioeconomic, biophysical, naturalist, etc.), except those concerning a specific discipline such as models based solely on macro- or micro-economic theories or even purely environmental models. These data relate to socioeconomic, physical, and ecological disciplines such as vegetation structures, cultural practices, market trends, and climate. They can come from observation networks, in situ measurements, digital or paper maps, spatial interpolations, expert analysis, etc. The following figure gives an example of the parameters and data taken into account for the modeling of LUCC in tropical zones, in order to understand and simulate the dynamics of deforestation (Fig. 4.1).

The main modeling techniques for the study of LUCC correspond to MAS (Parker et al., 2001, 2002; Sanders, 2006), to CA for the simulation of spatial dynamics of land use (White and Engelen, 1997; Dubos-Paillard et al., 2003; Langlois, 2010), or probabilistic approaches based on macro- or micro-economic models (Veldkamp et al., 2001; Verburg et al., 2010; De Vries et al., 2000; Fekete-Farkas et al., 2005). Such models can be coupled together according to the integration levels of the complexity of the system. In this case, these are modeling environments characterized by modules which provide inputs for other environmental modeling. Understanding the algorithms for calculating these coupled models is sometimes difficult due to this coupling and these models sometimes appear as "black boxes."

These methods and tools are able to take into account the spatial distribution and the shapes of the main drivers of transformation of land use and landscapes such as the physical elements of the territory (slopes, altitudes, types of soil and rocks, slope exposure, etc.), biotic elements (vegetation formations and species, etc.), and socioeconomic elements

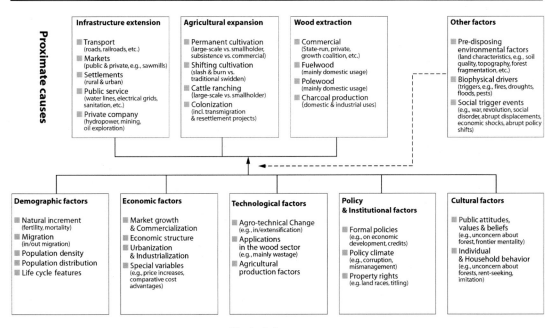

Figure 4.1
Parameters and data for modeling LUCC in the context of deforestation in tropical zones
(Geist and Lambin, 2001).

(buildings, infrastructure, activity zones, demographic trends, travels, etc.). The involvement of stakeholders in the development of development scenarios is necessary for the definition, the conception, and the evaluation of realistic scenarios.

4.5 Coupling prospective spatial modeling and decision aid tools

4.5.1 Companion modeling

The purpose of Companion Modeling (ComMod) is, as its name implies, to accompany a collective decision-making process as regards the sustainable management of a territory, by means of various types of models, some of them a-spatial, and others, more numerous, with a spatial reference. Models of the a-spatial type are used during the definition phase of the scope of the prospective approach and knowledge-sharing by the various actors forming the group. The ARDI (Actors, Resources, Dynamics, and Interactions) method devised by Etienne et al. (2011) is an example of method meant to serve as a guide in that phase. The ontological approaches based on a UML are often used to help clarify the terms used and to formulate in a common language and a class diagram the representations of the actors' knowledge and points of view (Caglioni and Rabino, 2007). The conceptual model jointly conceived by the group aims to collectively explicit the perceptions of the current dynamics and as projected into the future and also to reveal existing uncertainties. Statistical models and systems dynamics models can be associated in order to quantify the interactions of variables, to determine indicators, and to calibrate the conceptual model. Then comes the participative modeling phase consisting in simulating future evolution scenarios, assessing them, and discussing them. This modeling process based on the conceptual model is generally carried out using IT tools such as CA and MAS, coupled with GISs which provide the initialization data for the spatial configuration and the agents' characteristics (Gourmelon et al., 2013). Among the existing multiagent simulation platforms, CORMAS[1] has been designed by CIRAD (the French agricultural research and international cooperation organization) researchers for representing in a virtual spatial context the interactions between actors concerning renewable natural resources (Bommel et al., 2015; Bousquet and Le Page, 2004). Participative modeling relies on a medium called game board or simulation board, which is a schematic representation of the socioecosystem in the form of a regular plot plan. Each cell of the grid has a number of initial attributes that players modify through choices of actions at each round of the game. Fig. 9.4 is an example. This spatial medium then becomes the players' field of action as part of a role-playing game coupled with simulations. The spatial interface is generally associated to an IT tool that determines the modifications induced by the moves of the players and helps to follow them from simulation to simulation (D'Aquino et al., 2001; Etienne, 2012).

The interest of the ComMod modeling process is to combine several levels of organization, social entities, and spatial entities at different scales: plot, holding, terroir, commune, region, and country. This combination occurs via "maps" between which the players move depending on their needs and decisions of action (D'Aquino et al., 2012). This multidimensional virtual space is first of all a space for debating, an environment of resolution of environmental problems that facilitates the integration, in the players' reflection, of the complexity of processes operating at embedded levels.

4.5.2 Spatial prospective and companion modeling: toward a jointly constructed and informed decision

When spatialized prospective is combined with companion modeling, it ties up with the area of geogovernance (Masson Vincent et al., 2012; Masson Vincent and Dubus, 2013; Dubus et al., 2015). Geogovernance is an approach relying on the use of the methods and tools of spatial analysis, which is intended for putting within reach of all stakeholders concerned a pertinent territorial information, all along the construction chain of the territorial project. Its aim is to contribute, in addition to other methods, to making the territory's complexity intelligible and bringing out the latter's sociospatial stakes.

---

[1] CORMAS: COmmon-pool Resources and Multi-Agent Simulations. http://cormas.cirad.fr/indexeng.htm.

Figure 4.2
Classification gradient of the experiences.

This approach rests also on collecting the populations' spatial representations, their conceptions of the functioning of the territory, and their expectations as regards future development. The resulting knowledge is spatialized, according to various methods, and communicated to all the territory's stakeholders, be they residents, administrators, or decision-makers (Masson Vincent et al., 2012). The prospective modeling, companion modeling, and geogovernance triad is meant to put at the disposal of all the territory's decision-making players an easier access to the territory's spatiotemporal stakes, via approaches, tools, and innovative cartographies integrated into a process of to-and-fro between stakeholders and researchers. Maps, in particular, as results of prospective models, help stakeholders to navigate in the complexity and reach an informed decision (Dubus et al., 2010). So, this triad determines the outlines of a very fertile movement in geography since the beginning of the years 2000 although some of the tools used for companion modeling are anterior, such is the case, for example, with UML modeling, the first stable version of which was proposed in 1997.

All these approaches share the same project: to decide on the allocation of one or several spatial resources, in a collective and democratic manner. This process unfolds generally in a context of uncertainty about the future, tensions and even conflicts around the way these resources are divided and used. The tools and methods used in these approaches are extremely diverse. Christine Voiron-Canicio classifies them into two families: qualitative and quantitative (Voiron-Canicio et al., 2016). We can now add that the degree of integration of these methods into the overall mechanism for solving the project's support spatial problem shows a gradient that could be practical for classifying the real-life experiences on which this corpus of knowledge was built (Fig. 4.2).

For example, it can be considered that a model such as Cityscope[2] (Grignard et al., 2018) lies strongly to the right of the gradient whereas the ComMod approach would move a little more to the left. The issue here is not to measure the degree of computing complexity of the proposed solution but rather the degree of integration of the stakeholders into the process of appropriation/construction of the tools to help solving the support problem.

---

[2] Cityscope is a complete computation environment developed by the MIT's MediaLab enabling to visualize the complexity of a functioning city in an interactive 3D environment.

In our opinion, this is the best way to differentiate the vast number of coupling approaches between spatial prospective and companion modeling.

4.5.3 Introducing the modeling of the decision-making process into geoprospective

Observing these various experiments and results highlights a surprising phenomenon. Whatever the place of the experiences and models proposed on the gradient of Fig. 4.2, the stakeholders' decision-making process is implicit and always results from the man/machine interaction. It is difficult to verify whether this situation is the consequence of a thought-out and backed-up process or whether the issue was not really taken up clearly. However, it seems excluded that the researchers involved in that research might have considered that the topic was trivial. Generally, the aspects concerning the stakeholders' decision-making mechanism intervene at the level of map-making (Loubier, 2013). It is even the main premise of geogovernance. Therefore lots of works have concerned the development of visualization tools, in particular, with the use of 3D. The idea usually put forward is that a visual support facilitates dialogue among stakeholders for solving the problem. By doing so, the researchers adopt the following altruistic assumption. The stakeholders involved are capable to set aside their personal interests to the benefit of the common good, and even to forego some of the dividends of their social status such as the power to decide or influence the solution through a power relationship. Research work has shown that it was practically never the case (Emsellem et al., 2018). Actually, this criticism is often put forward as a limit to these approaches, in particular, within the framework of geogovernance.

This statement of fact leads us to focus on the modeling of the decision-making process in field experiences. This decision-making process is an intrinsic component of companion modeling. Yet, although the community has easily appropriated instruments stemmed from computing from most of them, it seems more hesitant to take on those of decision-making, in particular, those stemmed from operational research and utility functions. However, simple tools are available, for example, choice matrices and their criteria (Maximin, Minimax, Savage, Hurwicz, etc.). Yet they have the advantage to objectivize the solution fields or at least, to limit the ambiguities of each stakeholder by proposing a clear and measurable decision framework. Integrating decision modeling into a spatial prospective/ companion modeling coupling has recently been implemented in a highly conflictual context in Switzerland with very convincing results (Rojas and Loubier, 2017). Following a change of legislation at the federal level, Swiss communes were forced to reduce significantly their building areas. Therefore it had to be decided which cadastral plots would change from building to nonbuilding land. It is easy to imagine the stakes and the interpersonal conflicts generated by such decisions. The authors of the study showed that, beyond the rezoning, it is above all the feeling of lack of transparency in the decision-making process that generates most conflicts, with the feeling that decision-makers will protect their own interests by presenting a biased system, and this only because they have access to information that is not available to the average citizen. In such a context, it was deemed indispensable to develop an approach conducive to objectivizing the decision-making process. Then the researchers constructed a spatial modeling mechanism based on a multiple criteria decision method: the Analytical Hierarchy Process (AHP) developed by Saaty in 1980 (Saaty, 1980) coupled with a GIS. The coupling with a GIS was explored by Eastman and his team in the years 1990 (Eastman et al., 1995). The AHP approach makes it possible to give an objectivized and consistent weight to important spatial layers to meet the target, then a GIS raster is used for carrying out a weighted combination of these layers which, in the end, gives a score of adaptation to the initial target visualized in map form. In a conflictual context, this approach guarantees that there is no bias of powers because it is impossible to give weights that would favor a place, in particular, without being detected by an absence of consistence. Several communes, including Zermatt—one of the main tourist destinations in Switzerland (see Chapter 7: The Touristic Model of Valais Facing Climate Change: Geoprospective Simulations of More Environmentally Integrated Development Models)—, carried out their rezoning process successfully by using that approach. Therefore it seems that geoprospective modeling coupled with companion modeling would gain much by better integrating the subpart of the explicit modeling of the decision into the whole of the tools available within the prospective modeling, companion modeling, and geogovernance triad.

4.6 Uncertain causal models in the geoprospective approach

4.6.1 Rationale and options for uncertain causal models

According to Blecic and Cecchini (2008), decision-makers exploring possible futures for their territory need models of causal knowledge linking phenomena of concern among them. Clearly, these models must integrate uncertainties and will be partially subjective, that is, limited to their authors' knowledge. Nevertheless, decision-makers must take decisions based on projections of expected outcomes of their actions on the territorial system. Uncertain causal models (Pearl, 2000) meet this need. They couple a causal graph representing cause-to-effect relationships among the system elements to an appropriate quantitative formalism of uncertain knowledge (see Section 3.1). This modeling option is particularly interesting in geoprospective approaches where uncertainties must be integrated in the knowledge of complex territorial systems. Uncertain causal models are strategic aggregated models allowing us to understand big-picture trends in the spatial system, when the latter are known with uncertainty, for example, by using indicators and relationships among them to simplify more specific but poorly understood causal mechanisms. More particularly, geographers at ESPACE have been developing geoprospective uncertain causal models for several years (Fusco, 2008, 2010, 2012; Scarella, 2014; Dubois et al., 2016).

BNs (Jensen, 2001; Korb and Nicholson, 2004) are the most common option and use Bayesian probabilities to quantify uncertain knowledge. They can be produced by expert knowledge, through knowledge discovery in databases, and through a mix of these two approaches. The ease of probabilistic calculus after the key developments of the 1980s and 1990s (Gibbs sampling, junction trees, etc.) makes them an extremely powerful tool. Modelers can get closer to the complexity of real systems by integrating a large number of variables, multiple and entangled causal paths, as well as nonobservable variables and even the influence of omitted variables. Having a BN of the system under study, we can finally use Bayesian inference in a highly flexible manner (reasoning can be diagnostic, predictive, intercausal, etc.). Possibilistic networks (Benferhat et al., 2002) and credal networks (Cozman, 2004) are alternatives to BNs using more sophisticated formalisms of uncertain knowledge: possibility theory and imprecise probabilities, respectively. Although allowing more flexibility in modeling uncertainties, these formalisms are harder to operationalize. It is thus no surprise that, as of today, geographers mainly focused on BNs in geoprospective approaches. Within this section, we will present some examples of geoprospective models through BNs and we will conclude with an application of possibilistic networks, by highlighting their contribution to geoprospective thinking.

Whatever the employed formalism, uncertain causal models must be adapted to the specificities of spatial systems. A Bayesian, possibilistic, or credal network is an a-spatial model of relationships among variables. The modeler must thus find specific solutions to integrate spatial relations. A second difficulty concerns the modeling of feedbacks as uncertainty propagation algorithms can only be implemented on directed acyclic variable graphs. Possible solutions are either integrating time or delimiting the study domain to a part of the complex system where relationships present cause-to-effect asymmetries which are easier to model. Within geoprospective studies, these difficulties have been first tackled with BNs.

4.6.2 Bayesian Networks as geoprospective models

Fusco (2012) developed a BN from expert knowledge as a geoprospective model of metropolitan dynamics on the French Riviera (Fig. 4.3). The model identifies the key spatial components of the metropolitan system (the coastal area, the hinterland, and the whole metropolitan area) and uses probabilities to describe their specific spatial interactions (population spillover, interaction among economic activities, etc.).

This kind of model can describe the interactions among a limited number of spatial components and is not appropriate for fine-grained spatial modeling. However, it introduces important innovations for geoprospective modeling. First of all, it makes expert knowledge elicitation practically possible by using noisy logical gates (Noisy OR and Noisy AND, modeling sufficient and necessary causality, respectively). The number of model parameters that experts must provide is considerably reduced and their estimation becomes feasible.

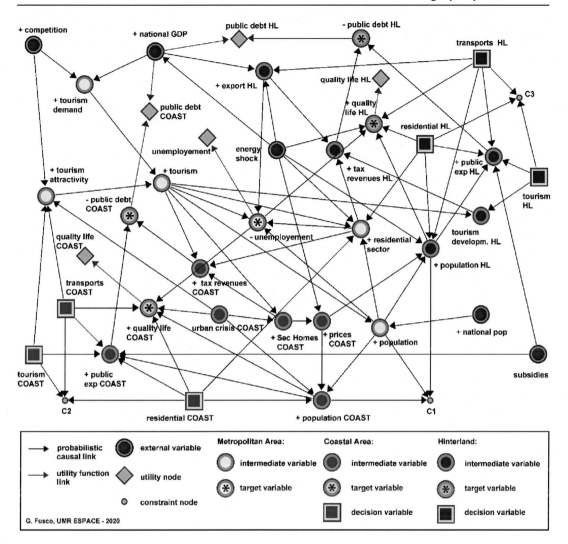

Figure 4.3

Geoprospective Bayesian Network modeling metropolitan dynamics on the French Riviera.

It even becomes possible to use panels of experts (for example, in a Delphi approach; Linstone and Turoff, 2002) or actors (in a participatory initiative). Second, it integrates explicitly new internal and external variables acting as wildcards within the geoprospective time horizon, it identifies target and decision variables, and transforms the BN in a decision graph (Jensen, 2001). Third, it proposes different model-based heuristics to build contrasted scenarios and uses the probability calculus to identify optimal policies for each scenario. The trend scenario is thus compared to two sets of alternative scenarios. The first are obtained through backcasting: after identifying goals to be achieved or situations to be avoided on the target variables, the BN determines the most probable causal antecedent events for these scenarios. The second are obtained by activating the wildcards (here an energy shock and an urban and social crisis in the coastal cities) and by propagating with probabilities the effects of these disruptions down the causal chain (forecasting).

This work also innovates by introducing variable feedbacks through dynamic BNs. A first model takes account of the relationships among the variables within a phase of the economic cycle (half cycle of 3−5 years). Within a phase, we can assume a certain stability for the modeled phenomena (economic activity, employment, and population either grow or stagnate). A static model can thus be built: the time variable is not considered, and loops are excluded. This model describes the territorial system for a time-slice that is short enough to identify causal drivers and their consequences within the time period of 3−5 years. This static model is the starting point for developing a dynamic model for long-term simulation. This is done using dynamic BNs, where variable loops and the time variable can be integrated. Dependent variables within a time-slice can become causal antecedent events for drivers in the following time-slice. The impossibility of modeling variable loops within BNs is thus solved, as variable B, depending on A at time t1, can influence A at time t2, and so forth, increasing the variety of possible relations in the model. Two hypotheses underpin dynamic BNs:

- The Markov hypothesis, according to which variables in a given time-slice only depend from a limited number of previous time-slices (the first-order hypothesis only considers the preceding time-slice, but higher order hypotheses can be used to model path dependency);
- Phenomena stationarity, that is, the invariance of the probabilistic parameters of the BN over time (this hypothesis can be partially released).

As far as spatial relations are concerned, a different modeling solution is to build a BN as a model of the interactions of each spatial unit with its neighbors within the territorial system. It is here fundamental to focus on the right neighborhood and on the most relevant spatial relations. In the geoprospective model of the evolution of regional urban networks (Fusco, 2010), the BN model describes the spatial relations of each municipality within neighborhoods of varying size, taking into account the attractive power of surrounding urban centers. Describing these relations is a preliminary phase to simulate the evolution of the regional urban networks over the decades, based on parallel models for the almost 1000 municipalities within the study area. The probabilistic parameters of the model are learned from past data and a trend scenario is then calculated with probabilities under the assumption of system stability. In this work, as in a previous one on the socioeconomic and urban dynamics of a coastal region (Fusco, 2008), time is integrated by coupling state variables (describing a spatial unit at time t1) with dynamic ones (describing the evolution between t1 and t2). Unlike the geoprospective model for the French Riviera (Fusco, 2012), these models are learned from geographical databases having a time depth of 20 years. In the coastal region model, for example, by using static and dynamic variables we can describe complex interactions: accessibility levels produced by the transportation networks are an explicative factor for the housing characteristics in each municipality, which heavily impact local demography (old-age ratio, household size, and their evolution over time). The latter is in its turn a driver for housing dynamics, creating the beginning of a loop which could be used to develop a dynamic BN model.

4.7 Beyond probabilities in geoprospective uncertain causal models

Two last geoprospective models will close this review and go beyond the limits of probabilistic modeling. These models deal with the dynamics of social polarization in the metropolitan area of Marseille (France) over the last decades. The study area encompasses the cities of Marseille, Aix-en-Provence, Toulon, and Avignon and has more than 3 million inhabitants. The models were then used within a geoprospective approach. They focus on the spatial distribution of two target populations, professionals and the unemployed, seen as structuring feature of social polarization in the metropolitan area (Centi, 1996; Fusco and Scarella, 2011). The overrepresentation of these populations is thus considered as an indicator of valorization or devalorization, respectively, for the municipal housing stock. However, knowledge of the factors inducing social polarization of the municipalities in the study area is uncertain. Several factors contribute to housing valorization/devalorization, but they are not deterministic in nature. This motivates the use of uncertain causal models. Scarella (2014) first proposed a BN model. Tettamanzi and Fusco (2016) later developed an equivalent possibilistic network model. The probabilistic model and its possibilistic counterpart share the same structure of causal dependencies among the 26 model variables, covering different aspects as the geographic position of the municipality within the metropolitan area, residential mobility flows, the presence/absence of environmental amenities, housing stock characteristics, urban planning policies, and path dependency of social specialization. Other shared characteristics are

- Both models describe the spatial relations of each municipality with its environment (surrounding municipalities, nearby metropolitan centers), whenever the latter can influence its residential valorization.
- Both models are based on expert knowledge. More precisely, BN parameters were elicited and validated with the statistical analysis of historic data. A least committing probability-to-possibility preference preserving transformation (Dubois and Prade, 1988) was used to obtain the possibilistic parameters.
- They include observable and nonobservable variables.
- Probabilistic and possibilistic relations are modeled through noisy and uncertain logical gates, respectively, which reduces considerably the number of parameters to be elicited.
- They model uncertain relations, include leak parameters (to take into account the influence of omitted variables), and produce uncertain results.

But the two models implement different formalisms of uncertain knowledge: probabilities, whose sum and product combinations are well suited for an exact knowledge or their values, and possibilities, whose max and min compositions are better suited to qualitative knowledge. One more difference is the ability of max/min compositions to keep track of the elicited parameter having led to a given prediction. This facilitates sensitivity analysis of results to the elicited parameters. The BN model was used to guide a whole geoprospective application: 10- and 20-year trend scenarios were calculated through probabilities; disruptive scenarios were built through forward propagation of wildcard events; scenario comparison finally allowed to determine which municipalities were stable and which were subject to possible transformations. Spatial interaction and time dynamics were modeled by coupling the BN model to a GIS. The most probable outcomes at 10 years are the input for GIS calculations of spatial relations within the metropolitan area and these are the new input for the BN model of the coming decade.

The possibilistic network was only used to build a 10-year trend scenario of social polarization in the study area. The goal was to identify analogies and differences with the same scenario built through the BN model. The outcomes of the two models were also compared through an uncertainty-based interactive geodataviz (Cao and Fusco, 2016). The knowledge of complex phenomena like social polarization within a metropolitan area is represented through a system of dashboards with dynamic links between maps, diagrams, and text (Fig. 4.4). Levels of uncertainty, whether probabilistic or possibilistic, can be explored interactively and their comparison is particularly instructive for geoprospective purposes.

Fig. 4.4B and C shows the most straightforward use of probabilities. For each municipality, the 10-year trend scenario indicates the most probable prediction of the BN model: valorized, devalorized, or other. Compared to the present state (Fig. 4.4A), the most probable predictions show valorization spreading around Aix-en-Provence and more stability in the rest of the metropolitan area. But these maps of most probable results give a false impression of certainty. Differences with second or even third most probable outcomes could be very small in probability. When a significance threshold of 0.1 is considered for these differences, two or even three outcomes can be considered equally probable in the trend scenario (Fig. 4.4D). These situations are represented in the map through a combination of two colors and through the use of the gray color, respectively. These can be considered as more uncertain situations. Uncertainty increases when significance thresholds increase for probability differences (Fig. 4.4E and F). This is a way of eliminating the impression of certainty given by the set of strictly most probable outcomes.

It is particularly instructive to compare the most plausible outcomes predicted by the two models for the 10-year trend scenario. In the possibilistic model, the discrete set of possibility levels and their max/min combinations in possibilistic inference produce equally possible outcomes without imposing any threshold on possibility differences (Fig. 4.4G). The outcomes

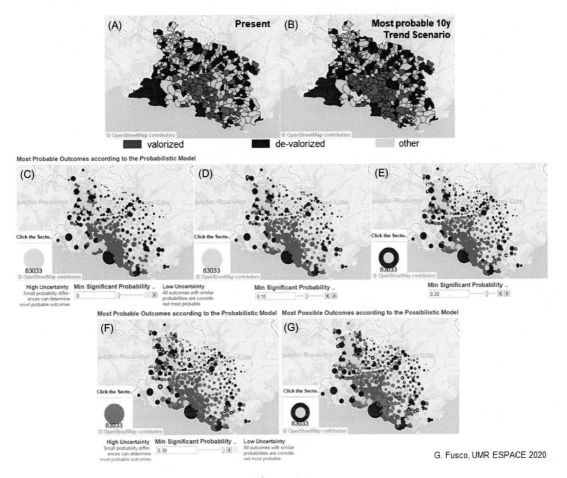

Figure 4.4

Exploring the 10-year most plausible scenario of social polarization (A: Present situation; B: Most probable 10-year trend scenario; C-F: Filtering the most probable 10-year trend scenario with different probability thresholds; G: Most possible10-year trend scenario).

of the possibilistic model seem in general more uncertain than those of the probabilistic one (Fig. 4.4C). When we filter the probabilistic outcomes by increasing difference thresholds, the BN prediction becomes more and more uncertain (the best agreement between the two models is obtained with a probability difference threshold of 0.25).

A sensitivity analysis of the model predictions was carried out using different values of the elicited parameters of the Noisy Max and Uncertain Max. These logical gates play a fundamental role in modeling the evolution of the valorization/devalorization of each municipality during the 10-year time span, which has direct consequences on the final predictions. The results of the analysis seem counterintuitive. Probabilistic predictions are less frequently impacted by perturbations in the parameter values, when compared to possibilistic predictions. At a closer look, though, BN predictions, when limited to the most probable outcomes, are more instable than possibilistic predictions. The former can become incompatible even in the presence of small parameter perturbations. The latter, by integrating more uncertainty, become either generalizations or restrictions of the original results, but are never totally incompatible with them.

The comparison also showed the differences between the Noisy Max and Uncertain Max logical gates, when evaluating the plausibility of rare events induced by the accumulation of concurring causes which could trigger them independently. In general, the possibilistic logical gate is more prudent and specific thresholds must be used to trigger the rare event. These technicalities may have important consequences when building disruptive scenarios.

Integrating uncertainty also allows a different understanding of the 10-year trend scenario. Either by filtering probabilities or by using possibilities, the most plausible outcome in 10 years is the continued valorization of the municipalities around Aix-en-Provence, two alternative outcomes for some municipalities and complete uncertainty in peripheral areas under concurring influence from several metropolitan centers. This is not just a different picture from the most probable predictions, but a more sincere one, highlighting the limits of our knowledge even in building trend scenarios.

To conclude, uncertain causal models can inform a geoprospective approach to social polarization in metropolitan areas. Methodologically, the use of probabilistic models should not be limited to identifying the most probable outcomes. Possibility theory is an interesting alternative for uncertain causal models. It produces more uncertain results reflecting the epistemic uncertainties in elicited knowledge. The results of the two models converge when probabilities are filtered through thresholds. The comparison with more widely used BN models opens the way to the use of possibilistic graphical models in geoprospective applications. However, further developments are needed to make possibilistic networks as versatile and powerful as BN models for scenario building in geoprospective applications.

The issue is not only to bring expert knowledge within reach of various audiences but also to make the complex aspects of the real understandable, and in particular spatiotemporal dynamics. To do so, geovisualization techniques are extremely effective.

4.8 Geovisualization—representing and sharing the knowledge gained from spatial models

Geovisualization designates the whole of the techniques available to visualize a phenomenon in order to help the user analyze and understand it, and which, for the most sophisticated, offer forms of visual exploration by interacting with the geolocalized objects.

There is a variety of visualization supports: map, graphic model, block diagram, 3D image, and film. Our purpose is not to inventory them, but to focus the attention on a number of methods particularly adapted to geoprospective. The first concerns graphic modeling of the choreme type which, applied to models and scenarios of the future, helps to better understand the spatiotemporal processes at stake in a given territory.

4.8.1 Graphic modeling and prospective choremes

The end-purpose of graphic modeling is to make visible spatial dynamics and the processes that generate them. This type of modeling is achieved using choremes. According to their designer, Roger Brunet (1980), choremes are the elementary graphic structures of the spatial organization of the geographic space. These elementary structures result from crossing four topological variables: point, line, area, and network, and seven configurations of space control: mesh, grid, attraction, contact, tropism, territorial dynamics, and hierarchy. These structures are represented by simple signs and forms—for example, a mesh, a front, an interface—and in limited numbers, around 30 initially (Brunet, 1986). New structures were subsequently added to the list within the framework of spatiotemporal choremes (Cheylan et al., 1997; Casanova, 2010). The table of elementary structures is used as a frame of reference in the modeling process.

Graphic modeling differentiates itself from graphic schematizing by the scientific methodology underlying behind its construction. An illustration is given with the Geoprospective project (ESPACE et al., 2013), carried out in 2012 for the Prospective Mission of the ministry of Ecology, Sustainable Development and Energy (MEDDE) as an extension of the Territoire Durable 2030 (Sustainable Territory 2030) program. This program is an expert prospective which presents in a narrative form the context of France's development by 2030 as well as four scenarios (BIPE, 2012). The program is presented in Chapter 13, How Do Public Policies Respond to Spatialized Environmental Issues? Feedback and Perspectives. The aim of the graphic modeling of prospective scenarios is to introduce the spatial dynamics likely to generate the change between 2010 and 30. The purpose is not only to represent in graphic form the dynamics of change in 2030 but also to highlight the processes involved, via a hypothetic-deductive approach based on the expert modeling phase. Fig. 4.5 shows the steps of the graphic modeling of one of the Territoire Durable 2030 scenarios: the trend prospective scenario, applied to the theme of pressure on the environmental resources (Casanova Enault and Chatel, 2017).

In that scenario, pressures on the environment increased between 2010 and 30. Territories restructure themselves around metropoles, and international openness favors the most connected regions. Urban sprawl continues while agriculture remains split into intensive production and an agriculture respectful of ecological balances (Casanova Enault and Chatel, 2017). The graphic modeling exercise breaks down the processes at work between

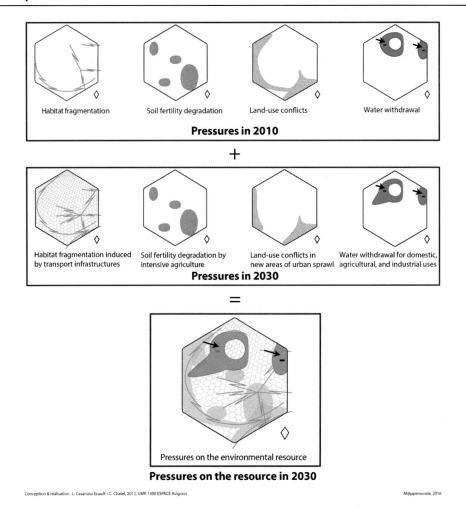

Figure 4.5
Example of a prospective choreme (Casanova Enault and Chatel, 2017).

the two dates. The environmental pressure is exerted through four processes: habitat fragmentation as a result of the multiplication of roads; contradictory dynamics generating conflicts of resource use, water in particular; resource degradation or erosion, and draining or withdrawal of water resources. Four graphic models depicting the pressures on environmental resources identified in 2010 are built and are used as a frame of reference for the evolution of spatial dynamics between 2010 and 2030. The breaking down of the four processes and their evolution between 2010 and 2030 helps to understand the "pressures on the resource" prospective graphic model in 2030.

Graphic models are at the same time vectors of communication of expert knowledge and supports for collective thinking. However, designing them necessitates learning the formalism of choremes and furthermore, the prior identification of the spatial processes at work in the prospective scenarios. For this reason, they are rather given for analysis in prospective workshops than given for constructing. On the other hand, 3D representations are thought to be the best way to facilitate participation.

4.8.2 3D representations and geoprospective

Various tools are available, going from block-diagrams to immersive modeling.

4.8.2.1 Prospective scenarios on landscape block-diagram

The support frequently used in participative approaches is the block diagram, which represents a landscape in 3D and enables participants to think spatially in simple and general standard situations. It is considered as one of the most effective dialogue tools (Michelin, 2000; Joliveau et al., 2000; Joliveau and Michelin, 2001). This is the type of representation chosen by the researchers of the MEDDE's Aqua 2030 prospective program to territorialize the prospective scenarios for 2030 (see Chapter 13: How Do Public Policies Respond to Spatialized Environmental Issues? Feedback and Perspectives). A block-diagram is representative of a type of French territorial system. That being so, the two block-diagrams of Fig. 4.6 concern the continental subsystem in wetland area.

They represent the initial situation in 2010 and the situation modeled in 2030, along the broad guidelines of one of the prospective scenarios envisaged. The directions of the scenario are drawn from visible, natural, and anthropic spatial elements. As certain directions are not visible—governance, fiscal policy, etc.—logos and labels are added. As an example, the scenario represented on the figure is characterized by two contradictory trends: the continuation of artificialization and rising awareness about environmental issues which is reflected notably in proactive policies. The visualization of the scenario's guidelines and specificity in relation to other scenarios is achieved through drawn objects, such as storage basins, solar farms, and rows of drip watering tape., as well as logos, such as that of on-site sanitation carried out thanks to a fiscal incentive policy.

4.8.2.2 3D geovisualization in geoprospective

3D in geoprospective is part of the latter's communication chapter. The approach consists in developing visualization tools enabling to erase the distance between the language of modeling specialists and the sensitive perception of stakeholders, some of whom do not have the expert knowledge needed to understand the scientific discourse. Among the approaches used to erase, or at least reduce, that distance, the one using immersion in a 3D environment has been developing significantly since computers offer a power compatible with the enormous needs in calculations required by this means of representation, and above all inexpensive.

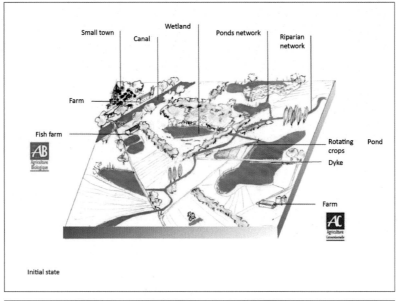

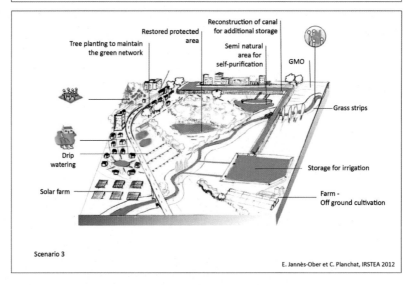

Figure 4.6
Example of prospective scenarios on a block-diagram. Source: Jannès-Ober and Lamblin, 2012, p.174 and p.177.

To tell the truth, 3D—it is hard to tell if it is a neologism or an acronym—groups together a vast amount of more or less interactive simulation environments of which we are going to give a thematic overview. The paradox is that 3D is only one of the properties of these environments. Two more have to be added, time management and atmospheres, to obtain a global map of the simulation tools used in geoprospective.

When speaking of 3D, it is generally assumed that it designates the whole of the computational environment, whereas factually, it is only the part relating to the geometric aspect taken into account in the system. Computationally, there are several kinds of 3D generation strategies aimed at representing a volume on a screen. De La Losa (2000) distinguishes three main families: scanning models, volume models, and boundary models. Translational, and above all hybrid scanning models, correspond to the extrusion functions that we know in modern GISs to give a volume to buildings, for example. This family consists in defining a trajectory via a function then applying it to a surface. The function can rotate around an axis. It is called rotational scanning (De La Losa, 2000). The best-known volume models are the voxels. They are the tridimensional extension of pixels used for representing an image in a computer. Voxels have been the first 3D models used to construct landscapes in video games. They were subsequently abandoned until the advent of the Minecraft game[3] which brought them back to the forefront. At the moment, this type of model is neither used in GISs nor in simulations for geoprospective. However, it is to be noted that trial spatial coconstructions have been carried out with the Minecraft game. Finally, boundary models are currently the most widespread in the case before us. These models have the advantage of allowing the use of the principles of topology in the 3D scene, as their boundaries are described by arcs and vertices. This approach is the one currently at work in 3D modeling tools for video games, the cinema and in the case of GISs, for the construction of field triangular irregular network models via the Delaunay interpolation.

The advantage of using topology in a 3D scene is that it can be made plastic, hence spatially and locally modifiable in real time (Loubier et al., 2017). This remarkable property opens the way to negotiation and convergence toward a coconstructed common solution in a spatialized problem, as well as to the exploration of possible futures. Then the 3D scene plays the role of point of representation common to all stakeholders and this helps to give a clearer picture of the stakes on the territory, and even to understand the stakeholders' conflicting views on a given situation.

---

[3] Minecraft is a sandbox game reminiscent of the Lego principle. Players entirely build and exploit their game environment with voxels. They can create complete worlds (rivers, cities, mountains, forests, etc.) and exploit the resources in relation with these environments. The aim is for players to "live" as long as possible in the worlds they have created.

4.8.2.2.1 Time: computed or real

Although it is central in the representation, the 3D in the strict sense is not sufficient to reduce the distance between stakeholders and scientists when it comes to making explicit the results of a simulation. It is also necessary to be able to address the mode of displacement and of interaction with the scene. Of course there are maps that use the dimension Z of the pixel as a means of measuring the intensity of a spatial phenomenon and are inert but although they make a territory visible efficiently, they are not suitable for coconstruction. Therefore introducing the temporal interaction with the scene is indispensable to ensure dialogue. At the moment, there are only two types of time: computed time and real time. Computed time corresponds to the construction of a film of images computer-generated from the scene. The advantage of this approach is that it enables to render a very high degree of verism. It is also possible to introduce the physical temporal dimension into the computed time, which helps to better imagine a landscape after a number of years, for example. The major inconvenient is that the displacement path is also precomputed. It is impossible to get rid of it, which prevents changes of scale and viewpoint. With real time, the scene can be displaced at will both heightways and widthways. On the other hand, despite the computers' increased computing power, the flow of these displacements is constrained by the complexity of the scene. That is to say by the number of faces of the boundary models displayed. Therefore it is absolutely impossible to obtain a degree of realism comparable to the one of computed time. In such context and in front of these so far irreconcilable limits, it is important to choose the right 3D visualization method and its role in the consultation mechanism.

4.8.2.2.2 The atmospheres

The volumes and the temporal interaction with the scene are not sufficient to provide a good perception of the territory or landscape analyzed in the context of a geoprospective process. A stakeholder will recognize himself fully in a known territory if he recognizes the atmospheres. In fact, it is mainly this last part of the simulation environment that leads to the immersion of stakeholders. Here the question is to deal, on the one hand, with the degree of quality of the textures on the volumes, and on the way the light and shadows are rendered on the scene. There are two families of methods to process atmospheres and they are available in rendering engines. The first family consists in exploiting the undulatory property of light. It is called an approach and illumination by radiosity. In that context, the algorithms compute the energy exchanges of the objects in the scene according to their absorption or restitution capacity. The technique consists in breaking down the scene into finite elements, then calculate the radiosity value for each of them. These two values are determined by the type of texture. For example, wood will not have the same values as glass. It is a global approach in so far as the calculation is carried out on the whole of the image. The second family exploits the particularity of the light. It is the ray tracing approach. Phong's algorithm is a textbook example of the ray tracing approach. It makes it possible to represent light inequalities on objects like brighter focal points. Of course, these two families are used jointly to obtain a realistic rendering of a 3D scene. In the end, the degree of verism will be directly correlated with the quality of the rendering of a scene's atmospheres.

Here too, the computing cost can prove prohibitive even though present-day computers are equipped with a graphic card the processor of which is meant for these operations. At the moment, there are only ersatz rendering engines in the GISs' 3D modules on the market. They cannot compare with those coming with the 3D modeling tools of the video games or of the computer-generated image software for the cinema. That is why the scenes proposed are generally rather cold.

4.8.2.3 The end-purpose: to make a territory visible

Introducing and building a 3D representation in a consultation process in geoprospective is therefore the outcome of an initial decision concerning the mechanisms of the consultation envisaged. The question is not to consider this stage as cosmetic and think that verist and real-time 3D is necessarily the best to achieve this. In fact, there are constraints that impose choices when it comes to selecting the visualization method. For example, if the spatial stakes concern a large area and the stakeholders are already engaged in a discussion, then the choice of the 3D scene in real time can be interesting in so far as it improves interaction between the different parties. On the other hand, a simulation presented in calculated time will help to better imagine the spatial consequences at local scales. Finally, field experiences (Loubier, 2013) have shown that paradoxically, total verism reduces the interactions of stakeholders in real-time 3D scenes. That is to say that they tend not to use the scene's plastic and topological aspects. It all happens like "in real life." For example, you cannot really cut down a tree in a square without having problems in real life. The verism of a virtual scene can produce this type of reaction in front of the scene. The stakeholder imports his real values "in silico" since he can see some territory (Chardonnel et al., 2003). It is therefore important to have a clear understanding of these cognitive mechanisms when setting the degree of realism of the consultation scene at the risk of interrupting the mechanism of exchange between stakeholders and scientists.

4.9 Conclusion

This chapter, without pretending to be exhaustive, illustrates the diversity of methods and tools available for developing a geoprospective approach, and, through them, the variety of ways to introduce the spatial dimension in scenarios, simulations, and collective thinking. All the modeling tools presented are used in other contexts than that of geoprospective. The novelty here is due, on the one hand, to the changes brought about to meet the requirements of spatial prospective—prospective graphic models, for example—, and on the other hand, to the introduction of methods opening up new perspectives—causal probabilistic models, decision-making process modeling, for example. However, the major novelty lies in the adoption of a new scientific modus operandi. Due to the omnipresence of uncertainty, the posture adopted is to "work with it", not only to erase or reduce it as much as possible, but also to take advantage of what it reveals about the systems studied. It is perceived as an opportunity and not as a handicap. Adopting this course of action leads to calling into question the conventional modus operandi, and more precisely to

- Considering that the uncertainty of the systems' evolution is one of the expressions of complexity and that including it in the scientific reasoning is at the same time a way to approach systemic complexity, and to prepare for the unforeseeable by trying to improve resilience, from an operational standpoint.
- Taking a different look at the qualitative, which is still too often depreciated: "qualitative is nothing but poor quantitative" (Ernest Rutherford).
- Setting free from the dependency on the past and observed data.
- Forgetting about the validation procedures that are inappropriate to complex system modeling and inventing new ones.

This raises two fundamental questions, on the one hand, the recognition of the specificity of geoprospective, and on the other hand, its validation. The recognition of the scientific nature of geoprospective, and incidentally of prospective, is still being debated. "Qualitative" scenarios are often considered nonscientific because the assumptions on which they are build are not based on formalized and reproducible protocols. As a consequence, their credibility is challenged. Conversely, quantitative scenarios based on assumptions corresponding to a partial viewpoint of the complexity of the future are considered credible, although the figures they give convey a misleading impression of accuracy (Mermet, 2005). The validation of the approach determines the confidence of potential users. Yet, the issue is not to assess the predictive competence of scenarios and projections (Sgard, 2008), but to ascertain that they meet the criteria of likelihood, coherence, pertinence, and usefulness. Such principles, together with the principle of rigor, must apply to the geoprospective approach as a whole. Therefore new validation protocols in line with the principles on which each method used is based should be established.

Applied in a geoprospective context, these various methods prove particularly appropriate to research work on the resilience of systems. They contribute in various ways to the prospective diagnosis. Anticipating a system's possible future dynamics is the level of knowledge bases from which its adaptive capacity to change can be studied. Simulations play a key role. Depending on their end-purpose, they can explore the possibilities of rare events occurring and measure the spatial consequences of such events, spot the recurrence of changes in certain locations, detect critical points by diversifying the initial spatiotemporal situations, etc. In the case of integrated models, the modeled system's capacity to generate possible bifurcations is researched by carrying out multiple simulations. Researching the adaptive capacity of categories of stakeholders can also be done by coupling qualitative scenarios and field interviews in order to detect how the stakeholders would respond to the phenomena included in the scenarios. Determining the stakes and stake-laden spaces as regards resilience results from the fineness of this prospective diagnosis, but also from its analysis by the stakeholders concerned. In this phase of collective reflection and of coconstruction of resilience strategies, the spatial outlook represented using the tools described earlier proves to be an extremely useful mediation tool.

Acknowledgements

Authors would like to thank Françoise Gourmelon for reviewing this chapter and for her constructive comments.

References

Alcamo, J. (Ed.), 2008. Environmental Futures: The Practice of Environmental Scenario Analysis, Developments in Integrated Environmental Assessment, vol. 2. Elsevier.

Amalric, M., Anselme, B., Bécu, N., Delay, E., Marilleau, N., Pignon, C., et al., 2017. Sensibiliser au risque de submersion marine par le jeu ou faut-il qu'un jeu soit spatialement réaliste pour être efficace? Espaces du jeu, espaces en jeu 8 | 2017 (in French).

Becu, N., Amalric, M., Anselme, B., Beck, E., Bertin, X., Delay, E., et al., 2016. Participatory Simulation of Coastal Flooding: Building Social Learning on Prevention Measures with Decision-Makers. International Environmental Modelling and Software Society, Toulouse, France, pp. 1−14 (in French).

Benferhat, S., Dubois, D., Garcia, L., Prade, H., 2002. On the transformation between possibilistic logic bases and possibilistic causal networks. Int. J. Approx. Reason. 29 (2), 135−173.

BIPE, 2012. Territoires durables 2030: un exercice de prospective. Rapport d'étude pour le Commissariat Général au Développement Durable, Délégation au développement durable, Mission Prospective du MEEDDM. Rapport d'étude, 100p. (in French).

Blecic, I., Cecchini, A., 2008. Design beyond complexity: possible futures—prediction or design? Futures 40, 537−551.

Bommel, P., Bécu, N., Le Page, C., Bousquet, F., 2015. Cormas, an agent-based simulation platform for coupling human decisions with computerized dynamics. In: Kaneda, T., Kanegae, H., Toyoda, Y., Rizzi, P. (Eds.), Simulation and Gaming in Network Society. Springer, Singapore, pp. 387−410. Volume 9 of the series Translational Systems Sciences.

Bousquet, F., Le Page, C., 2004. Multi-agent simulations and ecosystem management: a review. Ecol. Model. 176, 313−332 (in French).

Bradfield, R., Wright, G., Burt, G., Cairns, G., Van Der Heijden, K., 2005. The origins and evolution of scenario techniques in long range business planning. Futures 37, 795−812.

Brown, D.G., Page, S., Riolo, R., Zellner, M., Rand, W., 2005. Path dependence and the validation of agent-based models of land use. Int. J. Geogr. Inf. Sci. 19 (2), 153−174.

Brunet, R., 1980. La composition des modèles dans l'analyse spatiale. L'Espace Géogr. 4, 53−265 (in French).

Brunet, R., 1986. La carte-modèle et les chorèmes. Mappemonde 4, 2−6 (in French).

Busch, G., 2006. Future European agricultural landscapes—what can we learn from existing quantitative land use scenarios studies? Agric. Ecosyst. Environ. 114, 121−140.

Caglioni, M., Rabino, G.A., 2007. Theoretical approach to urban ontology: a contribution from urban system analysis. Stud. Comput. Intell. 61, 109–119.

Camacho Olmedo, M.T., Paegelow, M., Mas, J.F., 2013. Interest in intermediate soft-classified maps in land change model validation: suitability versus transition potential. Int. J. Geogr. Sci. 27 (12), 2343–2361.

Cao, C., Fusco, G., 2016. Representing Uncertain Futures: Social Polarization in the Metropolitan Area of Marseille. <https://public.tableau.com/profile/fusco#!/vizhome/RepresentingUncertainFutures/Story1>.

Casanova, L., 2010. Les dynamiques du foncier à bâtir comme marqueurs du devenir des territoires de Provence intérieure, littorale et préalpine. Éléments de prospective spatiale pour l'action territoriale, thèse de doctorat de géographie, Université d'Avignon, Avignon (in French).

Casanova Enault, L., Chatel, C., 2017. La modélisation graphique de phénomènes émergents pour répondre aux besoins de la prospective. Mappemonde 119 (in French).

Centi, C., 1996. Le Laboratoire marseillais. Chemin d'intégration métropolitaine et segmentation sociale. L'Harmattan, Paris (in French).

Chardonnel, S., Feyt, G., Loubier, J.-C., 2003. La maquette virtuelle comme fond de carte: une vision commune du territoire? In: Debarbieux, B., Lardon, S. (Eds.), Les Figures du Projet Territorial. L'Aube (in French).

Cheylan, J.-P., Libourel, T., Mende, C., 1997. Graphical modelling for geographic explanation. Spatial information theory a theoretical basis for GIS. In: Proceedings, International Conference COSIT '97, Laurel Highlands, PA, USA, October 15–18, pp. 473–483.

Clarke, K.C., Hoppen, S., Gaydos, L., 1997. A self-modifying cellular automaton model of historical urbanization in the San Francisco Bay area. Environ. Plan. B 24, 247–261.

Costanza, R., Sklar, F.H., White, M.L., 1990. Modeling coastal landscape dynamics. BioScience 40 (2), 91–107.

Cozman, F.G., 2004. Graphical models for imprecise probabilities. Int. J. Approx. Reason. 39 (2005), 167–184.

D'Aquino, P., Bah, A., Aubert, S., 2012. L'approche participative, incrémentale et itérative en modélisation: un changement profond de cadre méthodologique. Rev. Int. Géomat. 22, 201 (in French).

D'Aquino, P., Etienne, M., Barreteau, O., Le Page, C., Bousquet, F., 2001. Jeux de rôles et simulations multi-agents: un usage combiné pour une modélisation d'accompagnement des processus de décision sur la gestion des ressources naturelles, in Trebuil, Le pilotage des agro-écosystèmes: complémentarités terrain-modélisation et aide à la décision. CIRAD (in French).

De Chazal, J., Rounsevell, M.D.A., 2009. Land-use and climate change within assessments of biodiversity change: a review. Glob. Environ. Change 19, 306–315.

De La Losa, A., 2000. Modélisation de la troisième dimension dans les bases de données géographiques, thèse, Université de Marne-la-Vallée (in French).

Derbyshire, J., Wright, G., 2017. Augmenting the intuitive logics scenario planning method for a more comprehensive analysis of causation. Int. J. Forecast. 33 (2017), 254–266.

De Vries, B., Bollen, J., Bouwman, L., den Elzen, M., Janssen, M., Kreileman, E., 2000. Greenhouse gas emissions in an equity-, environment- and service-oriented world: an IMAGE-based scenario for the 21st century. Technol. Forecast. Soc. Change (63), 137–174.

Drouet, I., 2016. Le Bayésianisme Aujourd'hui. Fondements et Pratiques. Editions Matériologiques, Paris (in French).

Dubois, D., Prade, H., 1988. Possibility Theory: An Approach to Computerized Processing of Uncertainty. Plenum Press.

Dubois, D., Fusco, G., Prade, H., Tettamanzi, A., 2016. Uncertain logical gates in possibilistic networks: theory and application to human geography. Int. J. Approx. Reason. 82 (2017), 101–118.

Dubos-Paillard, E., Guermond, Y., Langlois, P., 2003. Analyse de l'évolution urbaine par automate cellulaire: le modèle SpaCelle. L'espace Géogr 32 (4), 357–378 (in French).

Dubus, N., Helle, C., Masson Vincent, M., 2010. De la gouvernance à la géogouvernance: de nouveaux outils pour une démocratie LOCALE renouvelée. L'espacepolitique 10 (in French). <https://journals.openedition.org/espacepolitique/1574>.

Dubus N., Voiron-Canicio C., Emsellem K., Cicille P., Loubier J.C., Bley D., 2015. Géogouvernance: l'espace comme médiateur et l'analyse spatiale comme vecteur de communication entre chercheurs et acteurs.

In: Colloque international du GIS PPDDP; Chercheur.es et acteur.es de la participation. Liaisons dangereuses et relations fructueuses, Paris (in French). <http://www.researchgate.net/publication/298201282_Geogouvernance_l'espace_comme_mediateur_et_l'analyse_spatiale_comme_vecteur_de_communication_ entre_chercheurs_et_acteurs>.

Eastman, J.R., Jin, W., Kyem, P.A.K., Toledano, J., 1995. Raster procedures for multi-criteria/multi-objective decisions. Photogramm. Eng. Rem. Sens. 61, 539–547.

Emsellem, K., Dubus, N., Voiron-Canicio, C., Loubier, J.-C., Cicille, P., 2018. Spatialité et géogouvernance: conceptualisation, expérimentations, évaluations. BSGLg 71(2) (in French).

ESPACE, GAÏAGO, ECOVIA, 2013. Géoprospective: modélisation et représentation graphique des scénarios de prospective pour un territoire durable à horizon 2030. Rapport de recherche, MEDDE—Commissariat général au développement durable—Mission Prospective (in French).

ESPON, 2007. Scenarios of territorial future of Europe. ESPON project 3.2.

Étienne, M., Du Toit, D., Pollard, S., 2011. ARDI: a co-construction method for participatory modeling in natural resources management. Ecol. Soc. 16 (1).

Etienne, M., 2012. Companion modelling: variant of geoprospective approach. L'Espace Géogr. 41(2).

Fekete-Farkas, M., Rounsevell, M., Audsley, E., 2005. Socio-economic scenarios of agricultural land use change in Central and Eastern European countries. In: Paper Prepared for Presentation at the XIth Congress of the European Association of Agricultural Economists, The Future of Rural Europe in the Global Agri-Food System Copenhagen, Denmark, August 24–27, 2005.

Feng, Y., Tong, X., Liu, M., 2007. Extended cellular automata based model for simulating multi-scale urban growth using GIS. In: ISKE-2007 Proceedings. <http://www.atlantis-press.com/publications/aisr/iske-07/>.

Fusco, G., 2008. Spatial dynamics in the coastal region of South-Eastern France. In: Pourret, O., Naïm, P., Marcot, B.G. (Eds.), Bayesian Networks: A Practical Guide to Applications. John Wiley, New York, pp. 87–112.

Fusco, G., 2010. Uncertainty in interaction modelling: prospecting the evolution of urban network in South-Eastern France. In: Prade, H., Jeansoulin, R., Papini, O., Schockaert, S. (Eds.), Methods for Handling Imperfect Spatial Information. Springer, Berlin, pp. 357–378.

Fusco, G., 2012. Démarche géo-prospective et modélisation causale probabiliste. Cybergeo 613 (in French). <http://cybergeo.revues.org/25423>.

Fusco, G., Scarella, F., 2011. Métropolisation et ségrégation socio-spatiale. Les flux de mobilité résidentielle en Provence-Alpes-Côte d'Azur. L'Espace Géogr., 4, 319–336 (in French).

Fusco, G., Tettamanzi, A., 2017. Multiple Bayesian models for the sustainable city. The case of urban sprawl, Lecture Notes on Computer Science, vol. 10407. Springer, pp. 392–407.

Geist, H.J., Lambin, E.F., 2001. What Drives Tropical Deforestation? A Meta-Analysis of Proximate and Underlying Causes of Deforestation Based on Subnational Case Study Evidence. LUCC Report Series No. 4. LUCC International Project Office, Louvain-la-Neuve.

Gibson, C.C., Ostrom, E., Anh, T.K., 2000. The concept of scale and the human dimensions of global change: a survey. Ecol. Econ. 32, 217–239.

Godet, M., 1986. Introduction to la prospective: seven key ideas and one scenario method. Futures 18, 134–157.

Gourmelon, F., Chlous-Ducharme, F., Kerbiriou, C., Rouan, M., Bioret, F., 2013. Role-playing game developed from a modelling process: a relevant participatory tool for sustainable development? A co-construction experiment in an insular biosphere reserve. Land Use Policy 32, 93–107.

Grignard, A., Alonso, L., Taillandier, P., Gaudou, B., Nguyen-Huu, T., Gruel, W., et al., 2018. The impact of new mobility modes on a city: a generic approach using abm. International Conference on Complex Systems. Springer, pp. 272–280.

Harvey, D., 1969. Explanation in Geography. St. Martin's Press, New York.

Helmer, O., 1981. Reassessment of cross-impact analysis. Futures 13 (5), 389–400.

Hervé, M., 2018. Mieux conserver la biodiversité en intégrant l'agriculture et en explorant les changements globaux dans l'aménagement du territoire, Doctorat, Aix-Marseille Université (in French). <http://www.theses.fr/2018AIXM0088>.

Houet, T., 2015. Usages des modèles spatiaux pour la prospective. RIG, 1, 123−143 (in French).

Houet, T., Vacquié, L., Sheeren, D., 2014. Evaluating the spatial uncertainty of future land abandonment in a mountain valley (Vicdessos, Pyrenees-France): insights form model parameterization and experiments. J. Mt. Sci. 12 (5), 1−18.

Houet, T., Aguejdad, R., Doukari, O., Battaia, G., Clarke, K., 2016a. Description and validation of a non path-dependent model for projecting contrasting urban growth futures. Cybergeo 759.

Houet, T., Marchadier, C., Bretagne, G., Moine, P., Aguejdad, R., Viguié, V., et al., 2016b. Combining narratives and modelling approaches to simulate fine scale and long-term urban growth scenarios for climate adaptation. Environ. Model. Softw. 86, 1−13.

Jannès-Ober E., Lamblin V., 2012, Prospective Eau, Milieux Aquatiques, et Territoires Durables 2030, rapport d'étude pour la Mission Prospective, Ministère de l'Ecologie, du Développement Durable et de l'Energie, Commissariat Général au Développement Durable, Délégation au Développement Durable, France (in French).

Jaynes, E., 2003. Probability Theory: The Logic of Science. Cambridge University Press, Cambridge.

Jensen, F.V., 1996. An introduction to Bayesian Networks. Taylor and Francis, London.

Jensen, F., 2001. Bayesian Networks and Decision Graphs. Springer, New York.

Jetter, A.J., Kok, K., 2014. Fuzzy cognitive maps for futures studies. a methodological assessment of concepts and methods. Futures 61, 45−57.

Joliveau, T., Michelin, Y., 2001. Modèles d'analyse et de représentation pour la prospective paysagère concertée; deux exemples en zone rurale. In: Lardon, S., Maurel, P., Piveteau, V. (Eds.), Représentations spatiales et développement territorial,. Hermes, Paris, pp. 239−266 (in French).

Joliveau, T., Molines, N., Caquard, S., 2000. Méthodes et outils de gestion de l'information pour les démarches territoriales participatives, un regard France-Québec, Saint-Etienne. CRENAM, Université Jean Monnet, Saint-Etienne (in French).

Julien, P.A., Lamonde, P., Latouche, D., 1975. La méthode des scénarios. Travaux et recherche de Prospective, (DATAR) 59 (in French).

Kok, K., 2009. The potential of fuzzy cognitive maps for semi-quantitative scenario development, with an example from Brazil. Glob. Environ. Change 19, 122−133.

Kok, K., Bärlund, I., Flörke, M., Holman, I., Gramberger, M., Sendzimir, J., et al., 2014. European participatory scenario development: strengthening the link between stories and models. Clim. Change 128 (3−4), 187−200.

Korb, K., Nicholson, A., 2004. Bayesian Artificial Intelligence. Chapman & Hall/CRC, Boca Raton, FA.

Lambin, E.F., Bauleis, X., Bockstael, N., Fischer, G., Krug, T., Leemans, R., et al., 1999. Land-use and land-cover change (LUCC): Implementation strategy. IGBP Report No. 48 and IHDP Report No. 10. Stockholm: International Geosphere-Biosphere Programme (IGBP); Geneva: the International Human Dimensions Programme on Global Environmental Change (IHDP).

Langlois, P., 2010. Simulation des systèmes complexes en géographie: fondements théoriques et applications. Lavoisier, Paris(in French).

Le Page, C., Abrami, G., Barreteau, O., Becu, N., Bommel, P., Botta, A., et al., 2011. Models for sharing representations. In: Etienne, M. (Ed.), Companion Modelling. A Participatory Approach to Support Sustainable Development. Springer, pp. 69−96.

Linstone, H., Turoff, M., 2002. The Delphi Method: Techniques and Applications. New Jersey Institute of Technology. <http://is.njit.edu/pubs/delphibook>.

Loubier, J.C., 2013. L'intérêt des maquettes virtuelles dans la compréhension des enjeux spatiaux. In: Masson Vincent, M., Dubus, N. (Eds.), Géogouvernance: Utilité sociale de l'analyse spatiale, Quae, pp. 119−127 (in French).

Loubier, J.-C., Voiron-Canicio, C., Genoud, D., Hunacek, D., Sant, F., 2017. Modélisation Géoprospective Et Simulation Paysagère 3d Immersive, Revue Internationale De Géomatique. Volume 27 − N°4/2017, 547−566 (in French).

Marks, R., 2007. Validating simulation models: a general framework and four applied examples. Comput. Econ. 30, 265−290.

Masson Vincent, M., Dubus, N. (Eds.), 2013. Géogouvernance, utilité sociale de l'analyse spatiale. Quae (in French).

Masson Vincent, M., Dubus, N., Bley, D., Voiron, C., Helle, C., Cheylan, J.-P., et al., 2012. La Géogouvernance: un concept novateur?, Cybergeo Eur. J. Geogr. 587 (in French).

Mermet, L., 2005. Des récits pour raisonner l'avenir. In: Mermet, L. (Dir.), Etudier des écologies futures—Un chantier ouvert pour les recherches prospectives environnementales. PIE Peter Lang (in French).

Mermet, L., Poux, X., 2002. Pour une recherche prospective en environnement, repères théoriques et méthodologiques. NSS 10 (3), 7−15 (in French).

Michelin, Y., 2000. Le bloc-diagramme: une clé de compréhension des représentations du paysage chez les agriculteurs? Mise au point d'une méthode d'enquête préalable à une gestion concertée du paysage en Artense (Massif central français), Cybergeo Eur. J. Geogr. 118 (in French).

Otterman, J., 1974. Baring high-albedo soils by overgrazing: a hypothesized desertification mechanism. Science 186 (4163), 531−533.

Parker, D.C., Evans, T.P., Meretsky, V., 2001. Measuring emergent properties of agent-based land-use/land-cover models using spatial metrics. In: Presented in the Seventh Annual Conference of the International Society for Computational Economics, New Haven, CT, pp. 28−29.

Parker, D.C., Berger, T., Manson S.M. (Eds.), 2002. Meeting the Challenge of Complexity. In: Proceedings of the Special Workshop on Agent-Based Models of Land-Use/Land-Cover Change. Joint publication of the Center for the Study of Institutions, Population, and Environmental Change (CIPEC) and the Center for Spatially Integrated Social Science (CSISS). CIPEC Collaborative Report No. 3. National Center for Geographic Information and Analysis: Santa Barbara, CA. Available from: <http://www.csiss.org/>.

Pearl, J., 1988. Probabilistic Reasoning in Intelligence System: Networks of Plausible Inference. Morgan Kaufmann, San Matteo.

Pearl, J., 2000. Causality: Models, Reasoning and Inference. Cambridge University Press, Cambridge.

Rojas, D., Loubier, J.C., 2017. Analytical hierarchy process coupled with GIS for land management purposes. In: Proceedings of the 22nd International Congress on MODSIM, 2017 (No. CONFERENCE), 3−8 December 2017.

Rothman, D., 2008. A survey of environmental scenarios. In: Alcamo, J. (Ed.), Environmental Futures: the Practice of Environmental Scenario Analysis, Developments in Integrated Environmental Assessment, vol. 2. Elsevier.

Saaty, T.L., 1980. The Analytic Hierarchy Process. McGraw-Hill, New York, NY.

Sajja, P.S., Akerkar, R., 2010. Knowledge-based systems for development. In: Sajja, P.S., Akerkar, R. (Eds.), Advanced Knowledge Based Systems: Model, Applications & Research, vol. 1, pp. 1−11.

Sanders, L., 2006. Les modèles agent en géographie urbaine. In: Amblard, F., Phan, D. (Eds.), Modélisation et simulation multi-agents; applications pour les Sciences de l'Homme et de la Société. Hermes-Lavoisier, pp. 151−168 (in French).

Scarella, F., 2014. La ségrégation résidentielle dans l'espace-temps métropolitain: analyse spatiale et géoprospective des dynamiques résidentielles de la métropole azuréenne, thèse de doctorat en géographie. Université Nice Sophia Antipolis (in French).

Schwartz, P., 1993. La prospective stratégique par scenarios. Futuribles176 (in French).

Schwartz, P., 1998. The Art of the Long View: Planning for the Future in an Uncertain World. Doubleday, New York.

Sgard, A., 2008. Entre rétrospective et prospective. EspacesTemps.net (in French). <https://www.espacestemps.net/articles/entre-retrospective-et-prospective/>.

Shafer, G., 1976. A Mathematical Theory of Evidence. Princeton University Press, Princeton, NJ.

Shafer, G., 1981. Jeffrey's rule of conditioning. Philos. Sci. 48 (1981), 337−362.

Tettamanzi, A., Fusco, G., 2016. Possibilistic Network. R-Geo-Soft Models. <https://zenodo.org/record/165857#.WCOPC_nhCDI>.

Turner II, B.L., Clark, W.C., Kates, R.W., Richards, J.F., Mathews, J.T., Meyer, W.B., 1990. The Earth as Transformed by Human Action. Global and Regional Changes in the Biosphere Over the Past 300 Years. Cambridge University Press (with Clark University), Cambridge, New York, Port Chester, Melbourne & Sydney.

Turner, B.L., Moss, R.H., Skole, D.L., 1993. Relating land use and global land-cover change. A proposal for IGBP-HDP core project. IGBP Report and 24/HDP Report; 5.

Turner II, B.L., Skole, S., Sanderson, G., Fischer, L., Fresco, L., Leemans, R., 1995. Land-Use and Land-Cover Change Science/Research Plan. IGBP Report No. 35 and IHDP Report No. 7. International Human Dimensions Programme on Global Environmental Change (IHDP), Geneva, Stockholm: International Geosphere-Biosphere Programme (IGBP).

Van Asselt, M.B.A., Storms, C., Rijkens-Klomp, N., Rotmans, J., 1998. Towards Visions for a Sustainable Europe, An Overview and Assessment of the Last Decade in European Scenarios Studies. ICIS, University of Maastricht, Netherlands.

Veldkamp, A., Verburg, P.H., Kok, K., de Koning, G.H.J., Priess, J., Bergsma, A.R., 2001. The need for scale sensitive approaches in spatially explicit land use change modeling. Environ. Model. Assess. (6), 111−121.

Verburg, P., van Berkel, D.B., van Doorn, A.M., et al., 2010. Trajectories of land use change in Europe: a model-based exploration of rural futures. Landsc. Ecol. 25 (2), 217−232.

Voiron-Canicio C., 2006. L'espace dans la modélisation des interactions nature-société, Colloque Interactions Nature-Société, analyse et modèles. La Baule (in French). <http://www.researchgate.net/publication/267631509_L'espace_dans_la_modelisation_des_interactions_nature-societe>.

Voiron-Canicio, C., 2009. Predicting the urban spread using spatio-morphological models. In: Murgante, B., Borruso, G., Lapucci, A. (Eds.), Geocomputation and urban planning, 176. Springer, pp. 223−236. Studies in Computational Intelligence.

Voiron-Canicio, C., Loubier, J.-C., Genoud, D., Hunacek, D., Sant, F., 2016. Intérêt de la simulation paysagère en modélisation géoprospective. Actes du Colloque SAGEO, Nice (in French). <https://www.researchgate.net/publication/322617695_Interet_de_la_simulation_paysagere_en_geoprospective>.

Wagner, D.F., 1997. Cellular automata and geographic information systems. Environ. Plan. B Plan. Des. 24, 219−234.

Walley, P., 1991. Statistical Reasoning with Imprecise Probabilities. Chapman and Hall, London.

Wang, C.F., 2001. Study on Urban Expansion Using Remote Sensing and Cellular Automata. Surveying and Mapping Press, Beijing.

White, R., Enguelen, G., Uljee, I., 2015. Modeling Cities and Regions as Complex Systems, from Theory to Planning Applications. The MIT Press, Cambridge MA.

White, R., Engelen, G., 1997. Cellular automata as the basis of integrated dynamic regional modelling. Environ. Plan. B 24, 235−246.

White, R., Engelen, G., 2000. High-resolution integrated modelling of the spatial dynamics of urban and regional systems. Comput., Environ. Urban Syst. 24, 383−400.

Withers, S., 2002. Quantitative methods: Bayesian inference, Bayesian thinking. Prog. Hum. Geogr. 26 (4), 553−566.

Woodwell, G.M., Hobbie, J.E., Houghton, R.A., Melillo, J.M., Moore, B., Peterson, B.J., et al., 1983. Global deforestation: contribution to atmospheric carbon dioxide. Science 222, 1081−1086.

Wright, G., Goodwin, P., 2009. Decision making and planning under low levels of predictability: enhancing the scenario method. Int. J. Forecast. 25, 813−825.

CHAPTER 5

Geoprospective approach for biodiversity conservation taking into account human activities and global warming

Guillermo Hinojos Mendoza[1] and Emmanuel Garbolino[2]

[1]ASES Ecological and Sustainable Services, Pépinière d'Entreprises l'Espélidou, Parc d'Activités du Vinobre, Aubenas, France, [2]Climpact Data Science, Nova Sophia—Regus Nova, Sophia Antipolis Cedex, France

Chapter Outline

5.1 Introduction

The landscape transformation has ecological implications (Lambin et al., 2003) that are necessary to evaluate, given that some transformations, like the development of artificialization and urbanization (we consider here that artificialization and urbanization processes are integrated into the same land use class), may affect the natural and rural areas and, in consequence, may induce a loss of biodiversity and ecosystem services These losses could provoke an increase of ecosystem vulnerability. "Ecosystem vulnerability" was

Ecosystem and Territorial Resilience.
DOI: https://doi.org/10.1016/B978-0-12-818215-4.00005-5

123

© 2021 Elsevier Inc. All rights reserved.

defined by Williams and Kapustka (2000), as the "potential of an ecosystem to modulate its response to stressors over time and space, where that potential is determined by characteristics of an ecosystem that include many levels of organization. It is an estimate of the inability of an ecosystem to tolerate stressors over time and space."

In order to assess the ecosystem vulnerability according to the Land Use and Land Cover Changes (LUCC), we consider that the simulation models help to assess the potential evolution trajectories of a territory according to spatial properties of the territory. Cellular Automata (CA) have been used to model landscape changes over the past years (Batty et al., 1999; Veldkamp and Lambin, 2001; Sui and Zeng, 2001; Lambin et al., 2003; Xie and Batty, 2003; Li et al., 2003; Barredo et al., 2003; Engelen et al., 2007; He et al., 2008; Aguilera Benavente and Valenzuela Montes, 2010; Sante et al., 2010; Feng et al., 2011; Basse, 2013; Wang et al., 2015, among many others), because these models simulate spatial complex phenomena like urban processes and their interactions with the different land use classes like natural areas, agricultural areas, transportation infrastructures, etc. CA represents the territory by a discrete lattice of cells and each cell corresponds to a specific land use/land cover state. Usually, the evolution of the cell state belongs to the state of each cell surrounding the considered cell, the evolution trend of each cell, and the intrinsic possibility of change of the considered cell. The analysis of the land use dynamic with the use of spatial and earth observation data allows determining the transition rules of the different categories of land use. The cells' evolution integrates these transition rules and it is calculated at each iteration of the model in order to simulate the dynamic of the territory.

Climate change also induces changes in biodiversity of ecosystems (Hassan et al., 2005; IPCC, 2014a,b) and those changes would be more and more significant according to the intensity of the climate warming until the end of the 21st century. Thuiller et al. (2006) have shown the potential impacts of global warming in Europe toward 2050. This climate change should affect the nature and structure of ecosystems (species composition) and, at the same time, the functioning of ecosystems. Garbolino et al. (2016, 2017, 2018) also presented potential impacts of climate change on some species and vegetation structures in France toward 2050 and 2100. According to these considerations, the conservation of biodiversity needs to be established not only on observed data but also on models that are able to forecast the potential changes of biodiversity level on the territory induced by climate change.

Our work is a first stage of a research which aims to assess the potential impacts of artificialization and climate change on the landscape units of the Alpes-Maritimes. This work is related to a social and regulatory demand: the need to promote the conservation of natural resources and biodiversity for the next decades. In the considered territory, the Regional Directorate for the Environment, Planning and Housing of the region Provence, Alpes, Côte d'Azur (DREAL PACA—Direction Régionale de l'Environnement, de l'Aménagement et du Logement in French) is the reference organization that have to define and apply two main regulatory instruments in order to reach the objectives of nature conservation: the Green and Blue Frames (TVB—Trames Verte et Bleue in French) and the Regional Scheme of Ecological Coherence (SRCE—Schéma Régional de Cohérence Ecologique in French). The aim of the definition of the TVB is to allow the interconnectivity between ecosystems in order to preserve the exchanges of species, energy, DNA, etc., in the different localities. The aim of the SRCE is to promote the development of the territory by taking into account the preservation of natural habitats and ecological networks. The main problems faced by the administration and the local authorities for the definition and implementation of such tools is that they do not really take into account the evolution of the elements of the territory, like the ecosystems and land use dynamics. In order to support these territorial decision-makers, we proposed a project financed by the region Provence, Alpes, Côte d'Azur named "Cassandre." This project finished in the early 2015 and produced a methodology and some results that underlined the need to simulate the evolution of land use and the ecosystems in order to support the territorial planning and make more resilient the Alpes-Maritimes to the global warming effects and the urban pressure. The results have been presented to the public authorities and stakeholders in charge of the application of the TVB and SRCE and they were integrated to a reflection document, edited by the Regional Economic, Social, and Environmental Council (CESER Conseil Economique, Social et Environnemental Régional in French), on the drivers to take into account for the territorial planning in the region.

In our research, we develop a simulation scenario of landscape transformation with a CA model that incorporates the transition rules that have been interpreted from the spatial analysis of the study area. In this model, we only focus our study by the use of landscape classes and we do not use socioeconomic parameters like the demography or housing marked: we suppose that it is implicitly integrated into some specific land use classes like urbanization or agriculture.

One of the main ideas of our methodology is to demonstrate that the contribution of landscape analysis allows identifying the main patterns to simulate the landscape dynamics, without ancillary data. In our model, the transition rules consider the constraints caused by the geomorphologic conditions, the natural protected areas, as well as the proximity of the transport network from urbanized areas, the closeness of the population centers, and the pattern of change between each land use category. These landscape patterns of change were identified using data from 1975 to 2011 in order to identify the spatial transition rules to project the future scenarios of landscape changes for 2050. We prefer to do not extend the simulation process of land use change due to the uncertainties that characterize the patterns of land use dynamics for a long duration.

The model takes into account the following four assumptions that are presented in the next section:

- Are we able to predict the landscape transformation according to the land use changes patterns?

- Is the landscape transformation influenced by the geomorphological factors?
- Are the changes of land use influenced by the urban infrastructure or the road networks?
- What is the dispersion level of the land use changes across the concerned territory?

We present the methodology developed in order to simulate the landscape dynamic that gathers three main steps: (1) the land use change analysis, (2) the land use transition rules definition, and (3) the model calibration and validation. We also present and discuss the results of the simulation of land use dynamics through 2050, taking into account the consistency of the results with the assumptions used to define the observed trends. After this part, we present the model of territorial biodiversity that we developed and the model of assessment of its evolution according to the potential impacts of climate change (Fig. 5.1).

We finally present the results of the expected impacts of land use changes and climate changes on biodiversity in the territory of the Alpes-Maritimes toward 2050. These results contribute to promote the definition and implementation of a dynamic and prospective approach of biodiversity conservation.

5.2 *Methodology to assess the land use transformations*

This part is divided into three main subparts (Fig. 5.2):

1. The land use change analysis (step 1): it is a presentation of the land use and land cover changes (LUCC) in the study area (French Riviera) based on a remote sensing data analysis since 1975;

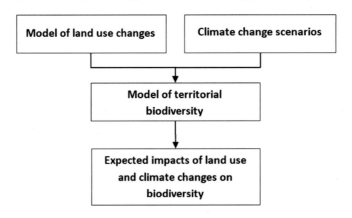

Figure 5.1
Methodology to assess the potential impacts of land use and climate changes on biodiversity.

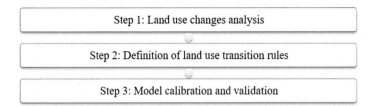

Figure 5.2
Methodology to simulate the present and future land use by Cellular Automata.

2. The transition rules definition (step 2) according to the identified land use change trajectories;
3. The calibration and validation of the simulation model of LUCC (step 3) performed by comparing the observed situation in 2011 and the simulated LUCC for the same period.

5.2.1 Land use change analysis

The study area covers the territory of the French Riviera which is considered as a biodiversity hotspot (Médail and Diadema, 2009). This context should lead the decision-makers to define priority conservation areas in the future (Myers et al., 2000) due to the potential urban development of this area. Urban development in this area has increased rapidly over the last 40 years and with a striking expansion, while agriculture areas have decreased systematically during the same period; meanwhile, the natural and seminatural surfaces are under constant pressure due to urban sprawl. Since 1970, this pressure has increased especially on the coastal strip that has been continuously urbanized and that is today completely occupied by artificial and urban structures. While the population only increased by one third between 1970 and 2000, urbanization on the coastal strip has significantly increased in the same period (DTA, 2008).[1]

In order to identify the main trends of land use changes, we did an analysis by taking into account the following parameters:

*   The surface of changes for each category expressed in hectares or cells, and the trajectory of the changes (e.g., *natural to agriculture or agriculture to artificial*).
*   The distance gradient of the different territorial elements that have an influence on the land change like transport networks, coastal strip, and established population centers.
*   The potential restrictions that represent the geomorphology of the development of urban/artificial surfaces.

---

[1] DTA is an urban and territorial planning policy developed by the state that fixes the objectives of development of infrastructure, transport, and sustainable development.

The analysis of these trends will be useful in order to develop the model of land use dynamics that will be implemented into the CA platform. They also contribute to define the transition rules of the different land use classes.

Land cover data were obtained from photointerpretation of the Landsat satellite imagery for the scenes of 1975, 1990, and 2011. This information was first of all processed by using IDRISI software to obtain a segmented thematic for each scene, which was classified by photointerpretation techniques with ArcGis software in the following classes: agricultural areas, natural and seminatural areas, and urban/artificial areas.

Fig. 5.3 shows the growth of urban areas. This figure was realized using data about the composition of surfaces in 1975 obtained by the photointerpretation process of Landsat images (Fig. 5.4). It was then compared with 1990 and 2011 surfaces obtained by the same procedure. The results of the photointerpretation in 1990 and in 2011 were validated with Corine Land Cover data and Local Development Plan in 2011. For 1975, it was not possible to validate the results because of the loss of ancillary data by the public administration.

The results of the photointerpretation (Fig. 5.4) show that the spread of urban/artificial surfaces along the coastal strip takes advantage of the assets provided by the agricultural surfaces. Outside of the coastal strip the trend is similar: the urban/artificial surfaces have also principally colonized the agricultural areas and also some natural areas.

In 1975, the agricultural surfaces occupied 40,398 hectares, decreasing to 31,135 hectares by 1990 (Table 5.1), so a loss of 9163 hectares. Meanwhile, the urban/artificial surfaces more than doubled between 1975 and 1990, going from 12,484 to 27,887 hectares. This change is remarkable for a relatively short period. The natural/seminatural areas occupied 376,548 hectares in 1975 and decreased to 369,811 hectares by 1990, showing a loss of

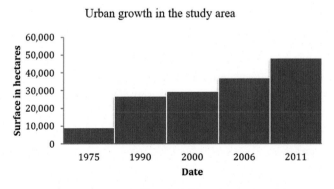

Figure 5.3
Urban growth in the French Riviera since 1975−2011.

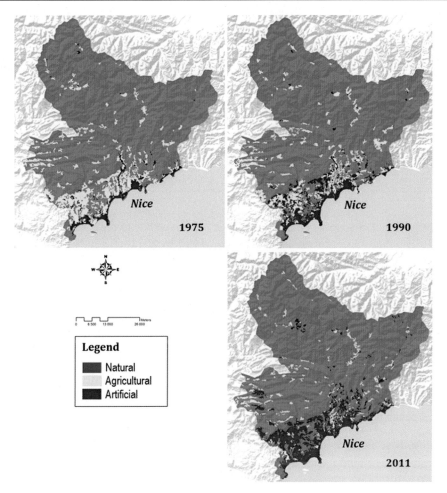

Figure 5.4
Spread of urban/artificial surfaces in the French Riviera since 1975.

6737 hectares. Table 5.1 shows that most of the changes provide the increase of the net urban/artificial surfaces.

In 1990, agricultural areas occupied 31,135 hectares, decreasing to 14,794 hectares by 2011 (Table 5.1), showing a loss of 16,341 hectares; meanwhile, the urban/artificial surfaces increased from 27,887 hectares to 46,213 hectares between 1990 and 2011, showing a gain of 18,326 hectares. The natural surfaces decreased from 369,811 hectares to 367,653 hectares in the same period (loss of 2158 hectares).

This analysis also shows that the changes are not completely homogenous throughout the study area because a few parts of these changes (3.7% between 1975 and 1990; 0.9%

Table 5.1: Changes of land cover observed between 1975 and 1990 and between 1990 and 2011 in ha.

Land use classes	1975	1990	Changes (ha)
Agricultural	40.398	31.135	− 9.263
Artificial	12.484	27.887	+ 15.403
Natural	376.548	369.811	− 6.737
Land use classes	1990	2011	Changes (ha)
Agricultural	31.135	14.794	−16.341
Artificial	27.887	46.213	+18.326
Natural	369.811	367.653	−2.158

Table 5.2: Trajectories of the main LUCC observed for two periods: from 1975 to 1990 and from 1990 to 2011.

LUCC trajectories	1975−90	1990−2011
Agricultural to artificial	62%	40%
Natural to artificial	34%	34%
Natural to agricultural	2%	25%

between 1990 and 2011) do not contribute to the increase of the artificial category. But, according to these results, the main changes show a significant contribution of agricultural and natural areas to artificial ones.

According to these first results, we also wanted to understand the trajectories of the changes. For the evaluation of land use change patterns, we used the module "Land Change Modeler" (LCM) of IDRISI Selva (Eastman, 2006). This evaluation consisted in identifying the principal change trajectories and the different patterns in the changes between the different spatial components of the territory. This evaluation was performed for the two periods 1975−90 and 1990−2011 by taking into account the surfaces of each land use class for two periods. In Table 5.2, we can observe that the intensity of the changes between the two periods is not the same for all the classes and we can also note the variation of pressure on each land use class.

We have identified several spatial patterns of the changes. In accord with Gomez-Delgado and Rodriguez Espinoza, 2012), the morphological patterns of land occupation are as follows:

- Aggregate: corresponds to traditional forms of urban growth adjacent to the consolidated core. This pattern is characteristic of the compaction process;
- Nodal: explains land use that comes close to major transport hubs (road crossings, linear infrastructure nodes, etc.);
- Linear: identifies all forms of land use which tend to preferentially occupy the spaces surrounding the communication ways and roads;

- Dispersed: they are mainly isolated stands centers points of infrastructure or communication nodes and usually far from consolidated urban centers (random behavior).

The analysis of the spatial patterns (amount of surfaces of spatial pattern classes between two periods) shows an increase in the spread of the changes (Table 5.3), which explains why the pure aggregate pattern has been decreasing in the territory. This dispersion does not occur at random; it mainly occurs at the proximity of the routes and along road networks. This explains why the nodal patterns and nodal aggregate have increased.

According to such assumption, we try to answer the following question: "Does the distance from a transport infrastructure have an influence on the territorial dynamics?" Recently, different authors have shown the role of transport infrastructure in the development of urbanized areas and take into account the influence of such structures in the simulation of the dynamics of the land use with CA in the Mediterranean area (Basse, 2013; Wang et al., 2015).

To evaluate such assumption, we measured the distance between the main road networks and the urban areas between two periods in order to identify the distance between the road networks and the main urban areas having changed. We discretize the distance into six classes (less than 1 km; at 1 km; at 2 km; at 3 km; at 4 and 5 km) and we calculate the percentage of each distance class from the urban center. These six classes were selected because, previously with the photointerpretation, we have identified that the main developments of urbanized areas are localized close to the road network and seem to remain in a radius of less than 5 km.

The road network and the localization of the principal population centers for the study area were obtained from the National Institute of Geography (2009). The thematic information concerning natural protected areas was obtained from Natura 2000. Table 5.4 underlines the influence of the distance of the changes from the population centers. It shows that the proximity of population centers is significant for the generation of a change because the main changes occur at a distance between 1 and 3 km from a population center (amount of 85% of changes for 1975−90; amount of 83% for 1990−2011). Localized areas between 1 and 3 km from a population center or village are the most likely to undergo a change, while localized areas less than 1 km and those more than 4 km are less likely to undergo a change. These results show that the shorter the distance, the higher the pressure of change.

Table 5.3: Spatial patterns of the changes.

Spatial patterns	1975−90	1990−2011
Pure aggregate	62%	57%
Nodal aggregate	23%	26%
Linear	7%	7%
Nodal	6%	9%
Linear aggregate	2%	1%

Table 5.4: Distance of the changes from the population centers.

Distance of changes from population centers	1975–90	1990–2011
<1 km	9%	9%
At 1 km	41%	39%
At 2 km	31%	32%
At 3 km	13%	12%
At 4 km	5%	6%
At 5 km	1%	1%
At 6 km	0%	1%

Table 5.5: Distance of the changes from the road network.

Distance of changes from road network	1975–90	1990–2011
<1 km	78%	76%
At 1 km	20%	22%
At 2 km	2%	2%

Table 5.5 shows that the distance from road networks and nodes of intersections is a factor that has a strong influence on the occurrence of change.

Most of the changes occur when the distance is less than 1 km from the roads or nodes of intersections (78% and 76%) for the two periods. The farther from the roads, the less significant the density of changes is.

Because the considered territory is mainly characterized by hills and mountains, we also tried to answer the question: "What is the relationship between the land use changes and the geomorphologic factor?"

The data concerning the geomorphology are provided by the vertical dissection model that was generated according to the method proposed by Priego-Santander et al. (2003). It provides information about the form, intensity, and the type of geomorphologic processes that act on the landscape (Bocco et al., 2010). We consider that geomorphology influences the land use changes in such a way that the changes are not random and do not occur everywhere. Table 5.6 shows the occurrence of the changes depending on the geomorphologic factors for each evaluated period.

Table 5.6 shows that most of the changes occur in "high hills moderately dissected" (50% and 48%), in "plains with hills slightly and strongly dissected" (21% and 19%), and in "highly dissected high hills" (16% and 18%) for the two periods of the study. This information explains the influence of the geomorphology on the land use changes: the frequency of the changes can be used as an indicator of this constraint. This analysis

Table 5.6: Relationships between geomorphologic factors and the changes.

Contribution of geomorphologic factors to changes	1975–90	1990–2011
High mountains strongly dissected	1%	1%
Highly dissected high hills	16%	18%
High hills moderately dissected	50%	48%
High mountains moderately dissected	3%	5%
Mountains slightly dissected	8%	8%
Plains with hills slightly and strongly dissected	21%	19%
Plains and hills slightly dissected	1%	1%

confirms that the dispersion of the changes follows a "regularity" depending on the accessibility provided by the geomorphology and that the idea that the dispersion in the concerned territory is a random phenomenon can be rejected.

In conclusion of this part, even if there is some spatial dispersion of the changes, it is important to note that this is limited by the presence of linear infrastructure, the proximity to population centers, and the topographic factors that prevent the indiscriminate dispersion through the territory of the Alpes-Maritimes.

5.2.2 Land use transition rules definition

The spatial transition rules explain the transformation of one state to another. This transformation depends on the force of life of each individual and the interaction forces, and the competition between the individuals who evolve in a given place with a particular environment. Each individual (cell, state, land use, etc.) is submitted to the environmental interactions, which can provoke their transformation into another state (Dubos-Paillard et al., 2003). The rules are interpreted in accordance with the knowledge base of the territory concerned, compiled from the precedent analysis. These rules will serve to simulate the potential change and territorial transformation with a CA platform.

According to Sante et al. (2010), a CA is a formal model composed by a set of cells which take determinate values. These values evolve in discrete steps according to a mathematical expression which is sensitive to the value of neighboring cells. A cell state is a description of the cell characteristics and the change in a state occurs according to a transition rule (Mitsova et al., 2011). Transition rules are defined according to a formal language which incorporates all the knowledge of the studied phenomenon (Dubos-Paillard et al., 2003). In this way, the transition rules must be defined according to a deep analysis of the studied system.

We used a developer of CA platform proposed by Langlois (2000), called SpaCelle (System of Environmental Cellular Automata Production) based on transition rules. The

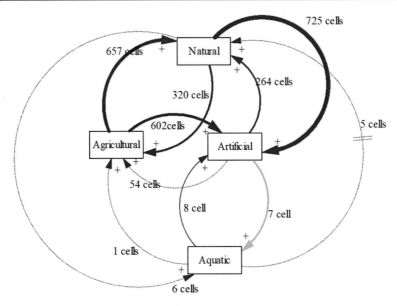

Figure 5.5
Causal relations of the 1990–2011 land cover changes.

advantages of this platform are principally that the user completely defines and handles the transition rules and defines all the components of the simulation model. Other advantages are the possibility to integrate the Geographic Information System (GIS) data multilayer in the platform interface; in addition, complex programming is not necessary. The principal limits of the platform are that the platform does not make the mobile objects, the distance factor can be introduced only in terms of the neighborhood parameter, and finally the language of the platform is French, a limiting factor for nonfrancophone users. However, we consider that the advantages of the platform outweigh any limits, which can be overcome by data preparation in the GIS.

According to the results of the assessment of the behavior of the territory during the two periods (1975–90 and 1990–2011), we selected the 1990–2011 period as the base of simulation because we consider that the pressures on agricultural areas presented during the 1975–90 period were a very important referential of the changes that occurred and the conditions of these structures differed greatly from the current condition of the agricultural areas. One of the reasons is that current distribution of this structure is much more fragmented than before, in such a way that the spatial behavior of the changes cannot be reproduced by the simulation. By contrast, all of the behavior identified for the second period of evaluation can be reproduced using the CA. The influence of the urban/artificial surfaces on the natural surface (Fig. 5.5) is the most remarkable while the agricultural areas are the most affected by the change, because these areas have lost surfaces from the synergic pressures of the urban/artificial and

Table 5.7: Spatial transition rules identification.

Trajectories	Cells	Hectares
Agricultural to natural	657	12,153
Artificial to natural	264	4883
Aquatic to natural	5	92
Natural to agricultural	320	5919
Artificial to agricultural	54	998
Aquatic to agricultural	1	18.5
Natural to artificial	725	13,411
Agricultural to artificial	602	11,136
Aquatic to artificial	8	148
Natural to aquatic	6	111
Agricultural to aquatic	10	185
Artificial to aquatic	7	130
Total	2659	49,351

natural surfaces and the areas that benefited the most from the changes are the urban/artificial ones. At the same time, it is possible to understand that this change is a very complex process.

Table 5.7 shows the number of cells that have changed for each trajectory and shows the equivalence of changes in hectares. The size of a cell is 18.50 hectares according to the spatial resolution of our analysis.

Table 5.7 underlines that the main changes are induced by the following three trajectories that are "natural to artificial" (27% of the changes), "agricultural to natural" (25% of the changes), and "agricultural to artificial" (23% of the changes). These results show not only that the urban/artificial areas increase by colonizing the agricultural and natural areas, but underline the contribution of the abandonment of the rural activities that allows the growth of natural areas in some parts of the territory.

5.2.3 Model calibration and validation

The model calibration and validation (Fig. 5.6) consisted in obtaining a model of 2011 from the simulation with CA using as baseline of simulation the land use thematic of 1990. A combination of three different methods of validation was used for the calibration in order to have complementary approaches providing an accurate level of information.

First of all, we used the information provided by the visual interpretation that consists in observing the most representative differences between both maps (real map vs simulated map). Then, in the second method of calibration, we used the statistical approach provided by «Validate» module of IDRISI Selva which is based on the use of the Kappa indices.

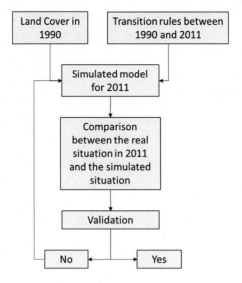

Figure 5.6
Model calibration diagram.

Then, we have evaluated the spatial pattern followed by the simulation by CA; that is why landscape ecology metrics (LEM) were used. These metrics can provide a vision of the similitude between both maps relating to the spatial pattern, the surface pattern of the patches, and the shape and the distribution of the elements of the map. This evaluation can provide more information than the kappa index application alone, because it enables us to know whether the transition rules follow the spatial pattern observed in reality. Even if the Kappa method was criticized (Pontius and Millones, 2011) other authors, like Basse (2013) and Mitsova et al. (2011), consider that this method can help to identify the locations of changes and the potential patterns that can emerge on the territory.

According to the identification of the transition rules of land use, it is possible to implement these rules into the CA platform. The transition rules in SpaCelle platform have the following structure:

$$E1 > E4 = AL(2,16)PV(E4,1)$$

where

E1: state 1 of the cell. For example, E1 is the natural area class.
E4: state 2 of the cell. For example, E4 is the urban/artificial area class.
AL(2,16): Transition probability. In this example, we can have two transitions in 16 years.
PV(E4,1): Interaction force. It depends on the radius of influence. In this example, the radius is 1 cell.

Applying this logic, the different transition rules were implemented in order to perform the simulation of the territorial dynamics from 1990 to 2011. In our simulation, we applied the Moore neighborhood which is a square nine cells with the considered cell in the center (radius of 1 cell). Fig. 5.7 shows two maps. The first map represents the observed situation for 2011 based on the photointerpretation of the 2011 Landsat image. The second map shows the simulated land use for 2011 from the 1990 land use model and the integration of the spatial transition rules. This simulated territorial representation was performed using 21 iterations in the SpaCelle cellular automata platform corresponding to 21 years of territorial evolution from 1990.

The visual interpretation allows confirming the coherence of the simulation by the spatial transition rules. The simulation reproduces the spatial patterns in a similar way of the reality. The main transformations are related to the increase of urban/artificial areas while the agricultural areas decrease. These transformations are mainly localized close to the coastal strip as it is underlined in the real situation in 2011. To complete these results, we have used a statistical approach to examine the quantitative coherence between both maps.

The IDRISI Selva "Validate" module offers a statistical analysis that answers two questions: (1) How well do a pair of maps that agree in terms of the quantity of cells in each category? and (2) How well do a pair of maps that agree in terms of the location of cells in each category? (Pontius, 2000).

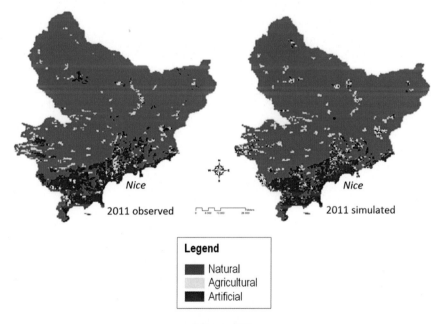

Figure 5.7
Results of model calibration.

The levels of information provided by the "Validate" module are defined by the next statistical expressions:

- $N(n)$: "is the agreement due to chance, which is the agreement between the reference and a map that has a membership of $1/J$ to each category in every grid cell, where J is the total number of categories in the analysis."
- $N(m)$: "is the agreement between the reference map and a map that has a distribution of m among the various categories in every cell, where m denotes a vector of the distribution of categories in the comparison map."
- $H(m)$: "is the agreement between the reference map and a modified comparison map, where the modification is to randomize the locations of the cells within each stratum of the comparison map."
- $M(m)$: "is the agreement between the reference map and the unmodified comparison map. It is the proportion of grid cells classified correctly, which is the most commonly used measure of agreement between maps."
- $K(m)$: "is the agreement between the reference map and a modified comparison map, where the modification is to rearrange as perfectly as possible the locations of cells within each stratum of the comparison map in order to maximize the agreement between the modified comparison map and the reference map."
- $P(m)$: "is the agreement between the reference map and a modified comparison map, where the modification is to rearrange as perfectly as possible the locations of cells within the entire comparison map in order to maximize the agreement between the modified comparison map and the reference map."
- $P(p)$: "is the perfect agreement, which is the agreement between the reference map and a map that has perfect information of both quantity and location. Therefore $P(p)$ is always 1."

The "Validate" module also provides the Kappa Index (Pontius, 2000). The range of this index varies from 0 (less than chance of agreement) to 1 (agreement almost perfect). This index consists of three main terms:

- The Kappa for no information (named Kno): it measures the gaps of information.

$$\text{Kno} = \frac{M(m) - N(n)}{P(p) - N(n)}$$

- The Kappa for grid-cell level (named Klocation) which measures the similarity of the position and distribution of the map elements:

$$\text{Klocation} = \frac{M(m) - N(m)}{P(m) - N(m)}$$

- The Kappa for stratum-level location (named KlocationStrata) which measures the coincidence between the number of cells:

Table 5.8: Results of statistical validation of the quantitative agreement between both maps.

Information of allocation	No [n]	Medium [m]	Perfect [p]
Perfect [$P(x)$]	$P(n) = 0.0012$	$P(m) = 0.9935$	$P(p) = 1.0000$
PerfectStratum [$K(x)$]	$K(n) = 0.0012$	$K(m) = 0.9935$	$K(p) = 1.0000$
MediumGrid [$M(x)$]	$M(n) = 0.0008$	$M(m) = 0.9466$	$M(p) = 0.9457$
MediumStratum [$H(x)$]	$H(n) = 0.0002$	$H(m) = 0.4377$	$H(p) = 0.4370$
No [$N(x)$]	$N(n) = 0.0002$	$N(m) = 0.4377$	$N(p) = 0.4370$
AgreementChance	0.0002	DisagreeGridcell	0.0468
AgreementQuantity	0.4375	DisagreeStrata	0
AgreementStrata	0	DisagreeQuantity	0.0065
AgreementGridcell	0.5089		
Klocation	**0.9157**		
KlocationStrata	**0.9157**		
Kno	**0.9466**		
Kstandard	**0.9051**		

$$\text{KlocationStrata} = \frac{M(m) - H(m)}{K(m) - H(m)}$$

- The Kappa Standard (named KStandard) which measures the proportion assigned correctly versus the proportion that is correct by chance:

$$\text{KStandard} = \frac{M(m) - N(m)}{P(p) - N(m)}$$

The results of these comparisons show an overall similarity (KStandard) very high (0.9051) with respect to the amount of coincident cells (KlocationStrata, 0.9157) and the similarity of the position (Klocation, 0.9157) (Table 5.8).

In order to evaluate the level of similitude between maps in terms of morphological conditions, we have also used the LEM in order to obtain information about the proximity, form, and surface. In this work, we used the following metrics (Table 5.9):

- Fractal dimension of the average patch (FD) measures the complexity of shapes; its value is between 1 and 2. Values close to 1 correspond to very regular perimeters, and values close to 2 correspond to very complex shapes.
- Patches Index (Shape Index): area or perimeter ratio adjusted to a square or a circle. When the form is compact, the index has a value of 1. This value increases with irregular shapes.
- Shannon Diversity Index: The "Shannon diversity" index determines in a neighborhood, the number of types of land present in approximately equal proportion. This index reflects the diversity of the landscape: the higher the index, the greater the number of types of land present in the same proportion.

Table 5.9: Landscape ecology metrics.

Map	FD	Patches index	Shannon diversity	Edge density
Real	1.44	7.64	0.48	0.19
Simulated	1.43	7.03	0.49	0.20
Difference	− 0.01	− 0.61	0.005	0.0003

• Edge Density: "Edge density" index calculates the length in a neighborhood of contours per hectare. This index reflects the complexity of the forms in the landscape: the higher the density contours, more complex is the landscape and the units are overlapped (Farina, 2011).

The application of these metrics between the real and simulated shows very few differences (0.01 for FD; 0.61 for Patches index; 0.05 for Shannon diversity; and 0.0003 for the Edge density). This result underlines that the simulated map has a high level of similitude to the observed map from the morphological point of view.

The results of the visual interpretation, statistical approach, and landscape metrics allow us to deduce that the transition rules applied within the CA have an acceptable approximation degree with the observed situation in the map of 2011 (around 90% of concordance). They also show that the model behavior follows an acceptable spatial and morphologic trend during the simulation. It is thus possible to make a territorial transformation model from simulation by CA.

5.3 Simulation results for 2050 and discussion

According to the previous results, we performed the simulation of the land use dynamic within 2050. Each iteration of the model represents 1 year. 2011 is the year of the beginning of the simulation and after completing 39 iterations the territorial transformation scenario is produced for 2050. The results show that the territorial transformation occurs in different ways (Fig. 5.8). While densification is constant near the coastal strip (plains and hills near roads and urban areas), in other parts of the territory (mainly in valleys and in low areas of mountains), the dispersion occurs, which can be explained by the influence of different territorial elements such as routes and road network, villages and population centers that together act as an asset of expansion.

The results presented by Fusco and Scarella (2008) support our results given that they mentioned the possibility of the urban expansion in the next decades because the territory has the suitability to host the population expected for 2030 by the demographic previsions, with the difference that our work gives a potential map of future expansion that was not presented earlier.

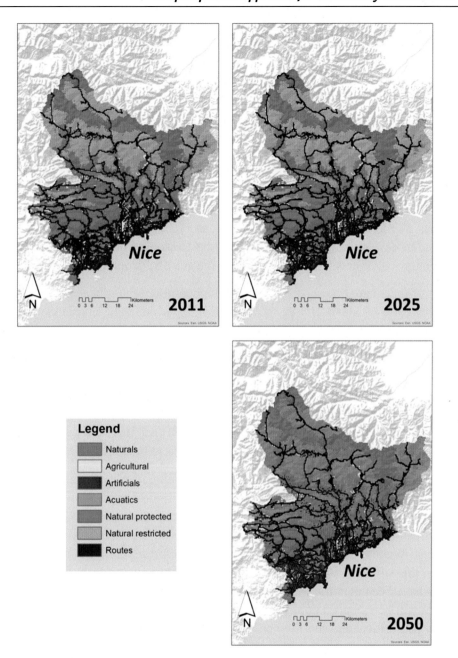

Figure 5.8
Landscape transformation scenario for 2050.

Our results are also similar to the results proposed by Basse (2013) concerning the simulation of urban expansion until 2040 in the coastal strip of the Alpes-Maritimes department. In this model, Basse demonstrated both a densification and a spread of urbanization in the whole coastal strip, with a significant increase of the rate of urbanization according to the potential implementation of a high-speed train network. The study proposed by Basse underlines the interest of the implementation of ancillary data like infrastructure projects, in order to assess the dynamic of land use. The absence of ancillary data in our model can be considered as a limitation of our model. But, in another hand, such ancillary data can also have a certain level or uncertainty: for example, due to problems of social acceptability, such infrastructure projects may be abandoned or delayed for many years or decades. In this context, we wanted to base our study only on observable spatial data in order to avoid assumptions that carry other uncertainties into the model. We also wanted to develop a methodology that can be transposed on other territories and requiring only spatial data like remote sensing images and geographic information.

The result of the simulation shows a possible duplication of the urban/artificial surfaces for 2050, which corresponds to 21% of the total territory, which may replace the natural and agricultural areas (Table 5.10).

From these results, it is possible to observe that the principal affectation of the spread of urban/artificial surfaces occurs on natural structures. In consequence, it is possible that biodiversity could be affected by this phenomenon of landscape transformation due to the fragmentation of natural areas. These affectations could occur potentially in the interior of the territory (middle and highlands) because the littoral strip has been very impacted by the urbanization since 1975 and there are few surfaces free to be built.

Using Corine Land Cover data and the photointerpretation of remote sensing images (landsat scenes), we have identified that the critical period of affectation for the coniferous forest will be between 2011 and 2025, where these structures will be most affected by a potential urban/artificial landscape. Moreover, this period is particularly critical for the mixed forest. We also know that the critical transition forest period has already started since 1990 when this structure was being affected. It is also possible to identify that natural grasslands have been particularly affected by urbanization since 1990 and that this assignment will be extended until the end of the century when these structures could be potentially in a critical and sensitive period from 2050, if current trends continue with the

Table 5.10: Land use dynamics between 2011 and 2050.

Land use classes	2011	2050
Agricultural	3%	1%
Artificial	11%	21%
Natural	86%	78%

same patterns of occupation and land use change observed over the past 40 years. Therefore it is possible that the urban/artificial areas augmentation will provoke the increase of the ecosystems vulnerability because they are more exposed than earlier to direct interactions with urbanized areas, and also because the agricultural areas have been totally affected and they are not more the "buffer" between urban/artificial and natural surfaces.

According to the previous results, we assess if the model retains the same behavior than the observed situation in order to validate our model.

Do the transition rules reproduce the observed spatial patterns?

The results of the comparison of spatial trend show that the change follows a continuous tendency, since 1975−2050, that enables us to validate that transition rules have complied with the behavior observed in the past (Table 5.11).

The exception is related to the dispersed patterns because the model does not consider a transition rule that takes into account only the random aspect that in reality happens with some changes. This omission is considered as a weakness of the model.

Do the transition rules reproduce the influence of the distance of road network on the changes?

The results of validation by the distance of road network confirm that the model reproduced the observed behavior (Table 5.12).

We can also observe that even if the dispersion phenomenon is progressive, most of the changes may occur close of the roads.

Do the transition rules reproduce the influence of the distance of population centers on the changes?

In the same logic, the validation of the results concerning the distance of the changes to the population centers confirms that the transition rules have complied with the tendencies of the observed behaviors (Table 5.13).

These results also confirm that urbanization is not random, given that the different elements of the territory have a strong influence on the existence of this phenomenon.

Table 5.11: Validation by spatial patterns.

Spatial patterns	1975−90	1990−2011	2011−50
Pure aggregate	62%	57%	49%
Nodal aggregate	23%	26%	28%
Linear	7%	7%	7%
Nodal	6%	9%	14%
Linear aggregate	2%	1%	2%

Table 5.12: Validation by road network distance measure.

Distance of changes from road network	1975–90	1990–2011	2011–50
<1 km	78%	76%	75%
At 1 km	20%	22%	22%
At 2 km	2%	2%	3%

Table 5.13: Validation by the distance of the changes from the population centers.

Distance of changes from population centers	1975–90	1990–2011	2011–50
<1 km	9%	9%	9%
At 1 km	41%	39%	37%
At 2 km	31%	32%	32%
At 3 km	13%	12%	13%
At 4 km	5%	6%	7%
At 5 km	1%	1%	1%
At 6 km	0%	1%	1%

This work demonstrates that it is possible to project landscape transformation studying late historical behavior and spatial pattern extracted from remote sensing images and geographical data. We have also been able to prove the strong influence of geomorphological factors and road networks above the dispersion of the land use changes and they can reject the idea that this dispersion is random and homogenous across the concerned territory.

As it is possible to observe, the territorial transformation occurs in a different way across the territory because near the coastal strip, urbanization is already consolidated and compact. This is due to the infrastructure and the strong attraction of the zone. In addition, the model shows the strong influence exerted on the transformation by road networks. It also shows that these factors provoke a significant dispersion of the urbanization, especially in hills and mountains of the department. Moreover, the analysis of changes shows that the transformation of the territory can provoke changes in the spatial organization and it is also possible to note that the changes follow a trend to dispersion influenced by the road networks. This dispersion could be a major change in the model of historical urbanization in the territory. Finally, these results can provide information about the needs of planning measures and about the most sensitive areas to change (the hills and mountains). The results can also support the ecological and conservation land use policies and prepare the policies to reduce the ecological vulnerability in the territorial system.

The principal limits of this work are related to that the spatial rules are stationary and depend on the level of knowledge of the observed phenomenon and in consequence, the emergence of the new urban/artificial areas is conditioned by the deterministic conditions of the spatial rules and the elements of the model. Moreover, the transition rules are identifiable only on relatively long time periods; the model has been tested only in a landscape scale, and it is possible for the model not to be optimal for intraurban simulation because this scale needs other data like infrastructure projects, economic and demographic trends, etc.

The landscape transformation of the concerned territory has ecological implications that must be evaluated. We consider that the changes may cause the biodiversity affectation because of the landscape fragmentation in the interior of the territory. This is why it is necessary to evaluate the alternative policies for territorial and biodiversity governance that enable the control of the urban expansion and dispersion as a measure of sustainable urban planning.

The next part is dedicated to the presentation of the methodology we used in order to characterize the biodiversity levels of the studied territory.

5.4 Model of territorial biodiversity

In order to assess the potential impacts of landscape transformation and climate change on the biodiversity of the Alpes-Maritimes, we developed a model of biodiversity based on landscape ecology. The first part is related to the definition of biodiversity we used in order to formalize our model. Then, we present the methodology to assess the biodiversity level of the studied territory.

5.4.1 What do we intend by "biodiversity"?

The term "biodiversity" is a contraction of the term "Biological diversity". It was proposed in 1985 by W. Rosen then taken up in 1988 by E. Wilson et al. in the book entitled "Biodiversity" (Wilson and Peter, 1988). Article 2 of the Convention of Biological Diversity edited by the United Nations Environment Programme (UNEP) on May, the 22nd of 1992, defines the "Biological diversity" as "the variability among living organisms from all sources including, inter alia, terrestrial, marine and other aquatic ecosystems and the ecological complexes of which they are part: this includes diversity within species, between species and of ecosystems." This definition can be considered as the official definition of biodiversity that integrates the notion of variability of the expression of life over the world in the different ecosystems.

According to Magurran (1988), the diversity of organisms seems to be "like an optical illusion. The more it is looked at, the less clearly defined it appears to be and viewing it from different angles can lead to different perceptions of what is involved." He explains why biological diversity is hard to define because it integrates two components: the variety and the abundance of species.

In 1991, Solbrig proposed a very complete definition of biodiversity considering that this concept is "the property of living systems of being distinct, that is different, unlike. Biological diversity or biodiversity is defined here as the property of groups or classes of living entities to be varied. Thus each class of entity—gene, cell, individual, species, community, or ecosystem—has more than one kind. Diversity is a fundamental property of every living system. Because biological systems are hierarchical, diversity manifests itself at every level of the biological hierarchy, from molecules to ecosystems." So, in this definition, biodiversity can be considered not as a result of life evolution but also as a property of living being that interact with their environment.

For other authors, biodiversity can be considered as a way, a criterion, to select the places according to their level of species variability. In this case, Sarkar (2005) explains that biodiversity "should be (implicitly) operationally defined as what is being optimized by the place prioritization procedures that prioritize all places on the basis of their biodiversity content using true surrogates. Thus biodiversity is the relation used to prioritize places."

According to the above definitions, we propose to consider the biodiversity as a property of ecological systems at all spatial, temporal, and functional levels. It is not a recognizable entity in space but a process that acts as a differentiator of life and its expressions. Biodiversity conservation is seen as preserving the ownership of ecosystems, landscapes, organisms, and all levels of ecological organization to differentiate from one another and maintain the heterogeneity of conditions but also the diversity of elements and relationships that maintain the dynamic equilibrium of natural systems.

This definition is the key element that defines the methodology we developed to characterize the biodiversity of the Alpes-Maritimes.

5.4.2 How characterizing the biodiversity of the studied territory?

Many works are dedicated to the assessment of biodiversity. They are often based on the definition of indices integrating field data and/or remote sensing images. Some approaches also combine DNA analysis in order to differentiate population structures among the species. There is a large bibliography on these subjects and we will not present here these methods. We suggest the readers to start with the book published in 1995 by Heywood and Watson: "Global Biodiversity Assessment" and to complete their study with the different indices proposed by Shannon and Weiner, Simpson, Jaccard, Pielou and McIntosh, Berger and Parker, Margalef, etc., and their applications in the field of ecology and landscape ecology.

To understand the biodiversity of the Alpes-Maritimes, we have developed a model of landscape units to integrate ecosystem functions. Landscape units consist of two major components: a physical component (stable component) that describes the sequence within the rock-relief-soil ensemble and has a low or very low rate of change over time and another component that describes the bioclimatic and land use dimension of the territory (dynamic component) and which is characterized by dynamism of several time scales. Both components can be handled separately in GIS and can be combined to obtain units connected by common features. This way of disaggregating the landscape according to the ecological regionalization makes it possible to enrich the knowledge on the geographical distribution of natural resources, ecosystems, and their dynamics over time. In addition, it helps to understand the level of tolerance to the changes that ecosystems can endure, the modifications of their services, as well as the risks, the geographical distribution of biodiversity, and the conflicts of land use (Bocco et al., 2010). The objective of this Model of Landscape Units is to understand the level of heterogeneity at different hierarchical levels due to the structure of the landscape and the patterns involved in determining the diversity of ecosystems.

The methodology is based on three main steps described in Fig. 5.9. It consists of the overlapping of four GIS data related to soil types, geomorphologic units, climate variables, and vegetation structures (or other land use structure like urban areas, water, and agriculture). These data are provided by public administrations in France and Europe. For the climate data, we used the outputs of the model ALADIN-Climat to locally evaluate the minimal and maximal temperatures trends in the middle of the 21st century. These data where regularizing by crossing environmental data that best describe the physical environment with temperatures

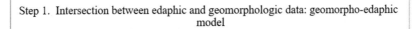

Step 1. Intersection between edaphic and geomorphologic data: geomorpho-edaphic model

⬇

Step 2. Intersection between geomorpho-edaphic model and climatic data: geomorpho-edapho-climatic model

⬇

Step 3. Intersection between geomorpho-edapho-climatic model and vegetation data: Landscape units model (biodiversity model)

Figure 5.9
Methodology to elaborate the model of territorial biodiversity.

from the ALADIN-Climat model output, several models of multiple regressions are established monthly for the period 2021–50. The obtained results give acceptable residues considering the scaling operated (from 12 km horizontal resolution to 90 m). This scaling was performed in the frame of the "Cassandre" project (Martin et al., 2013).

The result of this methodology is the characterization of 232 landscape units in the studied territory (Fig. 5.10).

Each landscape unit is characterized by all the modalities of variation of each data set. Photointerpretation techniques, based on remote sensing images and orthophographies, have been used in order to validate this model.

The climatic calibration of each landscape unit was performed by taking into account the criterion of ecological optimum.

According to Whittaker's "Ecological Continuum" theory, biotic communities are distributed in space along an environmental gradient, according to their ecological amplitude capacity so that less plastic or less resistant organisms grow in very specific areas where the environmental characteristics are very stable for these organisms. On the contrary, organisms with greater adaptability and plasticity may occupy larger extensions and distribute along areas with more variable conditions.

In line with IPCC predictions, it is expected that changes in vegetation and ecosystems will be different, since each species will have a different adaptation response. The composition

Figure 5.10
Model of territorial biodiversity showing the 232 landscape units.

of the various ecosystems will be affected at different speeds so that areas with human-induced impacts such as fragmentation will be affected more quickly than those with high natural features (Fig. 5.11).

This figure shows that trees and herbaceous plants will be very vulnerable to the climate change (for scenarios RCP4.5, 6.0, and 8.5) due to the difficulties for them to adapt and/or colonize other territories in order to face the kinetic of the climate evolution. If this

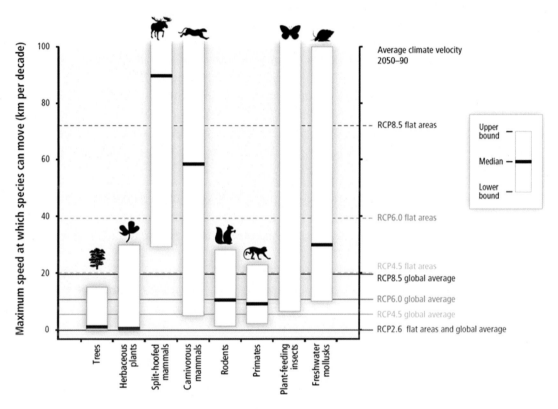

Figure 5.11

Maximum speeds at which species can move across landscapes (vertical axis on the left), compared with speeds at which temperatures are projected to move across landscapes (climate velocities for temperature; vertical axis on the right). White boxes with black bars indicate ranges and medians of maximum movement speeds for trees, plants, mammals, plant-feeding insects (median not estimated), and freshwater mollusks (IPCC, 2014a,b). (Figure SPM.5): Summary for Policymakers. In: Climate Change 2014: Impacts, Adaptation, and Vulnerability. Part A: Global and Sectoral Aspects. Contribution of Working Group II to the Fifth Assessment Report of the Intergovernmental Panel on Climate Change [Field, C.B., V.R. Barros, D.J. Dokken, K.J. Mach, M. D. Mastrandrea, T.E. Bilir, M. Chatterjee, K.L. Ebi, Y.O. Estrada, R.C. Genova, B. Girma, E.S. Kissel, A.N. Levy, S. MacCracken, P.R. Mastrandrea, and L.L. White (eds.)]. Cambridge University Press, Cambridge, United Kingdom and New York, NY, USA. (IPCC, 2014) assumption will be verified in the future, many plants will be in state of stress and may know an increase of their fatality rate (Fig. 5.12).

This graph also helps to identify the ecological range of a species. This parameter is its ability to reproduce in different ecological conditions. The species occupying very narrow zones on the gradient are specialized species, characterized by small ecological amplitude for the considered gradient. Others, on the other hand, occupy large areas because they have much wider ecological amplitude and are much less demanding (Dufrêne, 2004). The distribution of vegetation will be significantly influenced by climate change in the coming decades. At certain latitudes of the planet, changes in the distribution of vegetation are already occurring, and we can observe how temperate species have begun to move to higher elevation areas; it is also possible to observe how previously formed areas of tree species are now colonized by certain xerophilic and thermophilic species (MEA, 2005). Species with limited climatic thresholds are the most vulnerable in terms of extinction, given their limited ability to self-modify in the face of a change in temperature or precipitation. Among them are endemic mountain species and restricted island species. In contrast, higher tolerance species with long-spreading dispersal mechanisms and large populations have a lower extinction risk (IPCC 2007, 2014a,b).

For the determination of the ecological conditions which have an apparent influence on the distribution of these landscape units, the maximum and minimum temperatures of January, July, and November were taken into account, the precipitation for each month as well as the

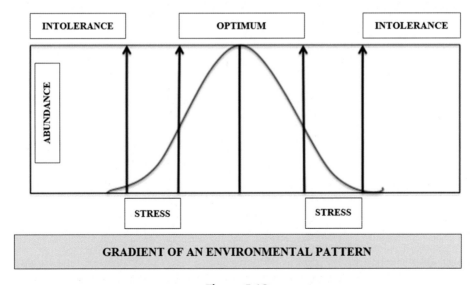

Figure 5.12
Distribution of abundances of a species among an environmental gradient and identification of its ecological optimum. Source: *Modified from Dufrêne, 2004.*

exposure, using the value of the statistical median for each of the variables. The objective of this calibration is to determine the climatic envelope (potential distribution) and optimal ranges of altitude and distribution of each unit of the landscape.

Based on these climate characterization results from 104 landscape units in the Alpes-Maritimes, which represent natural areas, we present the methodology for assessing the potential impacts of climate change on the distribution of these landscape units.

5.5 Methodology to assess climate change impacts on biodiversity

In order to assess the potential impacts of climate change on landscape units, we developed the CDS toolbox model which was promoted by ASES and Climpact Data Science (CDS) companies. For the determination of the potential ecological distribution of the landscape units according to the future climate, the A1B emissions scenario has been spatialized at 90 m for 2050. The procedure for assessing ecosystem dynamics in relation to climate change is to find the optimal set of conditions for each landscape unit according to the future climate, to evaluate potential spatial changes, and finally to observe whether the changes may or may not result in a landscape transformation and therefore whether or not they will have consequences for biodiversity.

The procedure for calculating the potential ecological distribution of landscape units consists in finding the reference conditions (maximum and minimum temperatures, precipitation, vertical dissection, and exposure) for each unit of the landscape, according to the future climate. Indeed, it is considered in the model that in some unit migration and displacement processes may cause adaptations to the new geomorphological conditions, since geomorphology is a relatively stable factor in the time scale of the model and therefore its dynamic is not noticeable. In other words, the model considers the possibility of finding one, several, or all the reference conditions in a different geomorphological unit from that of reference according to the future climate. If a unit will not find all the conditions in the same reference geomorphological unit, there will be three possible consequences: (1) the landscape unit adapts slightly to the new conditions and selects the locations with the conditions closest to its optimum of reference with a contraction of its populations. (2) The landscape unit adapts deeply to the new conditions. (3) The unit is not able to adapt to changes and disappears locally.

This procedure is carried out for all the conditions identified (11 variables[2]) and finally the linear weighting is used to identify the potential ecological distribution of the landscape unit according to the future climate, based on Fig. 5.13.

---

[2] Average of: Temp. max January, Temp. min January, Temp. max July, Temp. min July, Temp. max November, Temp. min November, Rainfall January, Rainfall July, Rainfall November. Vertical Dissection. Exposure.

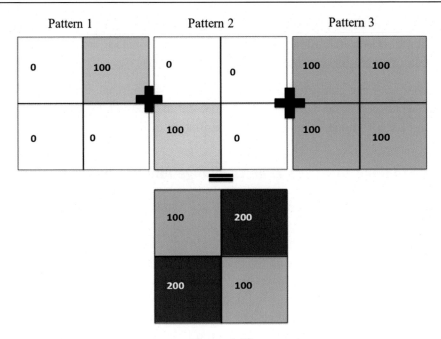

Figure 5.13
Potential ecological distribution of a species.

Potential Ecological Distribution = Pattern 1 + Pattern 2 + Pattern 3 + Pattern n...

The figure illustrates the spatial weighting of the variables. In the first frame (on the left), it is possible to observe the hypothetical presence of the maximum temperature condition represented in green and whose value is 100. This value is assigned to the localities that fulfill the expected condition for each of the variables observed. The second frame represents the locations where minimum temperature conditions are present; it is illustrated in light orange color. The third frame shows the locations where the precipitation conditions are in light blue. Finally, the bottom part represents the spatial weighting of the three variables. It illustrates in light blue the areas in which there is only one condition, and in dark blue areas of weighted presence (spatial convergence) of two out of three conditions.

After this step, it is possible to identify the level of similarity that will help the stakeholders into the decision process in order to identify the priorities of conservation for the different landscape units.

According to Eastman (2006), a decision rule is formalized from a series of arithmetic and statistical procedures that allow the integration of stabilized criteria into a simple index that helps to make comparisons of alternatives. In this case, the decision judgment for determining the distance to the reference point is based on the weighting previously

Table 5.14: Decision criteria, weighting, and global similarities.

Decision criteria										
1/11	2 /11	3 /11	4 /11	5/11	6 /11	7 /11	8 /11	9 /11	10/ 11	11/ 11
Extremely low	Very low	Low	Averagely low	Averagely high	Moderate	Hugh	Very high	Extremely high		Equal
Weighting										
100	200	300	400	500	600	700	800	900	1000	1100
Global similarity in %										
9	18	27	36	45	55	64	73	82	91	100

described. The interpretation of the expected weighting level for each landscape unit is provided by the scale in Table 5.14.

Areas where the value is 11/11 or 1100 are areas where conditions have not changed between assessment periods and therefore baseline conditions have been found fully in the climate change scenario. For areas with a value of 9/11 and 10/11, the changes are not very deep and, in principle, the landscape units are not subject to extreme pressure of change so that these areas are the places where adaptation will be potentially more likely. As the distance between the reference point and the results of the spatial weighting increases, the adaptation of the landscape units is less likely. Actually, the uncertainty of the model is higher in the areas between 6/11 and 8/11, and this uncertainty is more moderate toward the extremes because it is foreseeable that as the similarity increases, the possibility of adaptation is higher. On the contrary, as the similarity decreases between the reference point and the weighting, the possibility of adaptation decreases. Therefore, between the two extremes, the possibilities are much more uncertain.

The model considers three assumptions. Two of these are potentially predictable for model results that have an extremely high, equal, or even very high spatial weighting value. These assumptions correspond to the selection pressure and the adaptation pressure. For moderately high to high values, the uncertainty is high so that adaptation pressure or extinction is possible. This depending on the response capacity to the changes of the landscape units and the conditions external to the design of this model, such as ecological plasticity, phenotypic plasticity, or more complex and potentially ignored dynamics mechanisms.

In the CDS toolbox model, the selection pressure corresponds to the contraction mechanism of the distribution area of the landscape unit strictly on the localities where the changes are not very extreme for the whole of the landscape unit and the cost of the adaptation is received and assimilated to maintain the survival of the landscape unit. In this case, there is indeed a potential adaptation but there is not necessarily significant displacement and migration and this process mainly causes a spatial contraction of the landscape unit.

The adaptation pressure is even more complex because unlike the pressure selection, there are several possibilities. The first of the possibilities is also a spatial contraction of the landscape unit from an adaptation with resistance. For this first assumption, the changes are perceived negatively by the landscape unit and therefore it must select the characters most able to tolerate the changes in order to maintain the survival of the landscape unit, which would cause a migration to the localities having the characteristics closest to its ecological optimum. In spatial terms, this process would cause migration but at the same time a contraction of the populations, the result of which would be a necessary adaptation starting from the migration which could cause a fragmentation or erosion of the landscape unit. A second assumption for the adaptation mechanism is the possibility that changes are positively perceived by the landscape unit because these changes would favor the potential for colonization of new ranges, which could lead to an expansion of the landscape unit from adaptations to other conditions, for example, geomorphological, altitude, exposure, and soil.

For results with an extremely low to medium spatial weighting value, the local extinction assumption applies.

In the selection pressure hypothesis, the landscape unit reduces its distribution area to the localities where the conditions are close to its ecological optimum. In the case of adaptation, the landscape unit is able to change to occupy new areas of distribution.

5.6 Combined impacts of land use and climate changes on biodiversity

The Alpes-Maritimes have 55% of the alpine flora on 5.6% of the surface of the alpine arc, with about 3100 species of plants. It is one of the four main alpine endemism areas as the department has 125 alpine endemic species out of 417 endemic flora species, that is, 30%. According to Médail and Diadema (2009), the Alpes-Maritimes and Ligurian territory is one of the 52 largest refuges of Mediterranean plants, identified by genetic studies.

The final part of this research is to assess the potential impact of LUCC and climate change on the biodiversity of the French Riviera for 2050. In order to achieve this objective, we establish a model of landscape units for the whole territory. This model combines spatial data of geomorphology, soils typology, climatic means, and the vegetation structures (from CORINE Land Cover product). This model provides 232 landscape units for the whole territory.

Landscape units related to aquatic areas, urban areas, agricultural areas, and rocky areas are not taken into account in the analysis of results concerning the potential impacts of both land use changes and climate change. Only the most represented natural landscape units on the territory are taken into account in the following results.

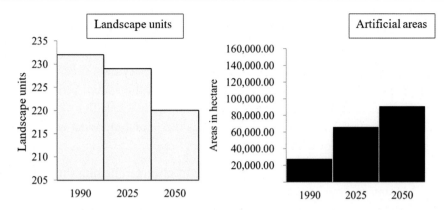

Figure 5.14

Evolution of the ecosystem richness according to the increase of the artificialization. Landscape units related to aquatic areas are not included in the assessment.

Table 5.15: Potential impact of landscape transformation (artificialization) on the landscape units.

Main landscape units	Landscape units (2011)	Landscape units (2050)	Landscape units loss	Variety loss (2050)	Variability changes (2050)
Sclerophylous vegetation	10	9	1	10%	0.12 nats
Meadows	19	18	1	5.26%	0.01 nats
Transition forests	21	21	1	4.5%	0.05 nats
Mixed forests	17	17	1	5.5%	0.04 nats
Conifers forests	21	21	0	0%	0.04 nats
Hardwood forests	15	14	2	13.3%	0.02 nats

Fig. 5.14 shows the relationship between the amount of landscape units and the dynamics of artificialized areas.

The results show an impact for 17 landscape units, which represent 7% of the biodiversity of the territory. These impacts will be generated by the landscape fragmentation due to the artificialization of the territory in natural and seminatural areas. The main fragmentations may occur in the hills and valleys of the territory, and in some mountain areas. Fortunately, these losses of landscape units should affect a few natural landscape units (six landscape units). The results of the assessment of the impact of artificialization on the diversity of the landscape units show that there is a risk of loss of variety and variability of ecosystems toward 2050 (Table 5.15).

Conservation policies and conservation tools such as nature reserves, nature or species management plans, and regulatory tools (TVB and SRCE) have the common characteristic of being focused on a fixed temporal condition or state of the attribute to be conserved and generally the objectives have not focused on ecological processes such as biodiversity, heterogeneity, or ecological continuity (Hannah et al., 2002; Bull et al., 2013). The static boundaries usually defined in nature conservation practices have a negative effect on biodiversity because they do not reflect the spatial influence of ecological processes (Múgica et al., 2002; Primack et al., 2012; Billet, 2011; Hannah et al., 2002; Bull et al., 2013), which often exceeds the poor functional administrative boundaries or cartography of natural systems.

Among the results, it was found that, in response to climate change, the expansion, geographic translation, and contraction of spatial distribution areas of landscape units are possible and that, in general, the results suggest that modification of the climate envelopes applies to the vast majority of landscape units. The results obtained in this work confirm that climate change could significantly affect and reduce ecosystem diversity by 2050. Thanks to the results obtained, it can be concluded that temperate landscape units will be very affected by the influences of climate change around 2050.

In general, the trend suggests that xeric and thermophilic landscape units will have the opportunity to expand and colonize new areas. However, the most important aspect is that although this trend seems clear and logical given the climatic dynamics (temperate ecosystems that tend to decrease and xeric and thermophilic ecosystems that tend to increase), the loss ecosystem diversity is observed in both cases. In other words, even if the trend shows a possibility of expansion for some landscape units able to adapt to future climatic conditions, there will be a loss of diversity because all the thermophilic or xeric landscape units are not favored by the same way and at the same level. As a result, some landscape units colonize more quickly and more areas than others, resulting in deep asymmetries in the spatial distribution of landscape units.

By observing on the maps the phenomenon of expansion compared to the reference spatial distributions of the landscape units, it is shown that even if some units will be able to adapt and increase their distribution areas, this surface increase will not represent an increase in diversity or variability. Therefore, in almost all cases, even if the trend is expansion or contraction, the general rule is a loss of diversity or a loss of ecosystem variability. Indeed, if the trend is a contraction of the units, it is due to the disappearance of one or more units and this causes a decrease of the variability and an asymmetry in the variability and the heterogeneity.

But if the trend is an expansion, this expansion is not proportional between all landscape units, causing a proportional dominance effect and consequently a decrease in variability and heterogeneity. As a result, biodiversity loss is inevitable even though climate change may favor certain landscape units or groups of landscape units.

Table 5.16: Potential impact of climate change on the landscape units.

Main landscape units	Landscape units (2011)	Landscape units (2050)	Landscape units loss	Variety loss (2050)	Variability changes (2050)
Sclerophylous vegetation	10	12	2	20%	0.39 nats
Meadows	19	15	4	21%	1.06 nats
Transition forests	21	21	0	0%	0.05 nats
Mixed forests	17	12	5	29%	0.75 nats
Conifers forests	21	15	6	29%	1.32 nats
Hardwood forests	15	9	6	40%	0.44 nats

Table 5.17: Potential impact of climate change and territorial transformation (artificialization) on the landscape units.

Main landscape units	Landscape units (2011)	Landscape units (2050)	Landscape units loss
Sclerophylous vegetation	10	7	3
Meadows	19	14	5
Transition forests	21	20	1
Mixed forests	17	11	6
Conifers forests	21	15	6
Hardwood forests	15	7	8

Table 5.16 presents the impact of climate change on landscape expressions for the 2050 horizon.

This table shows that for 2050 the Alpes-Maritime territory will face a potential impact of climate change on 48% of the natural landscapes. This does not mean that the territory will lose all these landscape units, but it means that these landscape units may face a significant loss of variety and variability due to the impact of climate change.

Table 5.17 finally shows the synergetic impacts of land use changes and climate change impacts on the landscape units toward 2050.

Fig. 5.15 represents the map of the main natural landscape units that are exposed to a risk of loss toward 2050 due to climate change and land cover transformation.

Evaluation of the impact of urbanization and climate change made from the definition of 103 natural landscape units characterized by various factors (geomorphology, climatology, pedology, vegetation formations) has shown that in total it will be 29 landscape units (28% of the diversity of the Alpes-Maritimes) that will be affected by these two phenomena by 2050. These landscape units present a risk of extinction, whether because of climate change or the dynamics of the artificialization of the soil. Seven landscape units are at risk of disappearing (major risk of extinction), as announced by all prospective scenarios for 2050.

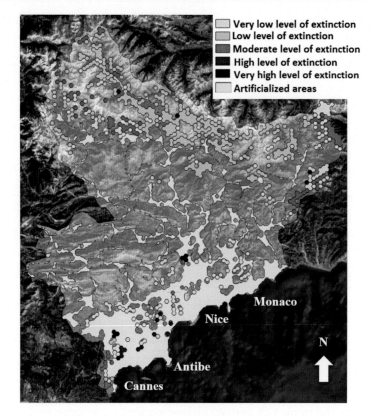

Figure 5.15

Map of the main landscape units having a level of extinction on the territory. The yellow polygons are related to the artificialized areas estimated for 2050. The colored scale represents the level of extinction risk of landscape units.

5.7 Conclusion

Biodiversity is considered as a characteristic or property of ecological systems at all levels of spatial, temporal, and functional organization, while taking into account that it is not a recognizable entity in space but a dynamic process that acts as a differentiator of life and its expressions.

Biodiversity conservation consists precisely in preserving the property of ecological complexes, ecosystems, landscapes, organisms, and all levels of ecological organization, differentiating from one another and preserving the heterogeneity of conditions, but also the diversity of elements and relationships that maintain the dynamic and chaotic equilibrium of natural systems.

This work demonstrates that it is possible to project landscape transformation by studying the historical behavior and spatial pattern of changes in past periods from the interpretation of the analysis and reproduction of spatial transition rules to the using simulation models by cellular automaton.

Through the results, it has been shown that with the CDS toolbox model, in response to climate change, the expansion, geographic translation, and contraction of the spatial distribution of landscape units are possible and that, in general, the results suggest that the modification of climate envelopes is a constant for the huge majority of landscape units because of the climate change in the study area.

This work shows the need to mobilize all mechanisms and means to conserve biodiversity despite the great uncertainties, because the forecasts are more and more pessimistic and for this reason, there is an urgent need to change the static vision of the conservation toward a dynamic and adaptive vision. The results are likely to be integrated into different regulatory documents and tools such as TVB and SRCE by the public administration and the territorial entities that have to manage the biodiversity conservation.

Finally, this work has made it possible to better understand the extent of the impacts that artificialization and climate change could induce on biodiversity at the Alpes-Maritimes scale.

Acknowledgments

We wish to thank the support provided by the administration of the local government of the Alpes-Maritimes Department and for the financial contribution provided by the Région Provence, Alpes, Côte d'Azur (France) for the "Cassandre" project.

References

Aguilera Benavente, F., Valenzuela Montes, L.M., 2010. Simulación de escenarios futuros en la aglomeración urbana de Granada a través de modelos basados en autómatas celulares (Simulation of future scenarios in the urban sprawl of Granada through models based on cellular automata). Bol.Asoc. Geógr. Esp. 54, 271–300.

Barredo, J.I., Kasanko, M., Mc Cormick, N., Lavalle, C., 2003. Modeling dynamic spatial processes: simulation of urban future scenarios through cellular automata. Landsc. Urban Plan. 64, 145–160.

Basse, Reine Maria, 2013. A constrained cellular automata model to simulate the potential effects of high-speed train stations on land-use dynamics in trans-border regions. J. Trans. Geo. Elsevier 32 (C), 23–37.

Batty, M., Xie, Y., Sun, Z., 1999. Modelling urban dynamics through GIS-based cellular automata. Comput. Environ. Urban Syst. 23 (3), 205–233.

Billet, P., 2011. L'évaluation environnementale, fondement de la prévention et de la réparation des atteintes à la biodiversité en droit français et communautaire. Approche critique. Rev. Jurid. de. l'environnement, numéro spécial 2011, 63–78.

Bocco, G., Priego, A., Cotler, H., 2010. The contribution of physical geography to environmental public policy in México. Singap. J. Trop. Geogr. 31, 215–223.

Bull, J.W., Suttle, K.B., Singh, N.J., Milner-Gulland, E.J., 2013. Conservation when nothing stands still: moving targets and biodiversity offsets. Front. Ecol. Environ. 11 (4), 203–210.

Dubos-Paillard, E., Guermond, Y., Langlois, P., 2003. Analyse de l'évolution urbaine par automate cellulaire: le modèle SpaCelle (Analysis of urban development by cellular automata: the SpaCelle model). L'espace Géogr. T32 4 (22), 357–378.

DTA, 2008.- L'essentiel de la DTA 06, DDE-SAET-CT-SIG.

Dufrêne, M., Réseau écologique - Structure écologique principale. Concepts - structure - stratégie d'élaboration. Version 1.0. MRW/DGRNE/CRNFB, p. 33.

Eastman, J.R., 2006. IDRISI 15.0. The Andes Edition. Clark University, Worcester, MA, USA.

Engelen, G., Lavalle, C., Barredo, J.I., van der Meulen, M., White, R., 2007. The Moland modelling framework for urban and regional land-use dynamics. In: Koomen, E., Stillwell, J., Bakema, A., Scholten, H.J. (Eds.), Modelling Land-Use Change, Progress and Applications. Springer, The Netherlands.

Farina, A., 2011. Ecología del Paisaje (Landscape Ecology). Publicaciones Universidad de Alicante, España.

Feng, Y., Liu, Y., Tong, X., Liu, M., Deng, S., 2011. Modeling dynamic urban growth using cellular automata and practical swarm optimization rules. Landsc. Urban Plan. 102, 188–196.

Fusco G., Scarella F., 2008. L'évolution de l'habitat dans les Alpes-Maritimes et dans l'Est Var (The Habitat Evolution in the Maritimes-Alps and the Est of the Var department). UMR ESPACE, Équipe de Nice dans le cadre du PREDAT des Alpes-Maritimes, France.

Garbolino, E., Sanseverino-Godfrin, V., Hinojos-Mendoza, G., 2016. Describing and predicting of the vegetation development of Corsica due to expected climate change and its impact on forest fire risk evolution. Saf. Sci. 88, 180–186.

Garbolino, E., Daniel, W., Hinojos Mendoza, G., Sanseverino-Godfrin, V., 2017. Anticipating climate change effect on biomass productivity and vegetation structure of Mediterranean Forests to promote the sustainability of the wood energy supply chain. In: Proceedings of the 25th European Biomass Conference and Exhibition, Stockholm, Sweden, 12–15 June, pp. 17–29.

Garbolino, E., Daniel, W., Hinojos Mendoza, 2018. Expected global warming impacts on the spatial distribution and productivity for 2050 of five species of trees used in the wood energy supply chain in France. Energies 11 (3372), 1–17.

Gomez-Delgado, M., Rodriguez Espinoza, V.M., 2012. Análisis de la dinámica urbana y simulación de escenarios de desarrollo futuro con tecnologías de la información geográfica (Urban Dynamics Analysis and Conception of Futures Scenarios from Geographical Information Technology). Editorial Ra-MA, Madrid, Spain.

Hannah, L., Midgley, G.F., Lovejoy, T., Bond, W.J., Bush, M.L.J.C., Lovett, J.C., et al., 2002. Conservation of biodiversity in a changing climate. Conserv. Biol. 16 (1), 264–268.

Hassan, R., Scholes, R., Ash, N., 2005. Ecosystems and Human Well-being: Current State and Trends, vol. 1. Millenium Ecosystem Assessment, Washington, DC, 917p.

He, C., Okada, N., Zhang, Q., Shi, P., Li, J., 2008. Modeling urban expansion scenarios by coupling cellular automata model and system dynamic model in Beijing, China. Landsc. Urban Plan. 86, 79–91.

Intergovernmental Panel on Climate Change, 2014a. Climate Change 2014: Impacts, Adaptation, and Vulnerability, Part B: Regional Aspects. Cambridge University Press, Cambridge, UK, p. 696.

IPCC, 2007. Climate Change 2007: Synthesis Report. In: Pachauri, R.K., Reisinger, A. (Eds.), Contribution of Working Groups I, II and III to the Fourth Assessment Report of the Intergovernmental Panel on Climate Change. IPCC, Geneva, p. 104.

IPCC, 2014b. Summary for policymakers. In: Field, C.B., Barros, V.R., Dokken, D.J., Mach, K.J., Mastrandrea, M.D., Bilir, T.E., Chatterjee, M.K., Ebi, L., Estrada, Y.O., Genova, R.C., Girma, B., Kissel, E.S., Levy, A. N., MacCracken, S., Mastrandrea, P.R., L.L. White (Eds.), Climate Change 2014: Impacts, Adaptation, and Vulnerability. Part A: Global and Sectoral Aspects. Contribution of Working Group II to the Fifth Assessment Report of the Intergovernmental Panel on Climate Change. Cambridge University Press, Cambridge, UK and New York, NY, USA, pp. 1–32.

Lambin, E.F., Turner, B.L., Geist, H.J., Agbola, S.J., Angelsen, A., Bruce, J.W., et al., 2003. The causes of land-use and land-cover change: moving beyond the myths. Global Environ. Change 11, 261–269.

Langlois, P., 2000. Automate cellulaire SpaCelle (Système de Production d'Automate CELLulaire Environnemental). Université de Rouen—Laboratoire MTG, France.

Li, L., Sato, Y., Zhu, H., 2003. Simulating spatial urban expansion based on a physical process. Landsc. Urban Plan. 64, 67–76.

Magurran, A.E., 1988. Ecological Diversity and Its Measurement. Chapman & Hall, p. 179.

Martin, N., Carrega, P., Adnes, C., 2013. Downscaling à fine résolution spatiale des températures actuelles et futures par modélisation statistique des sorties Aladin-Climat sur les Alpes-Maritimes (France). Climatologie 10, 51–72.

Médail, F., Diadema, K., 2009. Glacial refugia influence plant diversity patterns in the Mediterranean Basin. J. Biogeogr. 36, 1333–1345.

Millennium Ecosystem Assessment, 2005. Ecosystems and Human Well-being: Biodiversity Synthesis. World Resources Institute, Washington, DC.

Mitsova, D., Shuster, W., Wang, X., 2011. A cellular automata model of land cover change to integrate urban growth with open space conservation. Landsc. Urban Plan. 99, 141–153.

Múgica, M., De Lucio, J.V., Martínez-Alandi, C., Sastre, P., Atauri-Mezquida, J.A., Montes, C., 2002. Integración territorial de espacios naturales protegidos y conectividad ecológica en paisajes mediterráneos. Consejerla de Medio Ambiente, Junta de Andalucia, Sevilla.

Myers, N., Mittermeier, R.A., Mittermeier, C.G., Da Fonseca, G.A., Kent, J., 2000. Biodiversity hotspots for conservation priorities. Nature 403 (6772), 853–858.

Pontius Jr, R.G., 2000. Quantification error versus location error in comparison of categorical maps. Photogramm. Eng. Rem. Sens. 66 (8), 1011–1016.

Pontius Jr, R.G., Millones, M., 2011. Death to Kappa: birth of quantity disagreement and allocation disagreement for accuracy assessment. Int. J. Rem. Sens. 32 (15), 4407–4429.

Priego-Santander, A.G., Isunza-Vera, E., Luna-González, N., Pérez-Damián, J.L., 2003. Método para realizar mapa de diseccion vertical. Instituto Nacional de Ecología, SEMARNAT. <http://mapas.ine.gob.mx/website/metadato/cuencas/diseccion.html>

Primack, R.B., Sarrazin, F., Lecomte, J., 2012. Biologie de la conservation. Dunod, Sciences Sup, p. 384.

Sarkar, S., 2005. Biodiversity and Environmental Philosophy: An Introduction (Cambridge Studies in Philosophy and Biology). Cambridge University Press, Cambridge.

Sante, A., García, M., Miranda, D., Crecente, R., 2010. Cellular automata models for the simulation of real-world urban processes: a review and analysis. Landsc. Urban Plan. 96, 108–122.

Sui, D.Z., Zeng, H., 2001. Modeling the dynamic of landscape structure in Asia's emerging desakota regions: a case study in Shenzhen. Landsc. Urban Plan. 53, 37–52.

Thuiller, W., Lavorel, S., Sykes, M.T., Araújo, M.B., 2006. Using niche-based modelling to assess the impact of climate change on tree functional diversity in Europe. Divers. Distrib. 12, 49–60.

Veldkamp, A., Lambin, E.F., 2001. Predicting land-use change. Agric. Ecosyst. Environ. 85, 1–6.

Wang, Y., Monzon, A., Di Ciommo, F., 2015. Assessing the accessibility impact of transport policy by a land-use and transport interaction model—the case of Madrid. Comput. Environ. Urban Syst. 49, 126–135.

Wilson, E.D., Peter, F.M., 1988. Biodiversity. National Academy Press, Washington, DC, p. 521.

Williams, L.R.R., Kapustka, L.A., 2000. Ecosystem vulnerability: a complex interface with technical components. Environ. Toxicol. Chem. 19, 1055–1058.

Xie, Y., Batty, M., 2003. Integrated Urban Evolutionary Modeling. Centre for Advanced Spatial Analysis (University College London) Working Paper 68, London, England.

CHAPTER 6

Assessing the territorial adoption potential of electric mobility: geoprospective and scenarios

Christine Voiron-Canicio and Gilles Voiron

Université Côte d'Azur, CNRS, UMR ESPACE, Nice, France

6.1 Introduction

Transportation, with 13.41 gigatons of CO_2 emitted in 2016 worldwide, is the second contributor to greenhouse gasses behind energy and electricity production. Three-quarters of transportation-related emissions are due to trucks, buses, and cars. Yet, despite warnings issued for two decades, emissions are still very high. According to the International Energy Agency (IEA), road traffic generated 5.85 gigatons of CO_2 in 2016, that is, a 77% increase

Ecosystem and Territorial Resilience.
DOI: https://doi.org/10.1016/B978-0-12-818215-4.00006-7
© 2021 Elsevier Inc. All rights reserved.

since 1990. In the EU, during the 2005–15 period, GHG emissions dropped in all sectors, except in transportation (EEA data). In France, the contribution of transportation in CO_2 emissions went from 33% in 1990 to 38% in 2017 (from CITEPA, France), and the number of private cars on the road increased by 40%. In this worrying context, driving less and driving cleaner has become the leitmotiv of public policies. An example is the National Low-Carbon Strategy (SNBC). This road map drawn up by France in 2015 to reach carbon neutrality by 2050, caps emissions on successive 5-year periods until 2033, with objectives specific to each sector of activity. The strategy rests on four main lines: (1) improving the energy efficiency of light and heavy vehicles, (2) decarbonizing the energy consumed by vehicles, (3) controlling the growing demand for transportation by developing teleworking and car-sharing, and (4) encouraging the shift toward public transport and supporting active modes such as cycling.

From now on, mobility is considered as the main lever for action against climate change and pollution. The targets set are ambitious: 35% of sales of private electric or hydrogen vehicles by 2030 and 100% by 2040 and 30 million electric charging points against 28,600 at the end of 2019. Are these forecasts credible? They are perplexing because they are not based on any prospective study of the capacity of territories to change their mode of mobility so quickly and so drastically. The success of this ambitious policy depends on the way the population and companies will go along with the prescribed measures. But, due to the diversity of local contexts, the change to decarbonized mobility will not happen at the same pace everywhere, and the thresholds set will most probably not be attained by all communes at the planned dates. As stressed by Jacques Theys (Theys and Vidalenc, 2013), "The time horizon is well defined, what is uncertain are the possible courses which depend on the representations that stakeholders may have of urban inertias"—and rural, we could add. Indeed, the problem concerns the appropriation on the ground of measures decided at the national level, and more precisely, the local electric mobility adoption capacity. Yet, despite the stake involved in assessing such potential, no study or prospective study has been devoted to this issue in planning documents such as SRADDET (Regional Plan for Sustainable Development and Territorial Equality), SCOT (Territorial Coherence Plan), or PDU (Urban Transportation Plan).

Taking on this problem falls within a geoprospective approach. However, how to estimate the potential for adopting electromobility, bearing in mind that the latter is still not widespread in France? Indeed, on January 1, 2019, the number of electric cars amounted to 115,000, that is, 0.36% of the vehicle fleet. 30,944 electric vehicles (EVs) were sold in 2018, accounting for 1.45% of new registrations. The battery EV, and *a fortiori* the hydrogen type, is an innovation of which the spread in time follows a logistic curve, like any other innovation. On the other hand, its spread in space is still little known. This is because its knowledge comes up against several difficulties. The EV is a technology in progress, of which the most constraining element, autonomy, is improving progressively, thus constantly changing the situation. For all that, technological advances have difficulty in producing new practices on the ground. Moreover, decarbonized mobility remains vague. Information on the matter is still limited, partial, contradictory, and often obsolete. Reaching a sustainable mobility in all parts of a territory involves a dual scientific challenge. First, it is indispensable to improve our knowledge on the conditions and potentialities of electric mobility by precisely defining the electromobility system with its interacting components—technological, financial, socioeconomic, environmental, geographic, and political; then by measuring the possible adequation between the electromobility system and a given territory, a commune, a town, a region. The second challenge is to transfer this expert knowledge to local stakeholders. Then, the objective is to show how to take advantage of this geography of the potential of adoption of decarbonized mobility and implement action plans specific to the obstacles identified on the ground.

This chapter outlines the stages of the adopted approach. The latter has been elaborated with a view to being reproducible in any type of territory. The protocol has been established for a French Region, the South Region, located on the shore of the Mediterranean and including 953 communes. After diagnosing the current acceptance potential for electric mobility for each of the Region's communes, the possible evolution of the acceptance process is analyzed in a prospective phase aimed at assessing the potential by 2040, according to various scenarios.

6.2 The territorialized system of electric mobility

6.2.1 Electromobility, an innovation highly dependent on the territorial context

In the course of the last 10 years, numerous studies have been carried out on EVs and the spread of electromobility, in various disciplinary fields, each focusing on a particular aspect of the theme. Most of them concerned the technical specificities of EVs (Bady et al., 1999), services indispensable to its use, such as charging points (Chéron and Zins, 1997), as well as the economic and financial aspects (Kurani et al., 1995; CGDD, 2011; Emsenhuber, 2013). Studies concerned also the stakeholders system (Sadeghian, 2013) and conditions for the emergence of the EV (Bainée and Le Goff, 2012; Bainée, 2013). More recently, work was directed toward the perception and image of the EV (Lai et al., 2015; Walsh and Bingham, 2009), and the psychosocial dimensions of the acceptance of this type of vehicle (Poupon, 2017). Recurrent surveys are conducted on purchasing intentions per social group and age group, from a sociological or marketing point of view, and hence a-spatial. On the other hand, very few studies are carried out in a systemic framework and addressing the role of the territory in the emergence and spread of this innovation. However, two research works stand out. Jonathan Bainée, in his work on the emergence of the industry of the EV, considers the EV as a "territorialized composite system," that is, as a system lying at the crossroads of the joint evolution of the technology, the territory, and uses. He considers that the territory is not only the melting pot of innovations but that it is also capable to govern and direct their pace and form of emergence. As for Sadeghian (2013), this author points out that to each territorial configuration corresponds a mobility system and stresses that this relationship to the territory is more intense in the case of electromobility, because the EV, due to its limited autonomy and need to recharge, requires a different kind of relationship of dependency to the territory.

Therefore the electromobility system must be defined in its multiple components, and spatially, because the functioning of the system strongly depends on local structures and dynamics: on the spatial characteristics of territories, types of journey, kinds of habitat, and on the implication of local players in promoting decarbonized mobility.

6.2.2 Drawing up the territorialized system of electromobility

The protocol is in three phases. The first consists in listing the variables that fall within the electric mobility system, from existing literature and experts' statements. 25 variables were retained at the end of that phase. The purpose of the second phase was to determine the relationships between these variables. The Micmac method (Matrix multiplication applied to a classification) has been chosen; worked out by Michel Godet,[1] it enables one to describe a system using a matrix which connects all the system's components. The 25 variables identified are entered in a double-entry table. For each pair of variables, the direct influence relationship between variables i and j is discussed and filled in qualitatively. There are four choices: 0 (absence), 1 (weak relationship), 2 (medium), and 3 (strong). In the present case, 625 questions were put (25×25 variables). The following phase consists in identifying the key variables, those essential to the system's evolution. Three groups of variables can be derived from the results of this structural analysis (Table 6.1):

- The highly influential and not very dependent variables, also called "entry variables." They are seen as mostly explanatory of the system being studied, and conditioning the overall dynamics.
- The "relay" variables, highly influential and highly dependent. Any action on them will have concomitant repercussions on the other variables and a retroactive effect on themselves, thus deeply altering the system's overall dynamics.
- The highly dependent and not very influential variables, or "result" variables. Their evolution is explained by impacts mainly due to entry and relay variables.

---

[1] The Micmac method, created in 1971 by Michel Godet, belongs to the family of structuring tools of collective reflection.

Table 6.1: Classification in ascending order of the electromobility system's key variables (the relay influent and dependent variables are in italics).

The first 13 influential variables	The first 13 dependent variables
PEV fleet in use	*EV fleet in use*
Implication of the local political sphere	*Corporate EV fleet*
Share of individual housing	*Number of charging points*
Height difference	Operation of communication on EVs
Corporate EV fleet	*Purchase price and annual running cost*
Autonomy	Car ownership rate per household
Median income	*Local systems and electric self-consumption*
Existence of innovative companies	*Existence of innovative companies*
Local systems and electric own consumption	*Implication of the local political sphere*
Number of households	Number of service stations
Purchase price and annual running cost	Average daily professional mileage
Average daily professional mileage	Average driving speed
Charging time	Renewable energy production

6.3 Methodology: the conceptual model and the expert system

6.3.1 The conceptual model of the electric mobility's territorialized system

The conceptual model of the battery electric mobility's territorialized system has been drawn up within the framework of a research project[2] financed by ADEME (the French environment and energy control agency). Two research laboratories were in partnership, the EIFER laboratory and the ESPACE laboratory. Christine Voiron-Canicio and Gilles Voiron, who come under the ESPACE laboratory, were responsible for the workpackage titled "Regional potential of the battery and hydrogen electromobility." The conceptual model which they have drawn up was presented and discussed at a number of work meetings with ADEME, then submitted to a group of experts. The electromobility's territorialized system is structured around four main families of components that determine the capacity of a territory to adopt electric mobility: easy charging, EVs' adequacy to travel needs, interest and motivation for purchasing an EV, and local context (Fig. 6.1). Each component family is made up of geographic variables—framed by a thick line in the diagrams—and nongeographic [EV's autonomy, ICE (International Combustion Engine) and EV's purchase price, charging-up technical system, public aids, etc.].

The "charging" component:

---

[2] CATIMINI research project: Capacity of territories to integrate the innovations of electric battery and hydrogen mobility, carried out between 2017 and 2019.

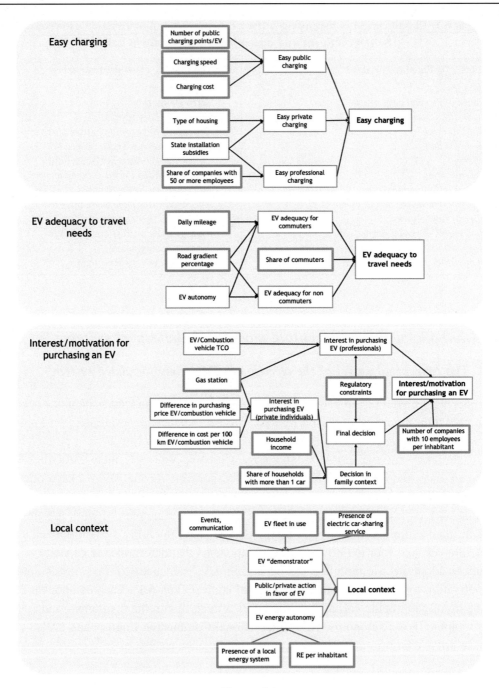

Figure 6.1
The electromobility's territorialized system.

It is the main element. The issues of access to charging and easy charging are essential for the spread of electromobility on a territory and remain an obstacle in a number of spaces. Three geographic variables contribute to ensuring easy charging:

- The number of charging points open to the public, on-street, in car parks, in commercial areas.
- The type of housing. This variable informs on the possibility to charge up one's EV at home at night. We posit that all those living in an individual house can charge their battery on a domestic socket. Home-charging is less obvious for people living in apartments. On the other hand, for new homes, fitting a charging point can be made easier, whereas in older apartment buildings, it will be more difficult (bad state of electrical installations, lack of room, difficulty to apply the "charging point right" in condos, etc.). The information taken into account in the analysis is the share of individual houses in the commune.
- The number of companies having 50 or more employees. We posit that small and medium-sized enterprises are progressively installing charging points for their fleet and their employees' vehicles.

The "electric vehicle adequacy to travel needs" component:

The EV adequacy to the commune inhabitants' travel needs component is determining in the uptake of electric mobility. Three geographic variables are taken into account:

- The share of the population working outside the commune of residence (commuters).
- The number of kilometers covered daily to commute to work.
- The topography of the place of residence, and more precisely the height difference. We posit that the more steep roads in the commune (gradient %), the more uphill drives will be disadvantageous for EVs.

The "interest/motivation for purchasing an electric vehicle" component:

For private individuals and professionals alike, the following elements are taken into account in the purchase decision: the difference in purchase price between an EV and a combustion vehicle, existing regulatory constraints, and notably in rural areas, the presence or not of a service station in the commune or in the immediate vicinity. We assume that the disappearance of service stations and the obligation of traveling long distances to reach a service station is an element favorable to EVs. Other variables are specific to each type of purchaser.

In the decision to purchase an EV, private individuals take also into account the difference of cost per 100 km, the household's median income, and the household's number of cars. A household with more than one vehicle will be more inclined to experiment the use of an EV as a complement to a combustion vehicle.

In the case of professionals, the comparison between an EV's total cost of ownership (TCO) and that of a combustion vehicle comes into play.[3] An EV's TCO is made up of two-thirds of fixed costs and one-third of variable costs, that is, the reverse of combustion vehicles. The maintenance of a combustion vehicle is estimated at 16% of the TCO against 9% for an EV. However, what is determining is the use that will be made of the vehicle. In a professional context, an EV is mostly adapted to dense tours, and notably to being used on short distances in an urban cycle. It remains unclear whether the average buyer, when choosing a vehicle, is able to identify precisely what determines the EV's economic performance in relation to that of a combustion vehicle for the use he envisages.

The "local context" component:

This component determines whether the local geographic context is more or less favorable to electromobility. Here, several types of variables come into play:

- Actions carried out by the public and private spheres to promote EVs in the communal territory. They include, on the one hand, the implication of the local political sphere in actions of various natures in favor of the development of electric mobility in its territory, and on the other hand, the presence of both public and private companies, innovative and exemplary as regards electric mobility—big groups such as Groupe La Poste, or startups.
- The EV "demonstrator" variable is a composite variable playing in terms of formal communication—events such as shows, fairs, France Electrique Tour, electric rally, and Riviera Electrique Challenge—, and informal. Can be added the regular and visible presence of private and corporate EVs circulating in the territory (cf. the condition "passing EVs on the roads on a regular basis—widespread solution" appearing in the study on growth factors for the market of electric private vehicles and light utility vehicles (ADEME, 2016), and the presence of car-sharing services.
- The "EV energy autonomy" variable is equally composite. It takes into account actions carried out by territories to appropriate their own energy dimension by producing renewable energy (RE), the presence of local energy systems, and electric self-consumption by private individuals.

This conceptual model is the framework of interpretation from which the characteristics of each commune will be analyzed in order to assess the adequacy between its mobility system and electromobility. The degree of compatibility between its geographic and socioeconomic structure and the conditions required by electromobility will enable to assess its capacity to adopt decarbonized mobility. However, this assessment is made difficult by the incompleteness of knowledge and uncertainties over the relationships between variables. Therefore it requires a particular methodology.

---

[3] The TCO includes in its calculation all of the direct and indirect costs generated by owning and using an EV.

6.3.2 The expert system

The method that we have developed has been guided by two principles: drawing on an expert knowledge emanating from a diversity of views both by experts and by stakeholders concerned by electric mobility—local authorities, private individuals, businesses—, and producing a protocol reproducible in all types of territory.

We use a rule-based system to address the issues of assessing future changes.

The system (knowledge-based system) consists of a knowledge base and an inference engine.

- The knowledge base is made up of a body of facts and inference rules. The facts are the information on the system's variables, most of it backed up by figures, and in some cases, in the presence/absence form. An inference rule comes as follows:

IF (rule premise) THEN (rule conclusion)

- The inference engine is a program which scrutinizes the rule premises in order to determine whether they are "right" or "wrong" taking into account the information included in the facts base.

Two software packages—NETICA[4] and BayesiaLab 9[5]—designed to work with Bayesian belief networks and influence diagrams, were used in turn.

We have built the inference rules of the territorialized electromobility system within the framework of tables concerning the variables of the four main component families. Each line of a table corresponds to a rule for a situation relating to specified conditions. The conditions—rule premises—are usually three, and expressed in linguistic form with two or three intensity levels—important/medium/weak; higher/equivalent/lower; a lot/little; easy/ uneasy; yes/no; high/low; favorable/unfavorable, etc. Some variables, such as the number of businesses, are expressed in classes of numerical values. Expert opinions are required for filling in the conclusion of each rule. These experts must assess whether the specified conditions are favorable versus unfavorable to EV, and with what intensity, by allocating a percentage. Opinions of 10 or so experts[6] have been gathered in the form of face-to-face interviews. Their opinions enabled us not only to complete the various tables constituting the knowledge-based system, but also to allocate a weight to the four components in the final calculation of the final overall capacity of acceptance of electric mobility. All experts give the highest weight to easy charging. Divergences appear in the importance given to the

---

[4] Netica is a Bayesian network software (a trademark of Norsys Software Corp.).

[5] BayesiaLab 9 is a French Bayesian network software.

[6] The administrator of the ENEDIS power grid, Smart Grid & Electric Mobility department; an energy syndicate of the South Region; people in charge of a car manufacturer's Electric Vehicles and Ecosystem department; the person in charge of travels and transport for a South Region's agglomeration community; an association promoting electric vehicles; a research firm specialized in charging infrastructures; a consultancy in sustainable mobility; ADEME.

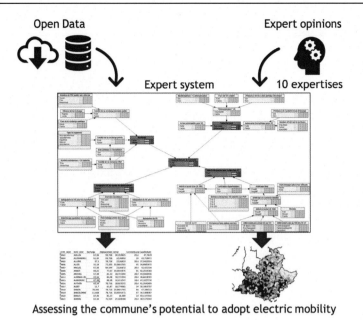

Assessing the commune's potential to adopt electric mobility

Figure 6.2
Structure of the territorialized system of electromobility (under Netica).

other components. A majority of opinions ranks second the interest for purchasing, and to a lesser extent, to the adequacy to travel needs and the local context.

For each rule, the synthesis of all "expert views" is carried out taking the median of opinions, a method used in fuzzy logic, and recommended for determining the final assessment of expert opinions expressed in linguistic variable (Bodjanova, 2005; Saneifard and Saneifard, 2012). From then on, for each of the four major components of the system, all propositions included in the tables become the rules of the knowledge base that the expert system's inference engine connects with the communal data—facts base—to deduce whether the commune's potential is favorable or unfavorable to electric mobility, and by what percentage (Fig. 6.2).

6.4 Assessing the adoption potential of electromobility in 2019

6.4.1 Evaluating the capacity of communes in the South Region to adopt battery electric mobility: analysis of the expert system's results

Most of the communes have good charging facilities (Fig. 6.3A). The worst off are of two types:

- On the one hand, the urban communes where the share of individual houses is low—under 25%: Marseille, Aix-en-Provence, Toulon, Hyères, Fréjus, Cannes, Gap,

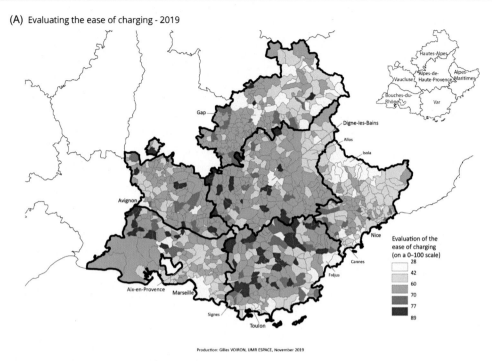

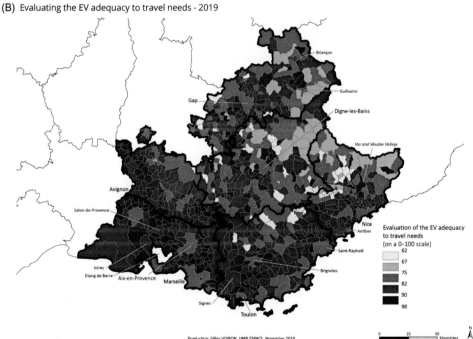

Figure 6.3
Evaluating the easy of charging (A) and the EV adequacy to travel needs (B)—2019.

etc. The differences depend on the number of charging points in relation to the number of private EVs and also on the presence of a car-sharing service with charging station, such as in Nice, unlike Marseille where the service is of the free floating type.

- On the other hand, mountain communes having a winter resort with housing almost exclusively made up of apartment blocks. The source of differentiation element is the number of charging points intended for the tourist clientele. As an example, in the commune of Isola (Alpes-Maritimes) where the easy-charging component is evaluated at 27%, there is only 9% of individual houses, and neither private EVs nor charging points, whereas in the commune of Allos (Alpes-de-Haute-Provence) where the percentage is 50%, the share of individual houses is 13%, there are no private EVs, but two charging points are available.

EVs seem to be adapted to the travel needs of the inhabitants of the South Region communes (Fig. 6.3B). Indeed, the lowest percentage is 62%. The class between 62% and 67% concerns only a small number of communes in the mountain hinterland of the three Alpine "départements" and in the Haut Var. Conversely, the communes with the highest adequacy to travel needs are in the outskirts of the urban poles which are employment pools for the nearby working population. This is confirmed whatever the size and location of urban centers; mainly around Marseille, Aix-en-Provence, and Istres, around the Étang de Berre, in the outskirts of Avignon; in the Var, between Toulon, the business park of the Signes plateau and Brignoles; in the Nice area, along the Var and Vésubie valleys. These belts of high adequacy are also perceptible in the two Alpine "départements," between Briançon and Guillestre, around Gap, as well as in the outskirts of Digne.

The interest in purchasing an EV for the commune's private individuals and professionals is very diverse according to the commune, depending on the average income, regulatory constraints, and aids for purchasing EVs introduced in some of the Region's areas—the Bouches-du-Rhône subsidy amounting to 5000€ for the purchase of an EV since November 1, 2018, for example (Fig. 6.4A).

The evaluation of the favorable role played by the local context reveals strong differentiations within the territory (Fig. 6.4B). Four groups can be identified:

- Extremely favorable local contexts (values between 90% and 100%), combining various types of actions having a ripple effect, going from producing RE to the implication of local authorities in setting up charging stations and/or car-sharing systems and in programs for rolling out intelligent electric systems. Some of the communes concerned are located in the coastal area—Fos, Marseille, Toulon, and behind, Le Castellet, the Nice Côte d'Azur Metropolis, Puget-sur-Argens in the Saint-Raphaël area, and others in the Alpine areas: Névache, Guillestre, Embrun, Les Orres, Sisteron.
- Favorable contexts (values between 70% and 90%) forming most of the time relatively homogeneous regional groups, such as nearly all communes in the Bouches-du-Rhône,

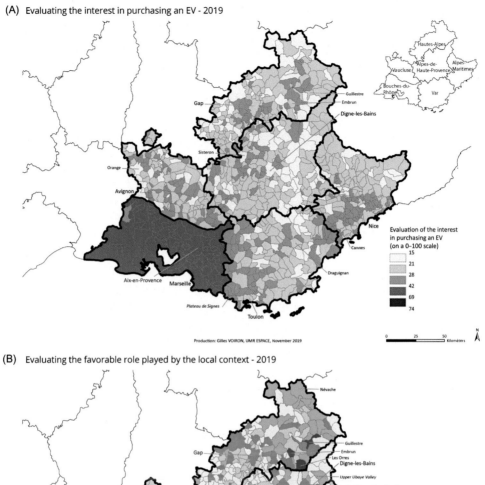

(A) Evaluating the interest in purchasing an EV - 2019

(B) Evaluating the favorable role played by the local context - 2019

Figure 6.4
Evaluating the interest of purchasing an EV (A) and the favorable role played by the local context (B)—2019.

the Var coastline and Centre Var, the east of the Alpes-Maritimes; in the Alpes-de-Haute-Provence, the Durance valley and its extensions toward Castellane and toward Digne, clearly becoming individualized. Further north, the upper Ubaye valley and the Gap area.

- Local contexts moderately favorable to the rolling out of EVs (values between 35% and 56%) are many in the Vaucluse. In the Alpes-Maritimes, they characterize communes in the lower Var valley the Vésubie and La Roya valleys.
- Hardly favorable contexts (20%−35%) are mostly found in a scarcely populated mountain central zone extending from north of Digne to the upper Var and Tinée valleys, in the Alpes-Maritimes, and in the far east, in the upper Roya valley.

Assessing the overall acceptance capacity for electromobility (Fig. 6.5):

First observation, the Region's main urban centers fall in the class with the lower acceptance potential: Toulon (34%), Avignon (33%), Marseille (40%), and Nice (38%). This can be explained by low or poor scores in each of the four components, and most particularly in the easy-charging component which carries significant weight in evaluating the overall capacity. Nice stands out among the other towns because of a slightly higher overall capacity (39%), which is due to better charging facilities. The communes located in peri-urban areas have a high potential (51%−61% class). Moreover, the mapping of results reveals the existence of extended zones with an acceptance potential higher than 51%: in the Bouches-du-Rhône, west Var (Signes plateau and Toulon peri-urban area), between Fréjus and Nice, and the peri-urban communes of the Côte d'Azur coastal back country, extending from the Grasse area to the hills of the Nice area. Conversely, the mountain communes in the Nice hinterland, Haut-Verdon, and the Barcelonnette area have a low potential (26%−38% class).

The model of EVs' territorialized acceptance capacity which has been devised for any territory, when applied to the South Region communes, shows an average adoption potential of 44.9%. The distinction between rural and urban communes shows substantial differences in potential. For example, the isolated towns stand out with the highest capacities for adopting EVs (48.6%), whereas in multicommunal conurbations, there is a significant gap between center-communes which have a low potential (41.5%) and neighboring peri-urban communes with a markedly higher potential (47%). On the other hand, in rural areas, the adoption capacity is low (42.7%). Now the significance of these estimates has to be evaluated.

6.4.2 Comparing the adoption potential and the electric vehicle registration rate

How to validate these spatialized potentials, knowing that the conventional validation procedures are inappropriate? Furthermore, EV sales, although they are increasing, are still

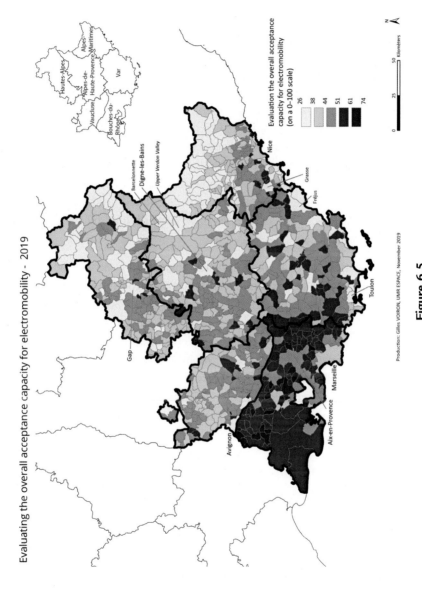

Evaluating the overall acceptance capacity for electromobility - 2019

Evaluation the overall acceptance capacity for electromobility (on a 0–100 scale)

26
38
44
51
61
74

0   25   50 Kilometers

Production: Gilles VOIRON, UMR ESPACE, November 2019

Figure 6.5

Evaluating the overall acceptance capacity for electromobility.

Table 6.2: Comparing adoption capacity and registration rate.

Types of commune	EV adoption capacity (in 2019)	EV registration rate (2016−19 period)
Multicommunal conurbations:		
Center-communes	0.92	0.92
Peri-urban communes	1.04	1.14
Isolated towns	1.08	1.10
Rural communes	0.95	0.84
Communal average = baseline	1.00	1.00

very few. 11,436 vehicles were sold in the South Region between 2016 and 2019, accounting for an average registration rate of EVs of 1.94%, that is, slightly above the national registration rate over the same period. The analysis concerns the comparison between the "electric mobility" acceptance capacity and the EV registration rate. As a reminder, the communal adoption capacity has been estimated based on the electric mobility territorialized system modeled in 2018, and applied on the most recent communal data available in 2019. The average registration rate is calculated on the total light EV sales from 2016 to 2019.[7] The values of the two series have been divided by their respective averages and shown in Table 6.2.

The values are either identical or very close in urban centers. In contrast, the adoption capacity estimated by the model slightly underestimates adoption in the Region's peri-urban area, and *a contrario*, overestimates the adoption capacity in the South Region's rural communes. The regional specificities thus brought forward are a valuable aid for local players wanting to define a strategy for promoting electric mobility that would match the various local contexts. However, in 2019 electric mobility is only at the first stage of spatial spread. Although there is no doubt that its progression will accelerate in coming years, there are uncertainties on the progression pace and the attractiveness of battery-type electric mobility versus new decarbonized energies such as hydrogen and biogas fuel.

6.5 Assessing the adoption potential of electric mobility in 2040

Based on the territorialized system of battery-type electric mobility, modeled in 2018 and applied on the South Region, scenarios have been drawn up of two plausible evolution variants by 2040, and discussed with ADEME. These are neither trend scenarios, nor breakup or normative scenarios. They correspond to two possible paths of development for decarbonized mobility, one concerning "all-electric" decarbonized mobility, and the other, diversified decarbonized mobility.

---

[7] Statistics: Sustainable Development Commission, data and statistical studies department (SDES), Ministry of Ecological and Social Transition.

6.5.1 Two plausible scenarios by 2040

Scenario 1: Toward an "all-electric" decarbonized mobility in 2040

This scenario is based on more favorable conditions for adopting EVs for both private individuals and companies. It can be summarized in three trends: technological progress that makes them attractive and increases connections with the local potential of renewable energies; democratization of the EV for private individuals and companies alike; strong support of the authorities at the national level—various aids and end of the sale of combustion vehicles in 2040, among others—a policy reflected at the regional and local levels.

In comparison with 2019, electric charging has been greatly improved: communes with over 10,000 inhabitants have a significant number of charging points per vehicle, and they are faster. All other communes have at least one charging point of the slow type, except for those which already had many charging stations in 2019, and those which had fast-charging stations. Small communes (under 200 inhabitants) have no charging points, except for those which already had some in 2019. Charging equipment is spreading fast in apartment buildings, with reduced installation time and costs.

The battery range of EVs has markedly improved, it has become higher. Communes with a steep relief are no longer disadvantaged as they were in 2019. This spatial homogenization combined with getting used to driving EVs makes the "adequacy to travel needs" component less determining than in 2019. Therefore its weight has been reduced in the expert system.

The interest-motivation for purchasing an EV has also evolved. Due to the now high range of EVs, households no longer need to keep a "spare" combustion vehicle for long week-end or holiday drives. Therefore the weight of the "share of households with two vehicles or more" variable has been reduced as compared with the 2019 model. Moreover, private individuals and companies alike are very much alive to pollution issues and are overwhelmingly in favor of decarbonized mobility.

The local context has changed between 2019 and 2040. On the one hand, the fleet of EVs has increased significantly, and the use of EVs has become commonplace. Furthermore, all communes over 10,000 inhabitants have an electric car-share service. On the other hand, energy autonomy has considerably developed. The production of RE has increased in close connection with the local potential of biomass and ground photovoltaic plants. All communes have local energy systems.

Scenario 2: Toward a diversified decarbonized mobility in 2040

This scenario is also in favor of decarbonized mobility, but differentiates itself from the first scenario with the introduction of the incidence of mature alternative energy sources—notably hydrogen—on adopting EVs. Competition is no longer from combustion vehicles but from hydrogen vehicles and, to a lesser extent, natural gas vehicles. Hydrogen is then strongly supported by the authorities that, correlatively, cease to back EVs.

Regarding the charging component, the number of charging points and charging speed is identical to scenario 1, the change concerns the end of public aid to installing charging points in apartment buildings, and so for various reasons: (1) previous aids have been successful, there is no need to extend them; (2) *a contrario*, subsidies allocated did not produce the expected effect and are abandoned.

The variables of the "EV adequacy to travel needs" component remain identical to scenario 1. The component's weight also remains identical to scenario 1.

Concerning the interest-motivation for purchasing an EV, the authorities have decided to subsidize hydrogen vehicles. Therefore the latter have become cheaper than an equivalent EV. Environmental constraints are identical to those in scenario 1. On the other hand, taxes on electric charging up have been either added or increased. In order to favor the development of hydrogen-fueled vehicles, subsidies are allocated to setting up hydrogen stations as well as to hydrogen as fuel. The TCO is unfavorable to EVs because they no longer benefit from aid, unlike other types of vehicles.

Some of the local context variables evolve differently from scenario 1. As an example, large urban areas with over 50,000 inhabitants retain their electric car-sharing service. In contrast, communes with over 10,000 inhabitants that did not have electric car-sharing in 2019 have opted for natural gas or hydrogen car-sharing. From then on, the "EV demonstrator" and "public/private action in favor of EV" variables are less important than energy autonomy. As in scenario 1, the RE potential plays a major role and local energy systems are omnipresent. By contrast, RE surpluses are no longer totally stored for EVs, but part is transformed into hydrogen to supply hydrogen vehicles.

In these two scenarios, both promoting sustainable mobility, the place of the EV is not identical. The first scenario establishes the supremacy of the battery EV. The combustion engine has declined significantly as a result of the combined effects of regulatory constraints, national and local public policies, and the environmental concerns of the civil society. In this "all-electric" context, alternative decarbonized fuels—hydrogen, biogas— have a marginal place. By contrast, they come into play in the second scenario. The diversified decarbonized mobility scenario is built on this new situation.

6.5.2 Simulation results from the two 2040 scenarios

The underlying questioning behind the geoprospective exercise remains the acceptance of battery-type electric mobility by communes. What will be the adoption potential of this electric mobility in each of the future contexts imagined for 2040? What will be the changes in the values of potentials, in the territorialization of these potentials as compared with 2019, and between the two scenarios for 2040? To be able to compare these potentials at the two dates, the territorialized system must retain the same structure. Therefore no variable is removed or added. However, the values of some parameters may be modified to take into account the impact of new phenomena, such as alternative fuels. Their incidence on the electric mobility territorialized system is expressed through public policies. For example, one of the incentives for diversifying decarbonized mobility is the end of subsidies allocated for battery EVs and related equipment. On the other hand, from then on, both private individuals and companies have a wider choice of decarbonized vehicles, etc. The weight of some of the components in the calculation of the global potential is also modified. Due to the technical advances concerning the battery's capacity, management, and efficiency, the range-related drawbacks disappear, and the "EV adequacy to travel needs" component weighs less. In contrast, the local context, being part of a logic of local production of RE part of which is interrelated with the rolling out of electric mobility, has a greater weight than in 2019. The weight of this component in the expert system is consequently increased.

The simulations of the two scenarios produce sharply contrasting results, especially as regards the "easy charging" and "interest for purchasing an EV" components.

In scenario 1—"all-electric mobility"—three-quarters of the communes have an easy-charging score above 60% (Fig. 6.6A). The gain is very high in towns, where aids for installing charging points in apartment buildings, combined with the massive rolling out of public charging stations in towns with over 10,000 habitants, make charging spectacularly easier.

The situation is the exact opposite in the simulation of scenario 2—"diversified decarbonized mobility" (Fig. 6.6B). The drastic drop of scores is mainly explained by the end of public aids for installing charging points in apartment buildings. Large towns where over 60% of housing is in apartment buildings are the most penalized. The recently installed stations, whether fast in large towns or slow in smaller ones, no longer suffice to make up for difficult charging. Communes with a high share of individual houses are less affected, because aids for installing charging points in individual houses were lower. The easy-charging scores are not only lower than in 2019 but also more homogeneous. Indeed, between 2019 and 2040, most communes have been equipped with at least one charging point so as to obtain a nearly complete regional network, thus filling "white areas."

The interest for purchasing an EV is logically boosted by the orientations of scenario 1—"all-electric mobility" (Fig. 6.7A). The simulation results show purchasing interests above 42%. As a result of State aids and further aids granted by local authorities, the purchasing price of a battery EV is equivalent to those of other types of vehicle,

(A) Evaluating the ease of charging - Toward an "all-electric" decarbonized mobility in 2040

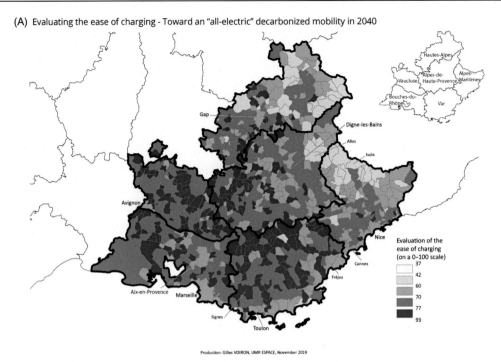

(B) Evaluating the ease of charging - Toward a diversified decarbonized mobility in 2040

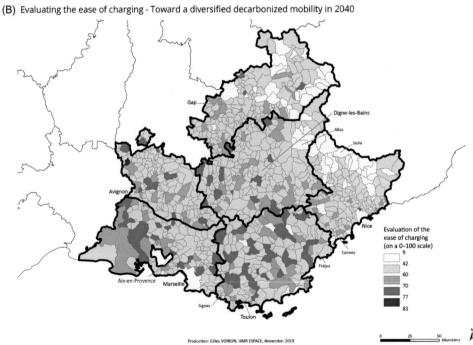

Figure 6.6
Evaluating the ease of charging in 2040: (A) Scenario *Toward an all-electric decarbonized mobility* and (B) scenario *Toward a diversified decarbonized mobility*.

(A) Evaluating the interest in purchasing an EV - Toward an "all-electric" decarbonized mobility in 2040

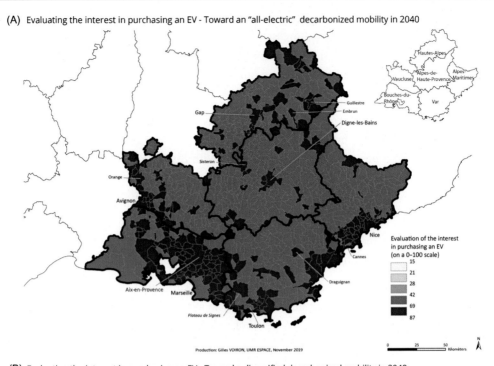

(B) Evaluating the interest in purchasing an EV - Toward a diversified decarbonized mobility in 2040

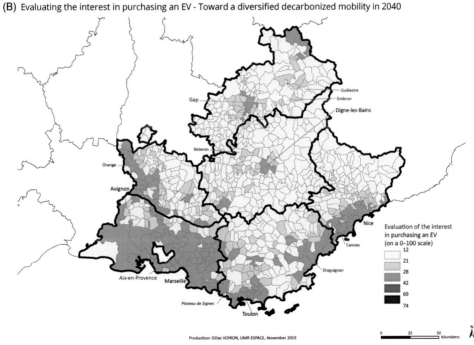

Figure 6.7

Evaluating the interest in purchasing an EV in 2040: (A) Scenario *Toward an all-electric decarbonized mobility* and (B) scenario *Toward a diversified decarbonized mobility*.

an advantage adding to a very competitive running cost. The purchasing interest is all the greater since environmental constraints ban the use of combustion vehicles in towns with over 10,000 inhabitants and in areas above the limits of air quality. It becomes attractive even in rural communes with median-income households.

The simulation results of scenario 2—"diversified decarbonized mobility"—are the opposite (Fig. 6.7B). EVs no longer benefit from specific aids, and as a result their attractivity drops steeply; for more than two-thirds of the communes, purchasing interest does not exceed 20%. However, two types of communes stand out with higher values—on the one hand, urban and peri-urban communes with high-income households enabling these to invest in a new EV, and on the other hand, urban areas having set up drastic environmental constraints for polluting vehicles.

The maps drawn from the simulations reveal possible future evolutions that raise questions. There is a striking contrast between the maps of overall capacities in 2040 (Fig. 6.8).

In addition to the differences in values between the two simulated scenarios, spatial differentiations peculiar to each scenario appear. The simulation of scenario 1—"all-electric mobility"—reveals a marked distinction between a large south and west zone, which is homogeneous and has high acceptance capacity (over 60%), and the north and east zone, which is mountainous and has an acceptance potential lower and more differentiated locally (Fig. 6.8A). The internal differentiations resulting from the simulation of scenario 2 are quite different (Fig. 6.8B). The highest potentials south and west zone is now highly differentiated, whereas the Alpine "départements" mountainous zone is characterized by an evenly low acceptance potential (38%). Furthermore, it is to be noted that the overall capacities resulting from this scenario are lower than those estimated in 2019. How to explain this phenomenon? This drop does not reflect a loss of interest for decarbonized mobility, and even less a return to combustion vehicles, but local situations open to various forms of decarbonized mobility. Indeed, a number of spatial contexts are more in adequacy with new energy sources, such as locally produced hydrogen and biogas. Battery EVs are no longer the only alternative to combustion vehicles, they have become an option among others.

Moreover, a number of communes seem to be more sensitive and reactive than others to the changes introduced in the scenarios. The possible future trajectories deserve attention.

6.5.3 Electric vehicles and communal trajectories between 2019 and 2040

The communal trajectories are analyzed by observing the class changes in the EV acceptance potential between 2019 and the two scenarios of 2040, which enables us to identify profiles of propensity for change. Only 20 communes have an unchanged position. They are small mountain communes the potential of which remains in the class of the lowest values. 14 small-size urban communes improve their acceptance potential between 2019 and

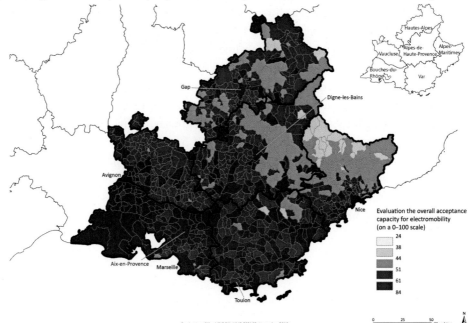

(A) Evaluating the overall acceptance capacity for electromobility - Toward an "all-electric" decarbonized mobility in 2040

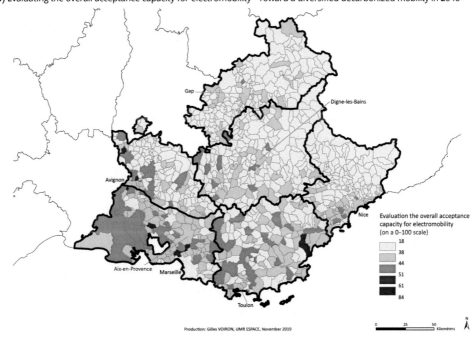

(B) Evaluating the overall acceptance capacity for electromobility - Toward a diversified decarbonized mobility in 2040

Figure 6.8

Evaluating the overall acceptance capacity for electromobility in 2040: (A) Scenario *Toward an all-electric decarbonized mobility* and (B) scenario *Toward a diversified decarbonized mobility.*

2040. All other communes in the region are characterized by fluctuations of potential with class changes. This type of profiles shows a strong sensitivity to the scenarios' diverging orientations. The acceptance potential of a first group made up of 473 communes increases markedly in scenario 1, but remains in the same class as in 2019 for scenario 2. As a reminder, this is not a sign of immobilism but of sensitivity to the new conditions of decarbonized mobility. A second group of 446 communes (47% of the total of the Region's communes) stands out for a lower acceptance potential than in 2019 in scenario 2. The potential of a number of these communes fluctuates greatly, going from class 4 in 2019, to class 1 in 2040 (15 communes), or from classes 4 or 5 to class 2 (127 communes). These fluctuations are mainly due to changes in easy charging and purchasing interest resulting from the orientations of scenario 2. The communes of this group are located in the peri-urban areas of the Aix-Marseille and French Riviera agglomerations. They also concern more rural areas in the hinterland—Center Var, Vaucluse, Durance valley.

6.6 Discussion

To this day, the volume of sale of EVs is very low, and no model of spatial spread, other than theoretical, can be drawn up. On the other hand, electromobility is conditioned by a number of indispensable, and even constraining elements, which are more or less present in territories. Electromobility is *de facto* a territorialized system. Once the structure of the system, its various components, and their interrelations have been determined, we have obtained the standard model of the territorialized system of electric mobility. Then, it becomes possible to evaluate if the conditions required by the model for practicing electromobility are met in a given commune, and at what level. It is with this in mind that the expert system has been designed. So the aim is not to estimate the number of people who adopt electric mobility in a commune but to assess the adoption potential—or acceptance capacity—existing in that commune.

What lessons can be learned from this geoprospective approach applied to the study of the territories' acceptance of electromobility? The first lesson concerns the reactivity to change of the territory being studied. The results of the scenarios show that the South Region's reactivity to the changes introduced in each scenario is high as a whole. It is especially the case with the communes located in the Region's most populated and active area. The second lesson concerns the resilience capacity, assessed from the drop in values of the overall potential between 2019 and 2040. A drop does not mean clinging to combustion vehicles of which France has planned to end the sale in 2040, but shows that there is room for other decarbonized energy sources. We can even make the assumption that the bigger the difference, the greater the possibility of a switch-over toward alternative energy sources. This opening of new possibilities can be interpreted as the sign of a potential adaptivity of the territorial system, which is a major resilience factor.

The presentations of the expert system and maps of results given for over a year aroused great interest among various audiences. From now on, the issue is for decision-makers to appropriate the tool and the underlying geoprospective approach. This expert system, like most prospective tools, is more aimed at decision-makers and people in charge of organizations than at administrators. It is adapted to the phase of reflective thinking in the course of which a strategy is devised and action plans are built.

The current phase is that of transferring the tool to professional and decision-making spheres. It requires actions of communication, awareness, and training of various natures: applications in workshops, training sessions, webinars, and discovering the tool's potentialities by means of a web simulator. This is a capital and at the same time critical phase which mobilizes the methods of geogovernance so as to transfer this expert knowledge to potential users as best as possible.

References

ADEME, 2016. Etude sur les relais de croissance du marché des véhicules particuliers et des véhicules utilitaires légers électriques. Rapp. d'étude (in French).

Bady, R., Biermann, J., Kaufmann, B., & Hacker, H., European electric vehicle fleet demonstration with ZEBRA batteries. SAE Technical Paper 1999-01-1156, 1999, < https://www.sae.org/publications/technical-papers/content/1999-01-1156/ >.

Bainée J., 2013. Conditions d'émergence et de diffusion de l'automobile électrique: une analyse en termes de "bien-système territorialisé," thèse, Université Panthéon Sorbonne, Paris I (in French).

Bainée, J., Le Goff, R., 2012. Territoire, industrie et "bien système": le cas de l'émergence d'une industrie du véhicule électrique en Californie. RERU 3, 303−326 (in French).

Bodjanova, S., 2005. Median value and median interval of a fuzzy number. Inf. Sci. 172, 73−89.

Chéron, E., Zins, M., 1997. Electric vehicle purchasing intentions: the concern over battery charge duration. Transp. Res. Part A Policy Pract. 31 (3), 235−243.

CGDD, 2011. Les véhicules électriques en perspective: analyse coûts-avantages et demande potentielle. Commissariat Général au Développement Durable, Etude n°41, Mai (in French).

Emsenhuber, E.-M., 2013. Determinants of the acceptance of the electric vehicles, Master thesis in Marketing, Aarhus University, Danemark.

Kurani, K., Sperling, D., Lipman, T.E., Stanger, D., Turrentine, T., Stein, A., 1995. Household Markets for Neighborhood Electric Vehicles in California (Report No. 462). The University of California Transportation Center, California.

Lai, I.K.W., Liu, Y., Sun, X., Zhang, H., Xu, W., 2015. Factors influencing the behavioural intention towards full electric vehicles: an empirical study in Macau. Sustainability 7, 12564−12585.

Poupon, L., 2017. L'acceptation de la voiture électrique: étude d'un processus, de l'acceptabilité à l'acceptation située, Thèse, Université de Lyon (in French).

Sadeghian S., 2013. Développer la mobilité électrique: des projets d'acteurs au projet de territoire, Architecture, aménagement de l'espace. Université Paris-Est (in French).

Saneifard, R., Saneifard, R., 2012. The median value of fuzzy numbers and its applications in decision making. J. Fuzzy Set. Valued Anal. 2012, 1−9.

Theys, J., Vidalenc, E., 2013. Repenser les villes dans la société post-carbone. ADEME/Ministère de l'Ecologie, du Développement Durable et de l'Energie (in French).

Walsh, C., Bingham, C., 2009. Electric drive vehicle deployment in the UK. Presented at the EVS24 International Battery, Hybrid and Fuel Cell Electric Vehicle Symposium, Stavanger, Norway.

CHAPTER 7

The touristic model of Valais facing climate change: geoprospective simulations of more environmentally integrated development models

Jean-Christophe Loubier

University of Applied Sciences and Arts of Western Switzerland, Sierre, Switzerland

Chapter Outline

7.1 Introduction

Valais is one of the 26 sovereign cantons of the Swiss Confederation. It is located in southwestern Switzerland, in the Alps, and covers the catchment area of the Rhône River from its source to its arrival in Lake Geneva. From a tourist point of view, the canton has

Ecosystem and Territorial Resilience.
DOI: https://doi.org/10.1016/B978-0-12-818215-4.00007-9

© 2021 Elsevier Inc. All rights reserved.

extraordinary advantages due to its landscape, cultural, and historical diversity. Valaisans have been able to exploit them since the first developments in tourism. The British came to the Alps to enjoy the fresh air from the mountains already in the end of the 18th century (Reichler, 2005). Historically, tourism in the canton has been developed through a luxury hotel built in the villages located in the high valleys, in order to offer recreational activities to their wealthy guests, hotel managers. It is therefore not an exaggeration to consider that Valais represents one of the places of origin for the development of tourism activities. This lasted for almost 150 years and helped to build the identity of Valais nowadays. From this point of view, tourism is a century-old activity in Valais and its notoriety extends far beyond the Swiss borders. Focused on mountain tourism, the question of climate change is an acute one.

This chapter provides a geoprospective analysis of the tourism issue in relation to climate change. To achieve this, we will describe in the first part, the socioeconomic mechanisms related to Valais tourism. We will also describe the natural and landscape environment in a dynamic context. In the second part, we will discuss about climate change and its potential effects on the canton's territory, with a particular focus on its links with tourism and the natural environment. Finally, in the last part, we will approach the geoprospective aspect through a simulation approach. We will then attempt to answer the following questions in a conclusion. What are the sociospatial consequences that climate change will have on Valais tourism system? Landscape and ecosystem balances are based on a system of slope use that has been developed over the past 150 years. Is this system resilient enough to withstand climate change? What would be the sociospatial consequences of an imbalance in economic, ecosystem, and natural risk terms?

7.2 The strength of tourism in Valais

In 2014, the canton carried out its last survey on the added value of tourism in Valais (Valaisan Tourism Observatory: 2014).[1] The report shows on page 8 that tourism contributed CHF 2.39 billion to Valais' added value in that year out of a total of CHF 16.5 billion. In full-time equivalent, it covers 24,058 jobs, which represents 18.6% of total jobs. Table 7.1 shows the percentage of dependence on tourism for the most affected economic sectors.

We observe with the first lines of this table that most of the tourism production apparatus is oriented toward ski resorts and related activities. This is confirmed by the fact that 40% of the value added (957 million) and 58% of full-time equivalents jobs (13,951 jobs) are directly produced by tourism service providers. This also indicates that tourism in the

---

[1] Available online: https://www.vs.ch/documents/303730/740702/
     Valeur + additionoftourism + in + Valais + 2014/ed696443-cbeb-409d-b697-da7d5ed33221.

Table 7.1: Economic branches and tourism dependency ratio (table constructed by the author on the basis of the Tourism Added Value document... p. 8).

Economic sectors	Dependency ratio (%)
Accommodation	91.6
Ski lifts	89.6
Culture, sport, and entertainment	59.4
Real estate	58.4
Restoration	50.4
Wood industry	26.4
Retail trade	26.2
Transport, warehousing	21.9
Households	20.8
Administrative services	18.4

canton is strongly oriented toward winter and mass skiing. It is almost a "monoculture" and the imbalance between winter and summer is extremely important. For example, a single sunny weekend in February brings in as much money for the Verbier ski resort as it does during its entire summer season (Scaglione and Doctor, 2011). Climate change is therefore perceived as a major risk for the canton's tourism system.

7.2.1 Tourism infrastructure

We see that Valais tourism is strongly oriented toward mass skiing. There is however an offer of well-being with thermal baths and a cultural offer but this is much less important than the winter offer.

The canton has 34 ski resorts for a total of 2.460 km served by 646 ski lifts. Unlike France, Valais' ski resorts (with the exception of Thyon) are all developed from preexisting villages. The most famous of them is Zermatt. Its size and worldwide reputation makes it a special case in terms of touristic power. Beyond the case of Zermatt, there is a great diversity of situations between stations in terms of market structure. Moreover, the resorts are only one of the components of Valais tourism area. For these reasons, the canton has developed its tools for monitoring the tourism market through the concept of a tourist destination. This approach makes it possible to integrate tourism into identified socioeconomic territories (Fig. 7.1).

7.2.2 Analysis of Valais tourism system

A multivariate and temporal statistical analysis was carried out in order to detect the general structure of the tourism market in Valais. The objective was to provide an image of the strength and dynamics of Valais tourism market in relation to demand. For this purpose, we have set up a database with the following variables for the years 2015, 2016, and 2017:

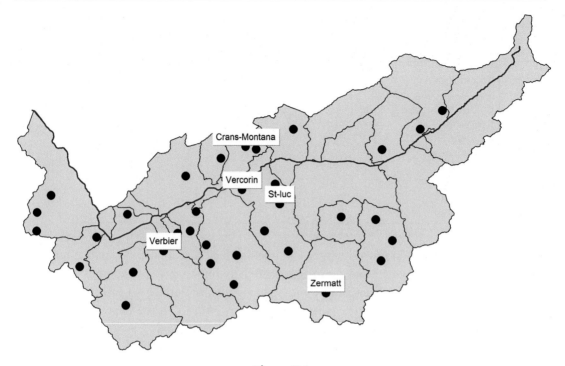

Figure 7.1
Subdivision of the canton by tourist destination and visualization of the ski resorts studied in the chapter.

- Hotel night stays;
- Self-catering accommodation nights;
- Cash register tickets—retail trade;
- Tax tourist tax;
- Number of hotels.

The analysis of the table shows that the destinations' situations in terms of business volume are very heterogeneous. There is therefore an activity gradient that makes it possible to rely on the possibility of establishing an effective typology of destinations. As a first glance, this heterogeneity also allows us to formulate the hypothesis that the effects of climate change will have different impacts on the tourism industry. As a result, the overall system will be affected in a different way. Indeed, although climate change affects all resorts, their capacity of adaptation is very different, depending on their financial resources on the one hand but also on the extent of the economic gap between tourism market size in summer and winter on the other hand.

Main component analysis is a factorial technique that formats the information contained in a quantitative data table into axes extracted one by one. Each axis retrieves a certain amount of information independently of the other axes' data. The process stops when 100% of the information contained in the table is retrieved in the axes. The number of axes necessary to achieve this is not limited. It can be seen that only five axes were sufficient to extract 100% of the information contained in the initial table (Table 7.2). This indicates that the data cloud is very elongated and thin. Half of the information is focused on axis 1.

We also observe that the eigenvalues greater than 1 are those of axes 1 and 2 and that the cumulative information of these two axes is 73.6%.

The analysis of the eigenvector table (Table 7.3) shows that all variables are well represented by axes 1 and 2, although axis 3 offers a better quality for the information of the cash register tickets variable.

According to these two analyses, we can consider that the factorial design (Fig. 7.2) would help us to build a typology of destinations. This could be possible since the information provides a 73.6% visual covering of all the information in the initial table and that all the variables are well taken into account in this design.

It can be seen that axis 1 corresponds to a gradient in the level of infrastructure. The further to the right, the destination has a better developed infrastructure. Axis 2 specializes in the volumes of use of these destinations. It is quite surprising that this axis opposes packages and cash register tickets, which seem to indicate that there are two models of destination use: that of second homes and that of day visitors.

In addition, it can be observed that the preferential direction of construction of tourist value is slightly oriented toward self-catering accommodation packages. It therefore seems that for Valais, the factor of wealth creation at the destination level is rather linked to a diffuse

Table 7.2: Table of eigenvalues.

	F1	F2	F3	F4	F5
Eigenvalue	2.554	1.129	0.899	0.368	0.05
Variability (%)	51.079	22.582	17.972	7.363	1.004
Cumulative (%)	51.079	73.661	91.632	98.996	100

Table 7.3: Table of eigenvectors.

	F1	F2	F3	F4	F5
Hotel nights	0.565	− 0.206	− 0.198	− 0.472	− 0.613
Para-hotels nights	0.53	0.196	0.061	0.796	− 0.209
Cash register tickets—retail trade	0.167	0.708	0.598	− 0.336	− 0.019
Tourist tax	0.072	− 0.639	0.764	0.05	− 0.003
Number of hotels	0.605	− 0.097	− 0.124	− 0.17	0.76

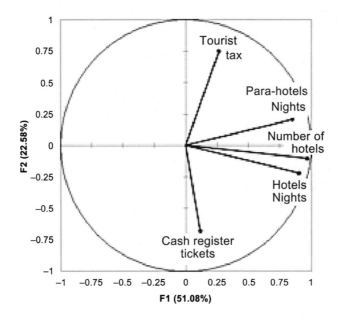

Figure 7.2
Factorial diagram (circle of correlations for variables).

environment with an *amateur* sector focused on self-catering accommodation. It would be interesting to conduct further studies on this point.

7.2.3 Temporal paths

As we have a time dataset, it is possible to study the time trajectories of the different destinations. First, a specific analysis by year shows that the underlying structure (the links between variables) does not change with the years (Tables 7.4—7.6). This shows that the overall tourism wealth creation system is stable at the canton level.

However, there are nuances in terms of intensity, such as the relationship between hotel nights and self-catering accommodation nights, which is increasing (0.51 in 2015, 0.59 in 2016, and 0.62 in 2017).

Analysis of the factorial design for the 3 years highlights three typical temporal behaviors at the destination level: a positive trajectory, a negative trajectory, and a stability.

It was found that 11 destinations out of 22, that is, half of the destinations move in factorial terms. We notice that the trajectories are essentially oriented in the vertical plane. This indicates that the changes mainly concern changes in the values of receipts and flat rates. In a way, this confirms the effect of specialization toward a type of customers segment that we have detected above. "Second home" destinations seem to be opposed to "day visitors"

Table 7.4: Correlation matrix (2015).

	Hotel nights 2015	Para-hotels nights 2015	Cash Register tickets—retail trade 2015	Tourist tax 2015	Number of hotels 2015
Hotel nights 2015	1	0.517	0.045	0.103	0.929
Para-hotels nights 2015	0.517	1	0.412	− 0.026	0.684
Cash Register tickets—retail trade 2015	0.045	0.412	1	− 0.027	0.13
Tourist tax 2015	0.103	− 0.026	− 0.027	1	0.087
Number of hotels 2015	0.929	0.684	0.13	0.087	1

Table 7.5: Correlation matrix (2016).

	Hotel nights 2016	Para-hotels nights 2016	Cash Register tickets—retail trade 2016	Tourist tax 2016	Number of hotels 2016
Hotel nights 2016	1	0.599	0.036	0.112	0.915
Para-hotels nights 2016	0.599	1	0.407	0.407	0.753
Cash Register tickets—retail trade 2016	0.036	0.407	1	− 0.063	0.146
Tourist tax 2016	0.112	0.407	− 0.063	1	0.088
Number of hotels 2016	0.915	0.753	0.146	0.088	1

Table 7.6: Correlation matrix (2017).

	Hotel nights 2017	Para-hotels nights 2017	Cash Register tickets—retail trade 2017	Tourist tax 2017	Number of hotels 2017
Hotel nights 2017	1	0.624	0.01	0.112	0.933
Para-hotels nights 2017	0.624	1	0.213	0.061	0.771
Cash Register tickets—retail trade 2017	0.01	0.213	1	− 0.124	0.138
Tourist tax 2017	0.112	0.061	− 0.124	1	0.105
Number of hotels 2017	0.933	0.771	0.138	0.105	1

destinations. Only two destinations (Val d'Hérens and Sasstal) have a combined movement (horizontal and vertical) from a factorial analysis perspective. Val d'Hérens increases its share of the offers but loses hotel and self-catering night stays. This specialization effect plays an important role in the type of infrastructure that is seen even in the landscape. Tourist destinations located near second homes have had an impact that tends to accelerate the urban sprawl in the area. This situation triggered a reaction from nature protection organizations which led to a popular vote in 2012 prohibiting municipalities to have more than 20% of second homes. As a result, the infrastructure development process stopped. Tourism stakeholders have changed their approach from construction tourism to supply tourism. More modern facilities have begun to replace the aging fleet of ski lifts. But above all, we have seen the development of new business models. It was, in particular, the development of an annual subscription at a discounted price by the Saas-Fee ski resort that launched the movement. Consortia of ski resorts now offer this kind of subscription (Magic Pass) and allow you to ski in several ski resorts. The revenue distribution is established a posteriori according to an internal stakeholders' agreement. With this new business model, the GDP generated by this part of the canton remains at a good level. Notwithstanding, the actors are concerned and wonder when climate change will condemn this activity in a large part of the ski resorts.

7.3 Nature in Valais

Valais is not only a touristy area. It also has an undeniable flora and fauna value as well as an important landscape diversity due to the altitudinal succession between the valley of the Rhône river on the one hand and the glaciers located on the highest peaks on the other hand. The federal Inventory of Landscapes has acknowledged 25% of the total surface area of the canton of national landscape importance by the federal inventory of landscapes, sites, and natural monuments office. Moreover, 9% of the total area occupied by biotopes of national importance in Switzerland are in Valais. These include dry meadows and pastures, marshes, amphibian breeding sites, and alluvial areas.

The canton is very active in the process of preserving its natural heritage. In 2019, the following paragraph was available on the official website of the Canton of Valais:

> The destruction of natural habitats, combined with the fragmentation of living spaces, is of particular concern for the Rhône valley. To address this problem, the Canton develops regional concepts aimed at ensuring ecological links and balances, based on the national and cantonal ecological networks. In September 2015, the council approved the regional concept of nature protection on the Rhône valley between Brig and Salquenenen.
>
> **Source: https://www.vs.ch/web/sfcep/biotopes.**

In this very rich natural context, there are a large number of plant species, some of which are rare such as the "Hierochloe odorant" and some are protected such as the "Sabot de Venus." In terms of fauna, 150 of Switzerland's 200 breeding birds are found in Valais. In addition, 24 of the 28 bat species are present in the canton. More generally, 12 of the 50 endemic species in Switzerland are found only in Valais. This is the case, for example, of the Simplon's Moiré, a butterfly whose area is only south of the Simplon. Finally, there are 47 species that are only present in the canton and not in other places within Switzerland. The wolf's process of conquering the Alpine space passes through the canton where several packs of wolves have settled. This regularly causes tensions with sheep farmers who are regularly attacked. The question of the future of this species is the subject of intense political battles that go beyond a cantonal framework.

One-third of the canton's surface area is covered by forests whose functions are to enhance the landscape and recreational, but also as an instrument of protection against natural hazards such as avalanches, landslides, and debris flows that regularly threaten homes.

Forest management is carried out in such a way that the protective role cannot be impacted by the natural life cycle of a forest. To this end, a sustainable management concept has been defined. It precisely describes the types of care to be provided to forests and the order in which they should be carried out (Frehner et al., 2005). It should be noted that Valais forest is gaining surface area at the expense of open spaces. The current forest cover rate is estimated at about 1000 ha per year (Institut fédéral de recherches forestières, 2010). This is due to the change of the use of agricultural land and the variation of the distribution and gradient of limiting factors in the forest, in particular, temperature and precipitation. Fig. 7.3 shows the effects of this evolution between 1950 and 2010 for the village of Vercorin.

1950

2010

Figure 7.3
Evolution of the Vercorin landscape between 1950 and 2010.

These dynamics, taken into a context of climate change, propose an ambivalent situation. On the one hand, this is beneficial in the field of CO_2 capture, protection against erosion and natural hazards and groundwater regime. On the other hand, the increase of forest area leads to a decrease in biodiversity, a reduction of agricultural potential and landscape value of the territory.

7.4 Climate change in Valais

Meteosuisse conducts important and recognized research on climate change in the Alps. It has had developed metrics since 1864. Thanks to that, it is possible to visualize the general trends. The year 2018 proved to be the warmest since the beginning of the measurements in Switzerland. The continuous series of maps below provides a clear overview of the climate trend (Fig. 7.4).

Fig. 7.4 is tragically explicit and there is a tendency for temperature to increase everywhere. In addition, precipitation volumes are also impacted (Fig. 7.5).

The detailed study of the sliding average curve shows a very slight upward trend in precipitation. MétéoSwiss also offers a map showing the distribution of precipitation trends. It can be seen that the Alps are more concerned by a precipitation decrease that can be significant on the southern slopes. However, it should be noted that the degree of statistical significance is not achieved for this part of Switzerland.

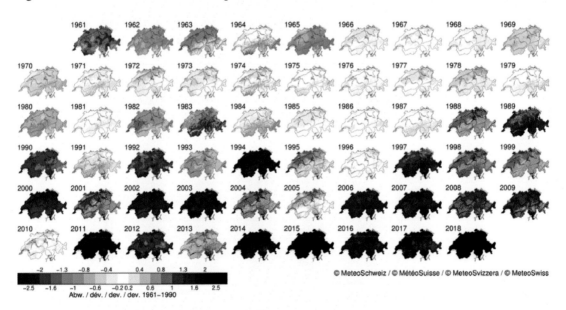

Figure 7.4
Temperature evolution between 1961 and 2018.

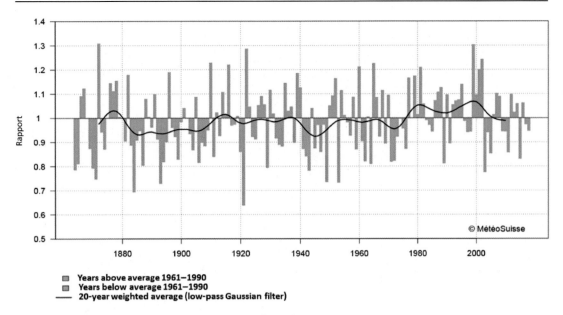

Figure 7.5
Evolution of annual precipitation in Switzerland (1864–2018).

Precipitations and temperature are two essential control factors for the maintenance of Valais' social and spatial systems. Indeed, for ski resorts, it is the combination of precipitations and low temperature that produces the snow necessary for this economy. Using the Swiss data cube (Giuliani et al., 2017), we carried out an analysis of the evolution of snow cover based on the comparison of Landsat images over a period from 1995 to 2017 (Mueller et al., 2016). The map below (Fig. 7.6) shows the overall evolution of this coverage.

When we overlay this map with the position of the ski resort surface we see that all resorts are affected by a significant decrease in snow cover. This implies that the stations will have to strongly develop artificial snowmaking systems to resist this decrease. In 2018, the resort of Champery invested CHF 11.5 million to install a brand new 300-gun snowmaking system that will cover all the resort's slopes in 2 days. Of course, such an investment will pay for itself in more than 10 years and is a major commitment for the station. The challenge is risky in the context of climate change because the climatic conditions necessary to produce snow (temperature/humidity coupling) are very precise and there is no guarantee that they will not be affected by climate change. MeteoSwiss projections on the evolution of the number of days of fresh snow show that this number is decreasing at all altitude levels (Fig. 7.7). The 1500 m × 2500 m areas should be observed with interest because it covers most of the canton's ski resorts.

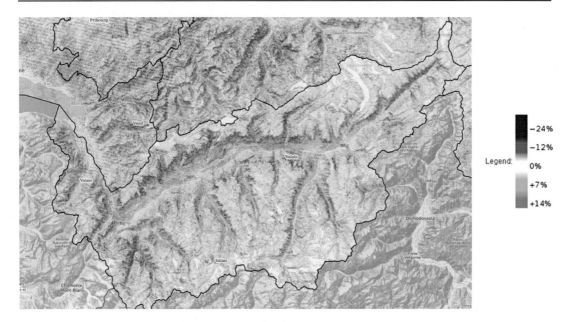

Figure 7.6
Evolution of snow cover in Valais (1995–2017).

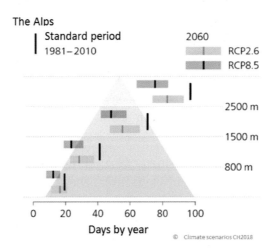

Figure 7.7
Number of days of fresh snow.

(RCP = Representative Concentration Pathways (IPCC) with: RCP8.5 is the worst-case scenario and RCP2.6 is the best-case scenario).

As a corollary to the snow issue, the use of water for tourism purposes raises a new environmental issue. Indeed, the volumes of water required to make a snow cover ready to ski are considerable (Magnier, 2016). Current calculations estimate that about 4000 m^3 of water is needed to cover 1 ha of a piste ready for skiing. Les portes du soleil, a Franco-Swiss cross-border ski resort with several resorts located in Valais, has 367 ha of artificially snow-covered slopes, which, by extrapolation with the volumes of water used to produce 1 m^3 of snow, corresponds to 1,468,000 m^3 of water (Magnier, 2016). In Valais, there are very few water reserves dedicated to this type of use. Pumping takes place either in the dam lakes, initially intended for electricity production, or in the rivers and reserves used by the municipalities for drinking water purposes. This levy, which was previously widely accepted by the population and the authorities, is beginning to be called into question because of the risk of a shortage of drinking water (as is the case of "La Clusaz" in France in winter 2018/2019) and because of the economic impact it has on other water-consuming sectors. For example, during an interview with the Radio and Television Suisse Romande (RTS), the operations manager of the Tseuzier dam from which the Crans Montana station draws its water for snow production estimates that this pumping process generates a shortfall of 1.2 million kW/h of electricity. This amount will not be produced, or the annual consumption of 240 households.[2] It therefore seems that the issue of water sharing and use will also be at the heart of the issues raised by climate change.

7.5 Conclusion of part one: issues and conflicts

With regard to what has been described above, it appears that climate change will weigh more and more heavily on Valais' tourism systems. The sector appears to be under threat and the authorities are considering a reduction of the number of inhabitants in the lateral valleys and also the number of jobs resulting from mass tourism. Under these circumstances, the entire sociospatial system would be affected. Indeed, the reduction of the population also leads to question local services such as the post office and even road maintenance. For the latter aspect, the canton is considering the maintenance of its current network and also to stop using certain roads on the basis of the abandonment of some urbanized areas (Département de la mobilité, du territoire et de l'environnement, 2018). As we can see, a positive feedback loop could be set up in the dynamics of the sociospatial system and lead to a recomposition of the territory.

In terms of the natural environment, we have seen above that the wide open spaces are conquered by the forest. We also observe that new species that were not initially present in Valais are emerging. This is the case, for example, of the Cardinal, a butterfly species whose range extends from the Canary Islands to Spain and which has been regularly observed since 2005 in the Rhône valley (Padfield et al., 2014). In terms of ecosystem value, irreversible losses are already underway. Indeed, the value of biological diversity is

---

[2] https://www.rts.ch/info/suisse/10036758-la-secheresse-menace-la-production-des-canons-a-neige-en-station.html.

much more important for mountain pastures than for forests. The simple fact of the forest increasing its area via the main limiting factors (temperature and precipitation) will mechanically act on the decrease of this value. In this context, the heritage value of the canton's natural environment will decrease overall. However, rare or unknown species (butterflies, wolves) could benefit from this change to establish themselves in the long term. Nevertheless, their contribution will probably be insufficient to compensate for the decline in biodiversity. In addition, harmful and unknown species could also establish themselves and take the place of local species. This is already the case for the impatient dwarf and the Asian Capricorn, which are considered to be serious threats to biological diversity.

7.6 Geoprospective: a multiparadigm simulation approach

The second part of our article proposes a geoprospective approach to address the question of the evolution of the tourist model of Valais in a sociospatial framework. The goal is to develop an instrument that can explore possible futures and represent them in order to help to develop a more responsible, sustainable, and integrated tourism policy. It is not a question of proposing ready-made solutions but rather of seeking to visualize the possible effects of a situation in order to enable decision-makers to assess what is at stake. It will eventually be possible to simulate the effects of one or more actions on the territory using this simulation tool.

7.6.1 General concepts

Multiparadigm simulation is an image used by the author that is inspired by the vocabulary used to describe computer languages. A computer paradigm is a way of approaching programming in a certain language. A multiparadigm language is therefore a language that allows the simultaneous use of several programming paradigms. Since we use several simulation tools with different approaches in our general model, we thought this word was appropriate to describe the geoprospective tool we have implemented.

This tool consists of a formal architecture that addresses different components of the sociospatial system with different mathematical approaches and adapted to the simulation of these components. This architecture is based on computer tools that exchange information, process it specifically, and produce results that can be used by humans.

To achieve an operational tool, it is also necessary to provide that the systems might be able to communicate with the general model as simple as possible. Not to mention the man-machine interface, it is important to make it easy to introduce scenarios into the system. That is, to be able to calibrate the general model with the minimum number of operations on the one hand and that these operations can be easily understood by decision-makers on the other hand in order to prevent decision-makers to be suspicious about the process given the complexity of the system.

These introductory remarks are essential for the development of an operational geoprospective approach. In this second part, we will only present the general formal aspects and simulations with their results. We will leave aside the exclusively IT component, which will be the subject of separate publications in specialized journals.

7.6.2 Description of the general model

It is very complex to develop a single model that addresses all the components of what we have described in the first part of our chapter. We have chosen to develop a specific simulation architecture by sections. These submodels simulate a particular component of the territory. Their results are used as parameters for calibrating the initial conditions in a particular spatial layer called the potential field. The latter serves as a basis for a Forrester simulation on the one hand and for the production of sociospatial maps of the canton on the other hand.

Fig. 7.8 describes the general architecture of the geoprospective tool we have implemented.

This approach allows us to visualize many of the mechanisms at work in a territory and to interact with them through the process of modifying a field of socioeconomic potential. It is

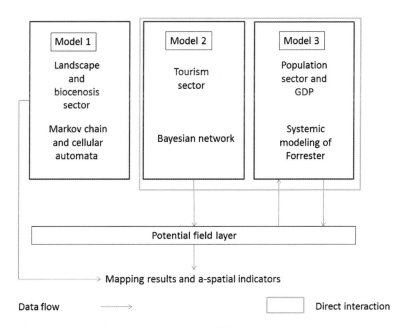

Figure 7.8
General architecture of the geoprospective tool.

this last operation that makes it possible to deal with the dynamics of human occupation with the corollary of an estimate of GDP on the economic and social level, the consequences of which will be measured in conclusion.

7.6.2.1 Potential field

The potential field is a particular spatial layer. The first part of the process was to work with a grid of georeferenced data of inhabitants and jobs in the canton with a resolution of 1 ha. This information comes from the Federal Statistical Office[3] (FSO) under the name Statpop (for the population) and Statent (for companies). This data therefore reflect the sociospatial situation of the canton. To create the potential field, we have added additional information to this data which is an attraction value that we have called "mass." This approach allows us, through a theory-based approach of spatial interaction, to make them exchange jobs and population. The shift process of these variables is controlled by two parameters: "mass" and distance. The idea is simple: the more "massive" a data point in the grid is, the more it will attract jobs and inhabitants at the expense of nearby points whose mass will not be sufficient to prevent this movement. In this way, we can simulate future developments in terms of population, employment, and GDP.

7.6.2.2 Construction of the mass for each data point

The mass of our potential field is a value P between 0 and 1 assigned to each data point of the "potential field" layer. To do this, we used an approach called the Huff model which is well known in the spatial interaction theory (Huff and Jenks, 1968). The equation below shows the version we implemented in the model.

$$P_{ij} = \frac{A_j^\alpha D_{ij}^{-\beta}}{\sum\limits_{j=1}^{n} A_j^\alpha D_{ij}^{-\beta}}$$

with

- A_j is the measure of attractiveness of one data point of the field;
- D_{ij} is the distance from i to j (1 ha);
- α is the attractiveness parameter (the number of inhabitants and jobs located at a data point);
- β is the parameter of decrease of the influence by distance (here a reverse Pareto function);
- n is the total number of locations.

---

[3] Available online: https://www.bfs.admin.ch/bfs/fr/home/statistiques/catalogues-banques-donnees/publications. assetdetail.6027955.html and https://www.bfs.admin.ch/bfs/fr/home/statistiques/industrie-services/entreprises-emplois/structure-economie-entreprises.html.

7.6.2.3 Advantages of using the mass concept

This technical architecture is very advantageous. On the one hand, we can spatialize the results of the systemic model and we can manage the local nuances of population evolution on the other hand. The feedback controls that we have performed show that our simulations are robust and effective.

Furthermore, from an operational point of view, if we introduce a change in the model, for example, to represent the spatial consequences of a policy or climate change on the sociospatial environment, simply means modifying the mass values of the grid. The complexity of updating the system is therefore outsourced, cutoff from the simulation process. This allows decision-makers to test their hypothesis through a quantification process via the mass. The new resulting mass of the data points will be used as an initial condition for the simulation section. The value of mass modification in the field of tourism is characterized by an expert approach treated by a Bayesian submodel.

7.6.3 Population sector and GDP

In this section, we present how population and GDP are simulated in the system.

7.6.3.1 The case of the population

The temporal evolution of the population is treated by a systemic model based on the same model as described in Kunte's paper (Kunte and Damani, 2015).

In this study, we have developed a transfer model with a breakdown of the population by 5-year age groups between 0 and 90 years old. The transition from one class to another is done by selecting one-fifth of the population in order to simulate aging. In addition, the following variables control the natural variations (birth rate by age group, mortality rate by age group). The movement part is controlled by two coefficients (emigration/immigration).

The simulation process is as follows.

For the 18 stages of age classes between 0 and 85 years old, the following algorithm is executed:[4]

- Recovery of one-fifth of the lower class population;
- Calculation of the number of deaths in the age category in question and the correspondent removal of this part of the population (deaths);
- Sending one-fifth of the population to the next level.

---

[4] Produced by Pascal Favre (pascal.favre@hevs.ch) as part of the sociospatial modeling project of the canton of Valais.

The last age group (85−90 +) contains only the calculation of the number of deaths for the volume of people present in it.

Step zero corresponds to the number of births for the population. It is a weighted sum of births by age group according to the birth rate assigned to it. In the end, the new population volumes are reallocated to the data points of the potential field.

7.6.3.2 The case of GDP

GDP is an indicator for measuring the economic performance of a country or region. This indicator is internationally comparable. It is widely used in political debates to measure the standard of living in a given region. In this project, we developed a cyclical analysis, that is, an analysis of the GDP growth rate over several years, that is, the period from 2012 to 2015. This indicator is calculated on the basis of the permanent resident population[5] and we have based our calculations on the FSO document: Revisions of GDP per capita (Office Fédéral de la Statistique, 2015). We have broken down the growth rate of GDP per capita using the following formula:

$$\left[\frac{GDP}{Pop\_tot}\right] = \left[\frac{\dot{GDP}}{HET}\right] + \left[\frac{Active\_\dot{employees}}{Pers\_actives}\right] + \left[\frac{Pers\_\dot{actives}}{Pop\_(15-64)}\right] + \left[\frac{Pop\_(15-64)}{Pop\_tot}\right]$$

With:

- GDP = Gross Domestic Product;
- Pop_tot = Total Population;
- HET = Effective Working Hours;
- Active_employees = Number of active people with a job;
- Pers_actives = Number of active people with a job + unemployed (as defined by the International Labour Office);
- Pop_(15−64) = the share of the population between 15 and 64 years of age.

It should be noted that the only variable that cannot be extracted from the population simulation model described above is the actual hours of work. This information is only known at the Swiss level (32.1 hours per week) and this is the information we used for our project.

Here too, the retro-predictions of Valais' GDP are working well. For example, the 2015 simulation showed that our GDP per capita was different from −0.59% against the real GDP. The least accurate simulation was in 2014 with an error of 3.9%.

---

[5] According to the most recent definition, the permanent resident population includes all persons of Swiss nationality whose main residence is in Switzerland, as well as persons of foreign nationality who have a residence or establishment permit for a minimum period of 12 months.

7.6.3.3 Tourism sector

This section of the model is treated by a Bayesian network. This is part of two main areas of research. The first area is in the field of strategic decision-making in an uncertain context. The second area focuses on the Bayesian network modeling approach. Although the literature is theoretically abundant in these two fields, there are currently no publications on experimentation in the field of strategic decision-making for a tourist resort. Furthermore, there are not experimentations of this kind particularly linked with the concept of decision support with the notion of the stress test which is at the heart of this part of the model.

7.6.3.4 Stress tests

The stress tests correspond to approaches mainly developed by the banking industry in order to assess the ability of organizations to withstand the effects of an economic crisis. In other words, to continue to meet their obligation (bank withdrawals, payment of debts, monitoring of market positions, and so forth). The criterion for success in these stress tests is to ensure that banks can stay above a certain value (8%) of the McDonough ratio (McDonough, 1998). Three types of risks are affected by this ratio:

- Credit risk;
- Market risk;
- Operational risk.

The relationship between these risks according to the banks' own capital capacities makes it possible to construct the following formula:

$$\frac{\text{Equity capital}}{\text{Credit risk} + \text{Market risk} + \text{Operational risk}} \geq 8\%$$

The application of this formula combined with a sensitivity analysis to the different risk configurations makes it possible to calculate a Value at Risk (VAR) for any bank. It is therefore a measure of default risk that corresponds to the fact that a bank can no longer honor its commitments, as we saw with the Lehman brothers bank in 2008. This figure must be greater than or equal to 8% for legal reasons (Basel II and III agreements) for the banking industry, but it can be generalized to other actors such as countries (Greece is an example of a characterized default), companies, or cities of significant size.

This VAR can be calculated using classical tools in the field such as JP Morgan's riskmetrik (Morgan, 1996) or extreme value theory (Boulier et al., 1998).

To our knowledge, this academic approach has never been applied to the tourism industry. Although very attractive, it is complex to implement and requires a considerable amount of time to achieve an acceptable result. We have therefore chosen to approach the establishment of a VAR using a Bayesian approach. This makes it possible to carry out comparable results in a simpler and more flexible way. This approach is based on the construction of a Bayesian network that can represent the global model of systemic risks in a tourist resort. The most suitable but not very known method for this construction is the GLORIA (GLObal RIsk Assessment) method developed by the research and development group of the national electricity supplier in France (EDF) (Naïm et al., 2011). This is an expert approach that help to define risks according to their relationship to defined objectives. At the first stage, we try to give a measure of quantitative objectives, for example, a turnover greater than or equal to a certain value. Then, in a second step, risks are identified that could prevent the achievement of these objectives. A Bayesian network is then built that describes the entire system. The section below describes the process for our project.

7.6.3.5 GLORIA for the definition of the notion of risk for tourist resorts

Risk is understood here as a potential event that could disrupt activity (Barthélemy and Courrèges, 2011). This disruption is understood as the difference between planned and actual objectives. In an extreme situation, the bankruptcy of a ski area can be considered as the failure to achieve the planned objectives.

It is quite delicate to manipulate the notion of risk. Formally, the probability of risks occurrence can be mitigated over time and its severity. It is easy to deal rigorously with the occurrence part of the risk considered. The gravity part is much more difficult to characterize for the following three reasons: its multicriteria nature, uncertainties about its effects and its interactions with other risks (Naïm et al., 2011).

In our approach, the definition of objectives is a crucial element. They were determined by a series of interviews with tourism stakeholders and a literature review.[6] These interviews made it possible to characterize the main objectives in the four specific areas recommended by the GLORIA method.

- Financial objectives;
- Technical objectives;
- Image objectives;
- Strategic objectives.

Notwithstanding, the definition of these risks, their occurrence, and their links remains empirical at this stage of our work since they concern only a panel of tourism stakeholders. They should be strengthened and validated by a quantitative survey of the entire tourism

---

[6] An experiment was carried out in this field via a Bachelor's degree work carried out last year under the supervision of Professor JC LOUBIER (Johan Reynard: Stress test pour les stations de ski: une vision quantifiée du risque systémique, 2016).

destination management industry in order to increase the scope of acceptance of the model. Nonetheless, this model remains entirely acceptable and functional.

Fig. 7.9 shows the Bayesian network in operation in the general geoprospective model.

The upper part of the network is the system of risks acknowledged as a potential impact factor to influence the objectives. It is at this level that expert action occurs where stakeholders can classify the risks based on their probability of occurrence with a qualitative scale (high; medium; low). Each of these occurrences has a "hidden" probability value which prevents that end users see it. This value is combined with the values obtained in the literature review and from interviews with experts (Reynard, 2016).[2]

As a result of this setting, the probabilities propagate through the network and eventually reach the VAR calculation area. This is presented in the form of a utility score. That is, a probability of achieving the overall objective combined with the impact of this likelihood

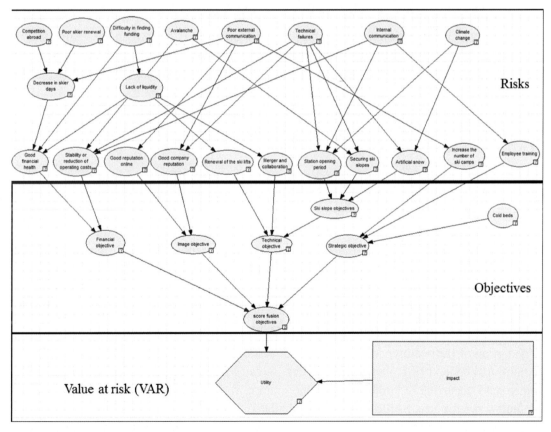

Figure 7.9
Bayesian network for the treatment of the economy of tourist destinations.

on the territory. Impact is defined as a preference. It is this value that will be added or subtracted on the initial mass of the data points of the potential field concerned by the analysis. From a simple point of view, utility corresponds to a preference score (which corresponds to a subjective point of view depending on the proposed problem) on a crossing between a state of the nature (e.g., climate change) and a series of possible actions (doing nothing; developing renewable energies, etc.). A table can be constructed with a value attached to each intersection that will quantify the "gain" if the situation occurs. There is a multitude of strategies to measure utility, but all are based on:

- The fact that the actions and states of the nature are countable;
- We can classify actions (i.e., establish preferences);
- We are rational (i.e., the ranking corresponds to a comparison of one action with another);
- We are consistent (we respect the transitivity of preferences. If we prefer apple over cherry and cherry over peach then we MUST prefer apple over peach).

In this context, the gain of an action can be negative. For example, in our utility function, if the objectives of the station are not achieved and the impact is high, then the "gain" is a value of -200 because the territory is strongly affected and loses jobs, population, etc. On the contrary, if the objectives are achieved and the impact is low then the "gain" is 1 because it corresponds to the current situation. The fact that it is not modified does not therefore bring any advantages to the territory but does not cause any losses either.

7.6.3.6 An understandable and empirical model

The Bayesian model is "easier" to understand for nonexperts. They can discuss or modify it because the influences of risks are visualized by the arrows on the graph. It is also more pragmatic because it relies not only on "hard" science but also on the empirical knowledge of the different stakeholders (e.g., to quantify the probabilities of a risk using scales). Classical work on the expert definition of risks has documented very well this aspect of the research (Fragnière et al., 2010). Indeed, it is common for real life problems to be vaguely defined and to have inexistent factual data. Therefore the use of an expert knowledge approach is common and is called probability elicitation. The most well-known tool is the probability scale (Fig. 7.10) of Druzdel and Van Der Gaag (2000).

The notion of the occurrence of several risks at the same time can be addressed in a conditional context (if a phenomenon occurs, it can lead to changes in the probability of other risks occurring by propagation in the graph). In the end, a VAR, that is, a risk of default score for a tourist destination, is calculated. This score is incremented on the mass value of each point of the potential field for the destinations concerned. For example, the Crans Montana score after analysis and propagation in the Bayesian network is -0.1314. The negative sign indicates that this value must be deducted from the mass value of the

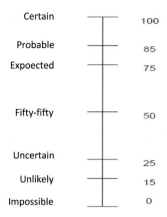

Figure 7.10
Probability scale.

points located in the destination "Crans Montana." The new mass value corresponds to the probability of attracting a point as a function of the expertise by the Bayesian network. We have recalculated the mass of 968 data points located at the destination Crans Montana this new mass then serves as the initial conditions for the temporal systemic simulation of population, employment, and GDP.

We also conducted a simulation on the following destinations: Crans Montana, Verbier, Zermatt, Vercorin, and St Luc. After setting up and calculating this new mass, we carried out a spatial analysis of the new population distribution. A Silverman smoothing mapping (Silverman, 2018) allows to visualize the changes. The map below (Fig. 7.11) shows the results of the simulation.

We can see that the impacts are stronger on the destinations of Verbier and Crans Montana. Vercorin and St Luc are also affected, but to a lesser extent. Zermatt is the destination that is least affected. However, visual appearances can be misleading because the study of absolute values shows that loss volumes are quite low. For Crans Montana, which has the highest values, the population losses are less than 2% per 6-year period and the GDP value is not affected. It therefore appears that, despite climate and societal change, the process of reducing the added value of the tourism industry will not be as strong as envisaged by the players in the field but will be spread over a long period. Population distribution changes for the lateral valleys also appear to be quite slow.

7.6.4 Landscape and biocenosis sector

The landscape and biocenosis part is treated by a well-known approach in ecology and modeling: Markov chains coupled with cellular automata. The literature is abundant on the subject and we will not develop further here the methodological aspects. Interested reader

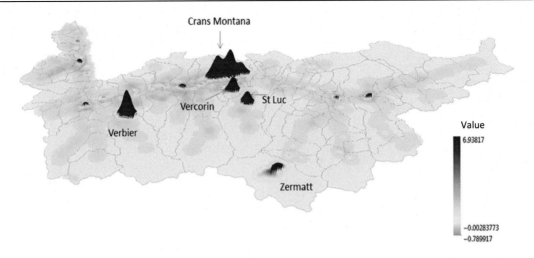

Figure 7.11
Results of the evolution of population in Valais' tourism destinations. Red peaks should be understood as intensities of density loss.

can consult Coquillard and Hill's manual for a seminal example in the field of ecology (Coquillard and Hill, 1997).

We used the Corine Land cover project's data of 2006, 2012, and 2018 to construct the models for the evolution of major land cover use, particularly those related to the surfaces of forest cover, agricultural areas, and wide open natural areas. We have seen in the first part of this chapter that agricultural open spaces are gradually being replaced by forests. The Markov model confirms this general trend. The detailed study of the global process of change dynamics allows us to build the following graph (Fig. 7.12).

It should be noted that this process is underway throughout the canton. Also landscape changes are slow and can only be noticed if a comparison is made among long periods of observation. However, the transformation of the landscape is well underway. In the case of wide open spaces gained by the forest (agricultural spaces), it is plausible that the mechanisms of the forest gaining surface from agricultural areas are accelerating. Indeed, the factors limiting forest expansion, in particular, temperature and precipitation levels, will increase upward and allow territorial gains to be made at the expense of alpine grasslands. This phenomenon will therefore necessarily reduce biodiversity in the long term according to two mechanisms: the replacement of grasslands by forests leads to a decrease in species diversity on the one hand and the modification of ancestral activities and the contribution of atmospheric nitrogen leads to a uniformity of biocenoses on the other hand. The biodiversity monitor in Switzerland developed an observation process on the biocenoses uniformity trend as it is shown in Fig. 7.13.

We are now able to produce a geoprospective analysis of the canton.

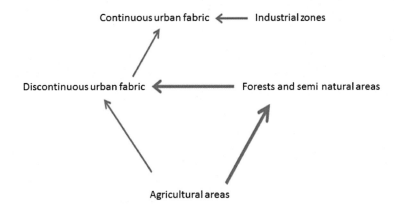

Figure 7.12
Graph of the dynamics of land use in Valais.

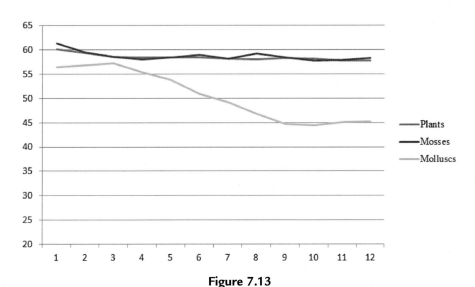

Figure 7.13
Diversity of species communities in meadows and pastures—index* of 0 (uniform) to 100 (diverse) of all paired comparisons of the sampling areas in percentage. *Mean values over a period of 5 years. Source: *From Office Fédéral de L'environnement, 2014. La biodiversité en Suisse (résumé). Available from:* <https://www.bafu.admin.ch/bafu/fr/home/themes/biodiversite/publications/publications-biodiversite/biodiversite-suisse.html> *(in French) (OFEV 2014, p. 13).*

7.7 The Valais of tomorrow: geoprospective panorama

Simulations have shown that all the sociospatial components of the canton are evolving. The main discovery is that these global changes are occurring slowly and continuously throughout the country. This can be a real problem because the absence of strong and spatially localized mutations may suggest that the current system is able to cope with all future situations, preventing the realization that new development policies must be considered. That said, the initial assumptions developed at the beginning of the chapter do not seem to be confirmed. In terms of population, there has been a slight decrease in population density at the heart of destinations as shown on the map in Fig. 7.11. In terms of jobs and added value, it seems that the phenomena in action are not yet strong enough to put under pressure the economic production apparatus. On the other hand, the gradient of global change of the whole system introduces to the discussion about water sharing in some destinations. As the forest rises, there is a risk that the local water cycle will be affected. However, it is currently impossible to give a picture of it. In our opinion, this issue will be the main factor disrupting the current sociospatial system and should crystallize the first strong societal controversies.

In terms of landscape, we can see that the process of forest expansion over wide open spaces is still in action. This leads to a loss of diversity, both in biocenoses and in biotopes. We also observe that the Rhône Valley is undergoing a diffuse urbanization to the detriment of agricultural territories which are mainly composed of orchards. For agricultural areas, two types of paths can be observed. Low altitude areas are gradually being transformed into residential buildings. The ones above have a variable future trajectory. The mountain pastures remain used by the herds and resist well even if we see that the combat zone is moving upwards. Some less accessible steep parts are clearly being overtaken by the forest. Finally, the wide open spaces located at intermediate altitudes and already surrounded by the forest are for the most part in danger of disappearing. To conclude on this panorama, it seems that a general impoverishment of the landscape and ecosystem value of the canton is slowly at work. The simulation shows that the lateral valley population appears to be more resistant to erosion than assumed in the original assumptions. This was found in the simulation due to the fact that despite the impacts of climate change, jobs remain little affected by a weakening of winter tourism activity. Finally, the general development of the population in Valais seems to be concentrated along the Rhône Valley. This situation indicates that the populations in the lateral valleys are only slightly affected by a population increase. In the long term, there is a risk of a population decline given that generational renewal cannot compensate for the mortality due to the aging of this population.

7.7.1 Focus on the tourism model

The general simulation suggests that the tourism system will not be impacted as quickly as expected. This seems to confirm that the strategy of pooling sky passes within destination consortia can significantly reduce the direct impact of climate change. It also seems that the general system is more resilient than assumed. In this context, annual climate variability would not be sufficient to really challenge ski tourism as proposed, as losses in 1 year can be recovered in the next year. Throughout the canton, GDP and induced employment are virtually unaffected.

This explains the current vision of stakeholders. It is common to hear among professionals in the sector that this model can still last 30 years. There is still time to continue with a classical approach. In our opinion, this is the most plausible explanation for the choice to invest in nondisruptive innovations with the traditional system. This is the case, for example, of Champery's snowmaking installation presented above. Indeed, although some low altitude stations close (Charmey in the canton of Fribourg), this has not yet happened in the canton of Valais with the exception of a small station (super St Bernard). For the latter, the explanations given are purely economic and linked to the replacement cost of lifts that have become obsolete in the face of the limited ski supply that this resort offered.

However, it is clear that the current tourism economic model will not be able to continue in this way for 30 years. Without even relying on the structural issue of climate change, the unsustainable approach and the significant exploitation of water resources for leisure activities are beginning to be discussed by tourists themselves. It will therefore be necessary to provide guarantees of good environmental conduct in order to attract skiers. Currently, a green washing speech is sufficient. However, preference surveys increasingly show that tourists are concerned about the issue and although their choice of holidays destination is not entirely determined by environmental considerations, it can make a difference in the final decision. Unfortunately, currently there is not a rigorous work under way in Valais which would be able to propose a more integrated economic model of tourism. It is feared that nothing will be done as long as the erosion process remains slow and diffuse.

7.8 Conclusion

Simulations have shown that the dynamics of climate change in the canton of Valais do not have yet a significant impact on the tourism system. It should be recalled that the simulation approach remains speculative since some parts of the general simulation model are based on an empirical scenario approach. The latter are characterized by the choices of actors recognized as competent. They set the realization values of the Bayesian network in a way that seemed obvious to them according to their belief. In this context, the geoprospective that we have conducted and that we present here does not meet the classical standards of a rigorous scientific approach. However, in a science of complexity such as man/territory interaction, it seems totally illusory to hope to control all the parameters involved. We would be then in the case of Laplace's demon, which is impossible. The path of an integration of actors and simulation can help to better understand the issue by avoiding the pitfalls of a prescriptive simulation that we now know almost never happens.

This chapter is therefore not intended to describe the canton's destiny in the distant future. It should be seen as a presentation of what could happen and why it could happen. The awareness process and the ease that the system provides allow us to explore possible responses. This process encourages the emergence of a new field of geography in relation to a more symbiotic approach to spatial analysis, of which geoprospective is one component.

References

Barthélemy, B., Courrèges, P., 2011. Gestion des Risques: Méthode d'optimisation Globale. Editions Eyrolles, Paris (in French).

Boulier, J.F., Dalaud, R., Longin, F., 1998. Application de la théorie des valeurs extrêmes aux marchés financiers. BanqueMarché 32, 5–14 (in French).

Coquillard, P., Hill, D.R.C., 1997. Modélisation et Simulation d'Ecosystemes. Des modeles analytiques à la simulation a événements discrets. Collection Ecologie. Masson, Paris, p. 273 (in French).

Département de la mobilité, du territoire et de l'environnement, 2018. Service de la mobilité. Concept cantonal de la mobilité 2040. Available from: <https://www.vs.ch/documents/529400/3859505/CCM + 2040 + - + Rapport/720702f6-36ad-4e41-8092-055bd5b74d59> (in French).

Druzdel, M.J., Van Der Gaag, L.C., 2000. Building probabilistic networks: "Where do the numbers come from?". IEEE Trans. Knowl. Data Eng. 12 (4), 481–486.

Fragnière, E., Gondzio, J., Yang, X., 2010. Operations risk management by optimally planning the qualified workforce capacity. Eur. J. Oper. Res. 202 (2), 518–527.

Frehner, M., Wasser, B., Schwitter, R., 2005. Gestion durable des forêts de protection. Soins sylvicoles et contrôle des résultats: instructions pratiques. L'environnement pratique. Office fédéral de l'environnement, des forêts et du paysage (in French).

Giuliani, G., Chatenoux, B., De Bono, A., Rodila, D., Richard, J.-P., Allenbach, K., et al., 2017. Building an Earth Observations Data Cube: lessons learned from the Swiss Data Cube (SDC) on generating Analysis Ready Data (ARD). Big Earth Data 1 (1), 1–18.

Huff, D.L., Jenks, G.F., 1968. A graphic interpretation of the friction of distance in gravity models. Ann. Assoc. Am. Geogr. 58 (4), 814–824.

Brändli, U.B., 2010. Inventaire forestier national suisse: résultats du troisième inventaire 2004-2006. Institut fédéral de recherches forestières, Birmensdorf, Suisse. Office fédéral des forêts et de la protection du paysage (in French).

Kunte, S., Damani, O.P., 2015. Population Projection for India—A System Dynamics Approach. In: 33rd International Conference of the System Dynamics Society, 19–23 July 2015, Cambridge, MA, USA.

Magnier, E., 2016. Les impacts hydrologiques de la production de neige dans un domaine de moyenne montagne. VertigO Rev. Électron. Sci. L'environ. 16 (1), (in French).

McDonough, W., 1998. Statement before the committee on banking and financial services, US House of Representatives. Fed. Reserve Bull. 84, 1050–1054.

Morgan, J.P., 1996. Riskmetrics Technical Document. Available from: <https://www.msci.com/documents/10199/5915b101-4206-4ba0-aee2-3449d5c7e95a>.

Mueller, N., Lewis, A., Roberts, D., Ring, S., Melrose, R., Sixsmith, J., et al., 2016. Water observations from space: mapping surface water from 25 years of Landsat imagery across Australia. Rem. Sens. Environ. 174, 341–352.

Naïm, P., Wuillemin, P.H., Leray, P., Pourret, O., Becker, A., 2011. Réseaux Bayésiens. Editions Eyrolles, Paris (in French).

Office Fédéral de L'environnement. 2014. La biodiversité en Suisse (résumé). Available from: <https://www.bafu.admin.ch/bafu/fr/home/themes/biodiversite/publications/publications-biodiversite/biodiversite-suisse.html> (in French).

Office Fédéral de la Statistique, 2015. Révision du PIB par habitant. Available from: <https://www.bfs.admin.ch/bfs/fr/home/statistiques/themes-transversaux/mesure-bien-etre/conditions-cadre/economiques/pib-reel-par-habitant.assetdetail.350183.html> (in French).

Reichler, C., 2005. Le bon air des Alpes: entre histoire culturelle et géographie des représentations. Rev. Géogr. Alp. 1, 9–14 (in French).

Reynard, J., 2016. Stress tests pour les stations de ski: une vision quantifiée du risque systémique, Bachelor Thesis, HES-SO Valais (in French).

Padfield, G., Baudraz, V., Baudraz, M., Chittaro, Y., 2014. Le Cardinal Argynnis pandora (Denis & Schiffmüller, 1775) s'est-il établi en Suisse (Lepidoptera, Nymphalidae)? Entomo Helv. 7, 99–111 (in French).

Scaglione, M., Doctor, M., 2011. The impact of inaccurate weather forecasts on cable-car use. In K. Weiermair, H. Pechlaner, A. Strobl, M. Elmi, & M. Schuckert (Eds.), *Coping with global climate change. Strategies, policies and measures for the tourism industry* (pp. 61-76). Innsbruck: Innsbruck University Press.

Silverman, B.W., 2018. Density Estimation for Statistics and Data Analysis. Routledge, London.

Valaisan Tourism Observatory (Observatoire Valaisan du Tourisme), 2014. Valeur ajoutée du tourisme en Valais : Analyse de l'offre et de la demande touristique, Available from: <https://www.vs.ch/documents/303730/740702/Valeur + addition of tourism + in + Valais + 2014/ed696443-cbeb-409d-b697-da7d5ed33221> (in french).

CHAPTER 8

Geoprospective assessment of the wood energy supply chain sustainability in a context of global warming and land use change within 2050 in Mediterranean area

Emmanuel Garbolino[1], Warren Daniel[2] and Guillermo Hinojos Mendoza[3]

[1]*Climpact Data Science, Nova Sophia—Regus Nova, Sophia Antipolis Cedex, France,* [2]*Plant and Ecosystems (PLECO), University of Antwerp, Wilrijk, Belgium,* [3]*ASES Ecological and Sustainable Services, Pépinière d'Entreprises l'Espélidou, Parc d'Activités du Vinobre, Aubenas, France*

Chapter Outline

8.1 Introduction

Wood biomass use for energy systems represents around 40% of the amount of renewable energy production in France (Cavaud et al., 2017). Almost 7000 wood-energy units are established for the collective and industrial energy production, and the development of such systems knows an acceleration since the beginning of the 2000s (CIBE, 2017). This phenomenon is mainly due to ensure France's commitments to reduce its greenhouse gas emissions for energy production.[1] This strategy is based on a regulatory and tax incentive

---

[1] DIRECTIVE 2009/28/EC of 23/04/2009 on the promotion of the use of energy from renewable sources. REGULATION No. 995/2010 of October 20, 2010, laying down the obligations of operators who place timber and timber products on the market. Council conclusions COM(2013) 216 final "An EU strategy on adaptation to climate change."

Ecosystem and Territorial Resilience.
DOI: https://doi.org/10.1016/B978-0-12-818215-4.00008-0
© 2021 Elsevier Inc. All rights reserved.

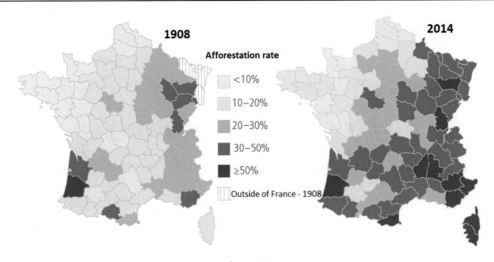

Figure 8.1
Afforestation rate of French territories in 1908 (left) and in 2014 (right) (IGN, 2018, modified).

context and on a considerable increase in forest areas (+6 million ha in 100 years, IGN, 2018; Fig. 8.1). Currently, this increase of forest cover is mainly induced by the abandonment of a part of rural activities, especially located in areas with hills and mountains (Talandier et al., 2016). The development of forestry in some territories also contributes to the increase of forest areas in France.

As shown in Fig. 8.2, wood biological productivity is not the same in the different places of France according to the bioclimatic patterns and the history of forest exploitation and dynamics. For example, the lowest productivity is located in the Mediterranean part of the French territory (less than 4 m^3/ha/year). This area knows an average growth rate of forest, due to a major abandonment of rural areas that compensates in part the small productivity level.

However, biomass power plants are developed in the south of France that may induce new pressures on forests and may cause social reactions against the biggest projects. The development of these projects is mainly induced by the growth rate of forest areas that is between 0.5% and more than 2% in the whole French Mediterranean area. The expansion of forest area is first located in the large Mediterranean arc and Corsica where most of the areas are pastoralism and agriculture abandoned land, which is very advantageous for tree development.

The fifth report of the Intergovernmental Panel on Climate Change (IPCC, 2014) has presented scenarios of increasing average temperature between 1.4°C and 3.1°C (baseline 2.2°C) over a period of 100 years. This climate change is expected to affect the nature and distribution of the constituent species of ecosystems and, at the same time, it would bring

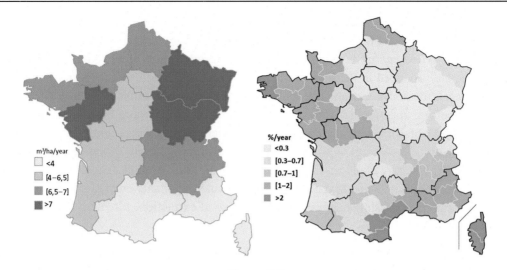

Figure 8.2
Wood biological productivity (in m^3/ha/year) and annual growth rate (in %/year calculated between 1985 and 2014) of forest areas in France (IGN, 2017, modified).

changes in the functioning of ecosystems like biomass productivity reduction, rise of wildland fire episodes, etc. Many researches show a potential impact of global warming on trees distribution (Badeau et al., 2004; Iverson et al., 1999; Thuiller, 2003; Thuiller et al., 2006; Garbolino et al., 2016) and ecosystem services like biomass production for bioenergy (Bellarby et al., 2010; Barney and Di Tomaso, 2010; Tuck et al., 2006; Garbolino et al., 2017, 2018) in Europe and America. Expected evolutions include changing the geographical boundaries of the current Mediterranean bioclimatic zone, particularly north of this area and in upland areas. Mediterranean species could hardly follow latitudinal displacement of their potential range at the rate advertised. They could possibly survive in the high thermal gradient hills, provided they are already present in the area (IGN, 2018). Due to the increase of heat waves in the next decades, other researches show that Mediterranean plants will experience (and some start to experience) a high level of hydric stress that will increase the mortality rate of species (Vennetier and Ripert, 2009; Ciais et al., 2005; Fink et al., 2004; Fischer, 2007; Niua et al., 2014; Reichstein, 2005; Charru et al., 2017).

The development of wood-based energy systems, for collective and industrial use, requires the involvement of at least the following stakeholders: the forest managers that are able to mobilize, prepare, and transport the wood biomass; the industry that delivers the energy system; the municipality that participates to the decision process for different activities (authorizations for wood taking, for the building of the wood boiler and housing, etc.); the public administration that controls these activities; and the customers that have more opportunity to choose its energy supplier. According to the roles of these entities and the life duration of a wood boiler that is usually planned for 30 years, it is relevant to plan the supply of wood biomass for at least three decades in order to manage the risk for the entire supply chain. Wood energy is mainly considered as a solution in order to remove woody wastes that are produced in the French Mediterranean area. It is also a mean for managing Mediterranean forests and scrublands that are often exposed to wildland fires. But, as mentioned before, uncertainties remain on the availability of the wood resource due to the potential impact of climate change on forest ecosystems. Other uncertainties are also related to the evolution of the energy demand and on the potential pressure of the development of the wood energy supply chain on the activities of using this resource (paper and construction industries, wood craft, etc.). Finally, some opponents emerge owing to the fear of landscapes and ecosystems alterations that the wood extraction, for energy demand, may provoke in the future.

Uncertainties remain concerning the potential development of wood energy supply chain in the South of France. The different stakeholders involved to the wood energy supply chain need to be supported on their decision process in order to ensure, as far as possible, the sustainability of this sector that is promoted by local, national, and European administrations.

In order to answer to a part of this problematic, we propose to assess the potential impact of climate change on the vegetation dynamics of 25 tree species toward 2050, by using the CDS toolbox model developed with ASES and Climpact Data Science (CDS) companies. These species stem from four main forest types mainly observed in the Alpes-Maritimes and used for the current and future bioenergy systems. This point is essential to ensure the sustainability of combustion systems that are built for a minimum life service of 25−30 years. In this frame, we also propose to assess the urban spread for 2050, in order to identify the most suitable areas to extract biomass for energy demand according to the distance of current and future housing areas.

8.2 A geoprospective approach based on spatial models

The aim of our study requires a Geographic Information System (GIS) platform in order to identify the best territories for the wood availability in 2050. This geoprospective study integrates the following spatial data and models (Fig. 8.3):

- The CDS toolbox model of vegetation dynamics due to the climate scenarios for 2050 (Garbolino et al., 2016, 2017, 2018; Garbolino, 2014) in order to assess the potential modifications of the vegetation structures that may affect the availability of biomass for its energetic use;
- A model of urban dynamics, based on the use of cellular automata (Hinojos et al., 2012), for the estimation of the urban spread toward 2050 in order to assess and identify the locations of the potential increase of the population and its energy demand.

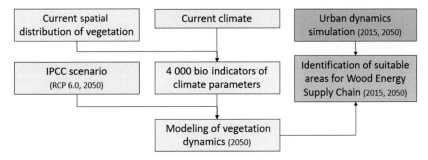

Figure 8.3

Methodology of the geoprospective approach defined in order to identify the most suitable areas for developing wood energy supply chain.

The following paragraphs introduce the assumptions, data, and the structures of each model. The results of each model are also presented. The integration of the two models is presented in Section 8.3.

8.2.1 Vegetation dynamics assessment

American foresters and French researchers (Iverson et al., 1999; Thuiller, 2003; Thuiller et al., 2006) provided models based on the overlap of plant observations and climate measures located in grids of 10−50 km of spatial resolution. The results are indicative plants of the climate based on the geometric model, like correlations.

In order to respect the intermittent nature of the field data, Garbolino et al. (2007) proposed a probabilistic method to quantify the relationship between 1.874 woody and herbaceous plants and 72 climatic variables measuring the average monthly climate over a period of 50 years in France. The results of this climatic calibration provided a set of bioindicators of climatic variables, the fundamental basis that should be reversed and complemented with new data for modeling the distribution of vegetation on French territory.

The climatic characterization of a taxon is based on a probabilistic calibration taking into account three main ecological assumptions (Garbolino et al., 2007, 2012):

- The effect of a factor on a plant's frequency follows a unimodal trend, defining an optimum frequency of the occurrences of the plant in a part of the range of a climatic variable;
- The effect of a factor on a plant is gradual, but the intermittence of the plant in the range of the climatic variable is possible;
- A plant is a better indicator of a factor if its occurrence is concentrated in one part of the range of the climatic variable, for example, if two plants are distributed in the same part of the range of a climatic variable, the most indicative plant is the one showing the highest frequencies at one or more levels of the range, even though the two plants may have the same optimum.

More recently, this calibration was applied on 4000 taxa in France (Garbolino, 2014) with the use of climatic data provided by Meteo France and distributed on the whole French territory at a spatial resolution of 1 km^2 in order to have the representation of the climatic behavior of each plant. The probabilistic calibration gives probability of occurrence of a plant into the range of a climatic variable. For this calibration, we used 36 climatic variables: Tmax—average of day temperatures for 30 years, Tmin—average of day temperatures for 30 years, and P—average amount of precipitations for 30 years, for 12 months.

The use of the inverse relation allows estimating the probability of occurrence of a plant in the climatic plots. The algorithm selects all the climatic plots that match the climatic range of a plant for all the 36 climatic variables. The next step is the calculation of the average probability of occurrence of a plant into a climatic plot according to the probability of occurrence of a plant into specific values of climatic variables. This calculation allows identifying the suitable areas for each plant by the calculation of the average probability of occurrence of a plant into the climatic plots. This algorithm is applied in order to assess the potential distribution of a plant in the territory for the current climate and for the climatic scenario for 2050. The comparison between the potential distribution of suitable areas for these two dates enables to identify the probable effect of the global warming on the spatial distribution of the suitable areas of forest types for the future.

We focus our study on four main forest structures that are very well represented in the Alpes-Maritimes (thermophilous sclerophyllous trees, thermophilous conifers, mesophilous conifers, mesophilous deciduous trees) that are characterized by 25 tree species (Appendix A).

8.2.2 Model validation

The validation methodology is based on the comparison of the spatial distribution of two datasets: the potential distribution of forest types, coming from the CDS toolbox model of plants distribution according to the current climate, and the maps of forest types provided by the French National Forest Inventory (IFN) managed by the French National Geographic Institute (IGN). IFN produces these maps with the combination of photointerpretation techniques and field observations of reference parcels.

Fig. 8.4 provides a qualitative validation based on maps of suitable areas for four forest types defined in our study and maps of observed forest types at the scale of the studied area (Alpes-Maritimes, France).

These maps show that the potential distribution of a vegetation type is always lager than the places of the observations, because the forest types do not colonize all the places where they should live due to the impacts of human activities like agriculture and housing, or other factors like the competition processes between them. These maps also underline that most of the plots of the observed vegetation frequently overlap the plots of potential suitable areas of each vegetation structure, which is a qualitative validation of the model.

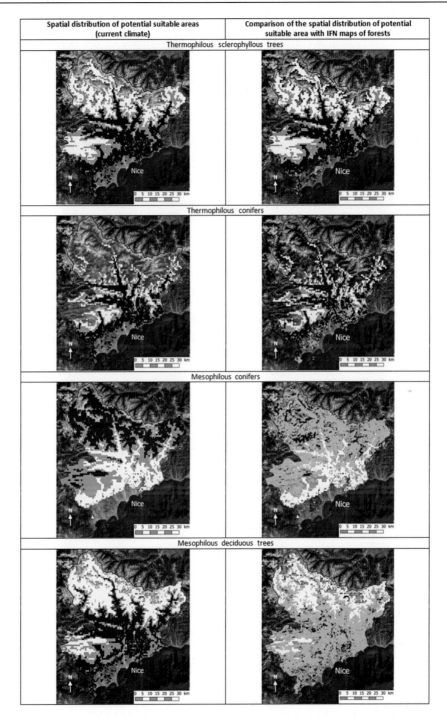

Figure 8.4

Comparisons of the probabilities (where low probability are in white, average probability are in orange and high probability are in red) to find the most suitable areas for the development of the four vegetation structures with the maps of their observation provided by the IFN (in blue).

Quantitative results in Table 8.1 show the amount and the percentage of plots of observed forest types of the IFN data according to the probability classes of the potential distribution of suitable areas for the current climate.

The quantitative results highlight that the plots of the high probability level correspond to the location for most of the observed forest types (between 54% and 87%), and the average probability level matches 13%−31% of the observed vegetation plots. The addition of the observed vegetation plots in the average and high probability classes of each vegetation type underlines that those two probability classes gather between 85% and 100% of the observed vegetation plots that overlap the vegetation plots assessed by the model. This result demonstrates that the average and high probability classes of the model can be considered in order to assess the potential suitable areas for the development of the forest types.

8.2.3 Assessment of suitable areas of forest types for 2050

By using the inverse relation between plants and climate, and the scenarios of climate evolution provided by the IPCC, it is possible to assess the probability of occurrence of plants. RCP 6.0 scenario was selected because this scenario is very close to the last average scenario (A1B) which was considered as one of the most probable scenario of climate evolution (IPCC, 2014).

Table 8.1: Statistical distribution of the forest types provided by the IFN-IGN with the probability classes provided by the model estimating the potential occurrence of suitable areas of the forest types according to the current climate.

Probability classes	Amount of plots	% of plots
Thermophilous sclerophyllous trees		
[0.01; 0.08]	0	0
[0.0.8; 0.38]	10	23
[0.38; 0.68]	34	77
Thermophilous conifers		
[0.11; 034]	0	0
[0.34; 0.50]	9	13
[0.50; 0.66]	60	87
Mesophilous conifers		
[0.01; 0.15]	157	10
[0.15; 0.40]	496	31
[0.40; 0.67]	932	59
Mesophilous deciduous trees		
[0.22; 0.30]	202	12
[0.30; 0.37]	538	31
[0.37; 0.49]	938	54

The following maps (Figs. 8.5 and 8.6) show the difference between the most suitable areas for the development of such forest types for the current period and for 2050. The probabilities of occurrence are discretized into three classes: low probability (white), average probability (orange), and high probability (red).

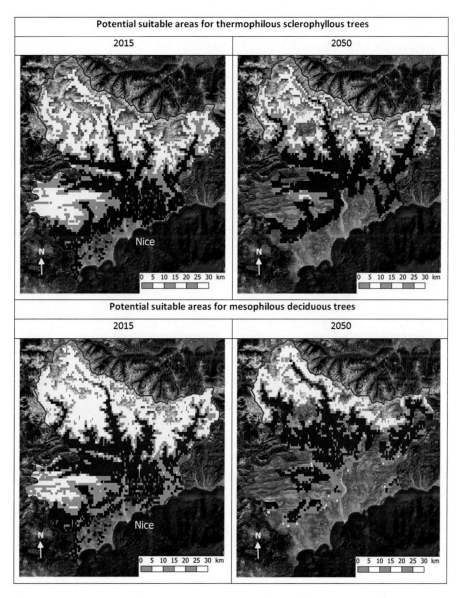

Figure 8.5

Comparisons of the probabilities to find the most suitable areas for the development of deciduous trees and evergreen trees in the French Riviera for the present time and 2050.

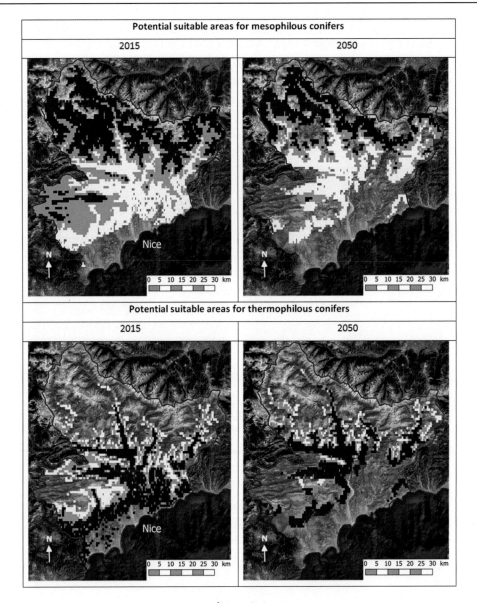

Figure 8.6
Comparisons of the probabilities to find the most suitable areas for the development of mesophilous and thermophilous conifers in the French Riviera for the present time and 2050.

Fig. 8.5 shows that deciduous trees have high probabilities of occurrence for the current period in coastal strip, south valleys, inner valleys, hills, and small mountains of the north part of the area. Their potential distribution in the future could be 150 m higher than today in average. This supposes that the deciduous trees may colonize more homogenous areas than today, located in inner valleys, hills, and small mountains. As shown in the maps, the potential ecological areas suitable for the deciduous trees should be located outside of the coastal strip in 2050.

The evergreen trees have suitable areas in the coastal strip and in the south part of the valleys. In the future, the suitable areas should be located further north, in hills, small mountains, and inner valleys. For those two forest types, climate change would affect significantly the populations that are currently observed in the coastal strip and the hills located in the south part of the territory, by the increase of stress and may be by the increase of the mortality rate.

Fig. 8.6 shows that the mesophilous conifers are mainly located in the north side of slopes of the mountains and in the valleys of the Alpes-Maritimes. They are also well represented in mountains and valleys of the north part. The potential distribution of suitable areas for 2050 shows a drastic reduction of these areas, which should be located in the extreme north part of the Alpes-Maritimes, especially in high mountains. These results underline a probable risk of stress increase for the populations of mesophilous conifers currently located in most areas of the territory.

The thermophilous conifers are currently well distributed in the coastal strip and in some valleys of the south part of the French Riviera. Their potential distribution in the future shows a decrease of ecological situations suitable for such forest types. The suitable areas could be located in higher areas, in some small mountains of the half south part of the Alpes-Maritimes.

The results of the modeling of the spatial distribution of suitable areas for the forest types show a potential decrease of these areas by 2050, for all forest types. This observation underlines a likely increase of stress for plants in 2050, which could induce a slowdown of trees growth and, in some places, an increase of mortality rate.

Fig. 8.7 shows the amount of the potential suitable plots of 1 km^2 in 2015 and in 2050 for the four different forest types. The comparison between the amount of plots according to the two periods underlines a potential decrease of the suitable areas for all the forest types.

Table 8.2 presents a loss of suitable plots in 2050 for each forest type.

This table shows an average of more than 1280 km^2 that will be less suitable for the different forest ecosystems in the Alpes-Maritimes that will represent 30% of the potential current suitable areas.

Fig. 8.8 presents the average altitudes of the potential suitable areas in 2015 and 2050 for the four forest types. This graph underlines that these areas will be distributed in preference in higher altitude in 2050 than in 2015.

Table 8.3 shows that the potential suitable areas for the different forest types in 2050 should be distributed between 105 and 204 m higher than the current situation.

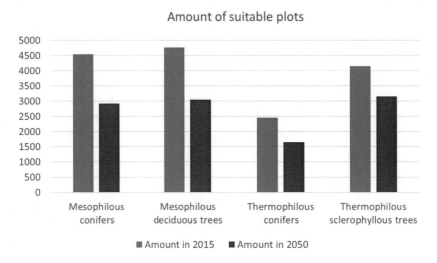

Figure 8.7
Amount of plots for the potential suitable areas of the forest types in 2015 and 2050.

Table 8.2: Amount of suitable plots in 2015 and 2050 and their difference.

	Amount in 2015	**Amount in 2050**	**Difference in plots amount**
Mesophilous conifers	4543	2924	− 1619
Mesophilous deciduous trees	4759	3045	− 1714
Thermophilous conifers	2458	1644	− 814
Thermophilous sclerophyllous trees	4152	3148	− 1004

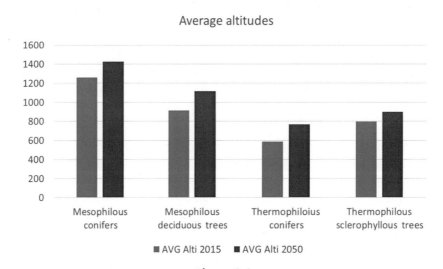

Figure 8.8
Average altitudes of the potential suitable areas in 2015 (AVG Alti 2015) and 2050 (AVG Alti 2050).

Table 8.3: Average altitude of the potential suitable areas in 2015 and 2050 and their difference.

	Average altitude 2015	Average altitude 2050	Difference in altitude 2050–2015
Mesophilous conifers	1259	1425	166
Mesophilous deciduous trees	914	1118	204
Thermophilous conifers	586	766	180
Thermophilous sclerophyllous trees	796	901	105

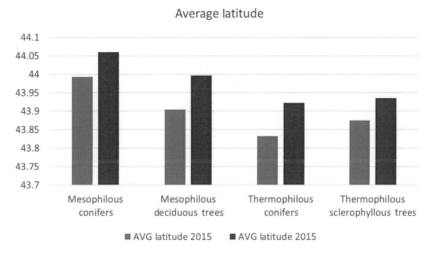

Figure 8.9
Average latitude of the potential suitable areas in 2015 (AVG Latitude 2015) and 2050 (AVG Latitude 2050).

Fig. 8.9 presents the average latitude of the potential suitable areas of the four forest types in 2015 and 2050. This figure shows that the potential suitable plots should be preferentially located in the northern part of the territory.

Table 8.4 underlines that the shift in the north part of the suitable areas in 2050 on the territory will be equal or less than 1.000 m.

Table 8.5 provides a more accurate evaluation of the spatial distribution of the suitable areas for the four forest types like the amount of new plots, the averages of altitude and latitude of the new suitable areas in 2050.

These statistics underline that the potential new plots in 2050 should be located in very high altitude and further north places, according to the current suitable areas. The natural potential spread of such species in altitude (more than 300m from current situation) towards 2050 seems to be very difficult according to the ability to tree species to colonize new

Table 8.4: Average latitude of the potential suitable areas in 2015 and 2050 and their difference.

	Average latitude 2015	Average latitude 2050	Difference in latitude 2050−15	Difference in meter
Mesophilous conifers	43.993	44.060	0.067	737
Mesophilous deciduous trees	43.904	43.997	0.093	1023
Thermophilous conifers	43.832	43.922	0.090	990
Thermophilous sclerophyllous trees	43.875	43.935	0.060	660

Table 8.5: Average altitude and latitude for the new potential plots of suitable areas in 2050.

	New plots in 2050	Average altitude	Difference in average altitude 2015	AVG latitude	Difference average latitude 2015	Difference in meter
Mesophilous conifers	209	2151	892	44.155	0.162	1782
Mesophilous deciduous trees	239	2114	1200	44.151	0.247	2717
Thermophilous conifers	409	1176	590	44.017	0.185	2035
Thermophilous sclerophyllous trees	289	1174	378	44.085	0.21	2310

Table 8.6: Amount of plots and average altitude and latitude for the potential loss of suitable areas in 2050.

	No probability in 2050	Average altitude	Difference in average altitude 2015	Average latitude	Difference in average latitude 2015	Difference in meter
Mesophilous conifers	1828	995	− 264	43.862	− 0.131	− 1441
Mesophilous deciduous trees	1953	917	3	43.84	− 0.064	− 704
Thermophilous conifers	1223	591	5	43.79	− 0.042	− 462
Thermophilous sclerophyllous trees	1293	803	7	43.795	− 0.08	− 880

Table 8.7: Amount of plots and average altitude and latitude for the suitable areas in 2015 that will remain suitable in 2050.

	Still observed in the same areas in 2050	Average altitude	Difference in altitude	Average latitude	Difference in latitude	Difference in meter
Mesophilous conifers	2715	1173	− 86	44.008	0.015	165
Mesophilous deciduous trees	2806	1158	244	44.003	0.099	1089
Thermophilous conifers	1235	712	126	43.907	0.075	825
Thermophilous sclerophyllous trees	2859	995	199	43.958	0.083	913

Table 8.8: Percent of probable plots in 2015 and still suitable in 2050.

	% Same probability	% Decrease of probability	% Increase of probability
Mesophilous conifers	3	92	5
Mesophilous deciduous trees	6	30	64
Thermophilous conifers	5	23	72
Thermophilous sclerophyllous trees	2	18	80

locations for long distances in few times (30 years). Planting campaigns could help such colonization in altitude if needed.

At the opposite, the potential loss of suitable areas in 2050 (Table 8.6) should be located in low attitude for the mesophilous conifers, but in the same average altitude for the other forest types.

This potential loss should occur in the south part of the current distribution of these forest types.

We do not present the results concerning the longitude because there are very few differences in the plots distribution in East and West parts of the territory between 2015 and 2050.

Table 8.7 shows the amount of the potential plots of suitable areas in 2015 that will remain suitable in 2050. It underlines that these plots should be located in slightly lower altitude for the mesophilous conifers, and in slightly higher altitude for the other forest types.

These probabilities of occurrence should be located in farther north part of the current distribution for the mesophilous deciduous trees, the thermophilous conifers, and for the thermophilous sclerophyllous trees.

Table 8.8 presents the percent of plots that should be still suitable in 2050 and the statistic distribution of these plots showing their trends (decrease, increase, or stability of probabilities of occurrence in 2050).

These statistics underline that, in the same suitable areas estimated both in 2015 and in 2050, the mesophilous conifers should mainly decrease (92% of the plots). At the opposite, the three other forest types should have an increase of their probabilities of occurrence, from 64% to 80% plots according to the type of vegetation structure. These latter show a slight trend of decrease of probabilities of the suitable plots, from 18% to 30%.

The four forest types show a very low proportion of stable probabilities in these plots (maximum equal to 6%), indicating that the impact of climate change may induce changes in all forest types' structure and functioning.

Fig. 8.10 underlines the differences of probabilities of suitable areas for the four forest types between 2015 and 2050. The white plots indicate a lack of probability in 2050, the orange plots correspond to a decrease of probabilities in 2050, and the red plots indicate an increase of probabilities in 2050.

For the mesophilous conifers, Fig. 8.9 indicates a potential increase of probabilities in few areas of the territory, and a potential decrease and lack of probabilities for many places. These results indicate that these species may face, in the future, a significant decrease in terms of probability of occurrence for almost all the places where currently they have a probability to find suitable areas for their growth.

For the thermophilous conifers, the results confirm a trend to a shift in the northern part of the territory in the valleys.

For the thermophilous sclerophyllous trees, the model shows a potential increase of the probabilities in the valleys and hills.

The model also indicates for the mesophilous deciduous trees a trend to find potential suitable areas in the northern part of the territory, in the valleys, and mountains.

8.2.4 Urban dynamics assessment

The assessment of the urban development is a complementary parameter in order to identify the areas that will require energy and their location in order to evaluate their proximity to the forest resources. This point is critical for the development of the wood energy supply chain because the distances have a strong influence on the cost of the resource and this is the reason why economical models take into account the distance in order to optimize the wood energy supply chain (Eliasson et al., 2017; De Meyer et al., 2014; Edel and Thraen, 2012; Kamimura et al., 2012; Alam et al., 2012; Viana et al., 2010; Kanzian et al., 2009; Ruiz et al., 2015; Grahama et al., 2000).

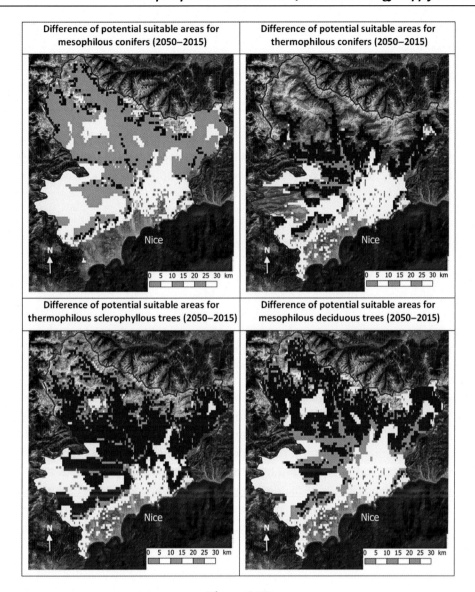

Figure 8.10
Difference of probabilities of occurrence of the suitable areas for the four forest types (in white: absence of probabilities in 2050; in orange: decrease of probabilities in 2050; and in red: increase of probabilities in 2050).

We use the results presented in Chapter 5, Geoprospective Approach for Biodiversity Conservation Taking into Account Human Activities and Global Warming (Hinojos et al.), that correspond to the assessment of the land use dynamic of the Alpes-Maritimes toward 2050. In the context of the development of the wood energy supply chain, this time range has been chosen due to the average duration of energetic systems life cycle (around 30 years of duration).

According to these simulations (see the next figures), the results show that the main location of the spread of artificial surfaces would occur on natural and agricultural areas. These changes could potentially occur in the inner part of the territory (valleys and hills) because currently the coastal strip has few surfaces free to be built.

These results have been integrated into our study in order to identify the potential surfaces and location of dense urban areas that may need to be supplied in wood in 2050 for energy purpose.

8.3 Models integration, results and discussion

Alpes-Maritimes territory gathers 36 wood biomass plants that are currently operating. Some projects are also planned in order to develop this industrial activity. Because such activity is established for a duration of 25−30 years, it is relevant to take into account the dynamics of wood resource and urban areas in order to assess the availability of wood resource until 2050 and its proximity from urban areas. This aspect is relevant because wood energy supply chain development belongs to a short supply chain in terms of distance.

The integration of the two models (model of potential suitable areas for trees and model of urban dynamics till 2050) was performed with geomatics tools allowing the overlap of the different maps of information and the extraction of the data. The scope of the model integration is to identify the location of the most potential suitable areas for trees development in 2015 and 2050, according to the spatial proximity of some urban areas in order to evaluate the feasibility of the development of the wood energy supply chain. In this frame, we also integrated the roads and pathways data for the purpose of assessing the potential accessibility for the stakeholders involved in the biomass extraction, preparation, and delivering.

Fig. 8.11 presents the comparison between 2015 and 2050 of the spatial distribution of the most suitable areas of forest development and urban areas (observed and potential).

Fig. 8.11 shows that for the current period, the most interesting areas for wood production are located close to the dense urban areas of the coastal strip, at a distance around 1−5 km which seems acceptable in terms of costs for the transportation and the preparation of the biomass. In 2050, the urban development should colonize the inner valleys but the most favorable areas for wood production should be located in the north part of the territory, which is farther from the dense urbanized areas localized in the coastal strip. Because some areas will be urbanized in the inner valleys of the territory, the supply chain should move close to these urbanized zones, near the wood production localities.

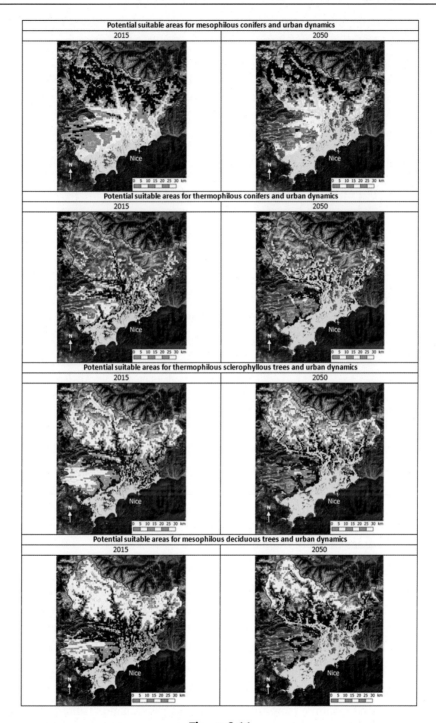

Figure 8.11
Comparison between the dense urban areas (in yellow) observed and their dynamic within 2050, and the suitability of the territory for the production of wood for the current period and 2050.

Table 8.9: Evaluation of potential suitable plots of trees with roads and pathways in 2015 and 2050.

	Plots in 2015	Plots with routes in 2015	Plots without routes in 2015	% Without roads in 2015	Plots in 2050	Plots with routes in 2050	Plots without routes in 2050	% Without roads in 2050
Mesophilous conifers	4543	4260	− 283	6	2924	2629	− 295	10
Mesophilous deciduous trees	4759	4476	− 283	6	3045	2739	− 306	10
Thermophilous conifers	2458	2408	− 50	2	1644	1589	− 55	3
Thermophilous sclerophyllous trees	4152	3977	− 175	4	3148	2967	− 181	6

Table 8.10: Evaluation of the distance from the potential suitable plots of trees with urban areas in 2015 and 2050.

	Urban 2015		Urban 2050		Difference 2015−2050			
	1 km	5 km	1 km	5 km	1 km	%	5 km	%
Mesophilous conifers	914	8450	897	3541	− 17	− 2	− 4909	− 58
Mesophilous deciduous trees	1358	11,300	966	3706	− 392	− 29	− 7594	− 67
Thermophilous conifers	1279	10,024	855	2698	− 424	− 33	− 7326	− 73
Thermophilous sclerophyllous trees	1349	11,077	1262	4668	− 87	− 6	− 6409	− 58

Table 8.9 shows that even if there should be a decrease of the potential suitable plots for trees development in 2050, these plots should stay slightly constant in terms of road and pathways inside the plots (only from 1% to 4% of plots more without roads and pathways in 2050). This point is relevant in order to maintain an efficient accessibility to the forest resources near the areas of energy demand.

Table 8.10 presents the amount of urbanized areas located at a distance of 1 km and a distance of 5 km from each of the four forest types. It underlines that it should be a significant decrease of forest resources (mesophilous deciduous trees, and thermophilous conifers, respectively, 29% and 33% of decrease) at 1 km from the urbanized area in 2050, and a more significant decrease at 5 km for the four forest types (from 58% to 73%).

This result underlines that the potential decrease of the suitable areas of trees development assessed in 2050 and underlined in Figs. 8.5 and 8.6 and in Table 8.7 will affect the proximity of the wood resource from the zones of energy demand. Table 8.8 has also shown that the future suitable areas will be more located in high altitudes, where the slopes are steeper than in lower altitudes in the Alpes-Maritimes.

There is another parameter to take into account for the development of such activities of wood extraction for energy purpose: the presence of protected areas like the National Park of the Mercantour (north part of the territory) and the Regional Park of the "Préalpes d'Azur" (south-west part of the territory). These natural parks get a regulation that constrains human activities inside their own territories, like forestry practices, in order to keep the natural dynamic of forest. Sometimes, the managers of protected areas allow the tree cutting in order to maintain some open landscapes or for phytosanitary objectives (Fig. 8.12).

Currently, the supply chain is not constrained by the presence of protected areas because the resource is located in other parts of the territory. But, according to the potential move of the potential suitable areas for wood production in the north part of the territory and in inner valleys where more protected areas are localized, and due to the reduction of potential suitable plots for 2050, the supply chain could face a potential reduction of resource availability in 2050.

Table 8.11 presents that the amount of potential suitable plots located in protected areas for the current climate is around 33% for all the forest types.

For 2050, due to the decrease of potential suitable areas, the amount of suitable plots should have an average of 27% inside the two main protected areas of the territory. This result also underlines that almost the same proportion of the potential suitable areas for wood production should be still located in protected areas in 2050.

Bellarby et al. (2010), Barney and Di Tomaso (2010), and Tuck et al. (2006) provided the assessment of climate change impact on bioenergy crops in Europe and North America. Their methodology based on the characterization of bioclimatic envelops in order to identify potential suitable or not suitable areas for the growth of species. Tuck et al. (2006) argued that European Mediterranean places will be less suitable for some bioenergy crops species due to climate change effects by the 2080s. Even if the development of wood energy supply chain is more related to the natural development of trees, we observe similarities in the impact of climate change on natural vegetation and its productivity.

The optimization of wood energy supply chain is largely based on the integration of parameters like distance from wood resource to the road network, physical parameters like slopes or costs of biomass transportation and processes to prepare the biomass (Eliasson et al., 2017; De Meyer et al., 2014; Edel and Thraen, 2012; Kamimura et al., 2012; Alam et al., 2012; Viana et al., 2010; Kanzian et al., 2009; Ruiz et al., 2015; Grahama et al., 2000). Multicriteria approaches in forest planning have been developed since the end of the 1980s (Mendoza and Spouse, 1989) and improved by using different mathematical technics and integrating more parameters (Diaz-Balteiro and Romero, 2008; Kangas et al., 2015).

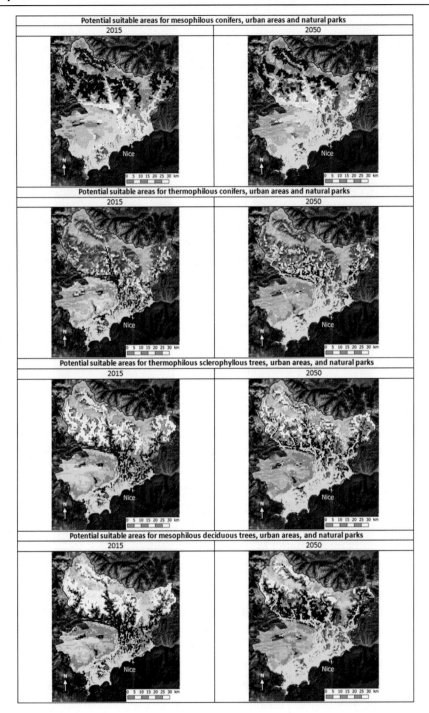

Figure 8.12
Maps of the most suitable areas for the four forest types and urbanized areas in 2015 and 2050 and the location of the National Park of the Mercantour and the Regional Park of the "Préalpes d'Azur" (light green).

Table 8.11: Amount of potential suitable plots for forest types in protected areas in 2015 and 2050.

	Plots in 2015	Plots inside protected areas in 2015	% Plots in protected areas in 2015	Plots in 2050	Plots inside protected areas in 2050	% Plots in protected areas in 2050	Difference 2050−2015
Mesophilous conifers	4543	1544	34	2924	792	27	− 752
Mesophilous deciduous trees	4759	1545	32	3045	835	27	− 710
Thermophilous conifers	2458	799	33	1644	459	28	− 340
Thermophilous sclerophyllous trees	4152	1354	33	3148	854	27	− 500

The originality of our study and methodology is based on the inclusion of dynamic processes of ecological and human parameters toward 2050 for the purposes of determining the potentiality of the development of the supply chain. First, we determine the potential dynamic of forest types by taking into account the effect of the climate change on trees spatial distribution. This approach is based on a climatic characterization of plants in France by using a probabilistic calibration which allows performing bioindicators of climatic variables in a very accurate way (75% of accuracy of calibration). Second, we integrate the potential distribution of the urban areas in order to identify the future areas where the energy demand will be located with a validation step giving 90% of accuracy of the model behavior to assess the potential areas suitable for urbanization.

This approach tries to encompass two major parameters for the development of the wood energy: the availability of the resource according to the potential impact of the global warming on forest ecosystems and the potential development of the urbanization in order to identify the potential areas of energy demand. The integration of the results of these two models underlines the need of a geoprospective approach in order to assess the feasibility of developing the wood energy supply chain in a changing world.

8.4 Conclusion

In the current context of global warming and according to other authors mentioned in this article, the results underline that climate change may affect the structure and function of Mediterranean forest types. This impact could affect the mortality rate of trees due to the water stress and it could also induce the substitution of mesophilous species by xerophilous and thermophilous species in many places in the studied area (Alpes-Maritimes, France). These changes could have an impact on the availability of biomass resource for the wood supply chain. ONERC (2015) also argues that this water stress may affect trees deemed adapted to the Mediterranean climate. This point emphasizes the need for forest managers to adapt their management methods to climate change (Lindner et al., 2014; Keenan, 2015) in order to promote a resilience strategy of Mediterranean forests. Although studies show that in some regions climate change may have an effect to stimulate tree growth and development, the majority of studies show a general increase in risks to forests (Lucier et al., 2009). Local studies in the East Mediterranean area of France (GREC-PACA, 2018) have shown that there is already an increasing decline in stands of fir (*Abies alba*), Scots pine (*Pinus sylvestris*), and pubescent oak (*Quercus pubescens*).

Our study also focuses on the potential dynamics of urbanized areas. The simulation of urban development shows a potential evolution in the North part of the territory until 2050. This result asks the question of the sustainability of energy systems based on woody biomass in such areas according to the potential availability of wood resource. It also raises the problem of the definition of adaptation strategies that have to be applied in order to promote the resilience of the Mediterranean forest close to these areas. The main private and public stakeholders affected by the development of wood energy sector can use some of these results for establishing their risk assessment and for defining the management strategy of their activities and territories.

References

Alam, Md.B., Pulkki, R., Shahi, C., Upadhyay, T.P., 2012. Economic analysis of biomass supply chains: a case study of four competing bioenergy power plants in Northwestern Ontario2014 Int. Sch. Res. Netw. 2012, 1–12.

Badeau, V., Dupouey, J.L., Cluzeau, C., Drapier, J., Le Bas, C., 2004. Modélisation et cartographie de l'aire climatique potentielle des grandes essences forestières françaises, Rapport final du projet CARBOFOR— Séquestration de carbone dans les grands écosystèmes forestiers de France, Tâche D1, Ecofor, 138.

Barney, J.N., Di Tomaso, J.M., 2010. Bioclimatic predictions of habitat suitability for the biofuel switchgrass in North America under current and future climate scenarios. Biomass Bioenergy 34, 124–133.

Bellarby, J., Wattenbach, M., Tuck, G., Glendining, M.J., Smith, P., 2010. The potential distribution of bioenergy crops in the UK under present and future climate. Biomass Bioenergy 34, 1935–1945.

Cavaud, D., Coléou, Z., Guggemos, F, Reynaud, D., 2017. Chiffres clés des énergies renouvelables, In: MEED (Ed.), Edition 2016, Service de L'observation et des Statistiques (SOeS), p. 76.

Charru, M., Seynave, I., Hervé, J.-C., Bertrand, R., Bontemps, J.-D., 2017. Recent growth changes in Western European forests are driven by climate warming and structured across tree species climatic habitats. Ann. For. Sci. 74, 33p.

Ciais, P., et al., 2005. Europe-wide reduction in primary productivity caused by the heat and drought in 2003. Nature 437 (7058), 529–533.

CIBE, 2017. Retour sur la saison de chauffe: bilan sur l'approvisionnement. Enjeux et perspectives: Recensement des installations, Bilan du Fond Chaleur, Valorisation des cendres. Séance plénière du Comité interprofessionnel du Bois Energie (CIBE), p. 22.

De Meyer, A., Cattrysse, D., Rasinmäki, J., Van Orshoven, J., 2014. Methods to optimize the design and management of biomass-for-bioenergy supply chains: a review. Renew. Sustain. Energy Rev. 31, 657–670.

Diaz-Balteiro, L., Romero, C., 2008. Making forestry decisions with multiple criteria: a review and an assessment. For. Ecol. Manag. 255, 3222–3241.

Edel, M., Thraen, D., 2012. The economic viability of wood energy conversion technologies in Germany. Int. J. For. Eng. 23 (2), 102–113.

Eliasson, L., Eriksson, A., Mohtashami, S., 2017. Analysis of factors affecting productivity and costs for a high-performance chip supply system. Appl. Energy 185, 497–505.

Fink, A.H., Brücher, T., Krüger, A., Leckebusch, G.C., Pinto, J.G., Ulbrich, U., 2004. The 2003 European summer heatwaves and drought? Synoptic diagnosis and impacts. Weather 59 (8), 209–216.

Fischer, E.M., 2007. The Role of Land–Atmosphere Interactions for European Summer Heat Waves: Past, Present and Future. ETH Zurich, p. 167.

Garbolino, E., Daniel, W., Hinojos-Mendoza, G., 2018. Expected global warming impacts on the spatial distribution and productivity for 2050 of five species of trees used in the wood energy supply chain in France. Energies 11 (3372), 2–17. Available from: https://doi.org/10.3390/en11123372.

Garbolino, E., Daniel, W., Hinojos-Mendoza, G., Sanseverino-Godfrin, V. 2017. Anticipating climate change effect on biomass productivity and vegetation structure of Mediterranean Forests to promote the sustainability of the wood energy supply chain. In: Ek, L., Ehrnrooth, H., Scarlat, N., Grassi, A., Helm, P. (Eds.), EUBCE-25th European Biomass Conference and Exhibition, "Setting the course of a biobased economy," Stockholm, Sweden, June 12–15, 2017, pp. 17–29.

Garbolino, E., De Ruffray, P., Brisse, H., Grandjouan, G., 2007. Relationships between plants and climate in France: calibration of 1874 bio-indicators. C. R. Biol. 330, 159–170.

Garbolino, E., De Ruffray, P., Brisse, H., Grandjouan, G., 2012. The phytosociological database SOPHY as the basis of plant socio-ecology and phytoclimatology in France. Biodivers. Ecol. 4, 177–184.

Garbolino, E., Sanseverino-Godfrin, V., Hinojos-Mendoza, G., 2016. Describing and predicting of the vegetation development of Corsica due to expected climate change and its impact on forest fire risk evolution. Saf. Sci. 88, 180–186.

Garbolino, E., 2014. Les Bio-indicateurs du Climat: Principes et Caractérisation. Presses des MINES, p. 129, Développement durable.

Grahama, R.L., English, B.C., Noon, C.E., 2000. A Geographic Information System-based modeling system for evaluating the cost of delivered energy crop feedstock. Biomass Bioenergy 18, 309–329.

GREC-PACA, 2018. Impacts du changement climatique et transition(s) dans les Alpes du Sud. Cahier thématique du groupe de travail « Montagne ». GREC-PACA, p. 48.

Hinojos, G., Garbolino, E., Godfrin, V., 2012. Urban dynamics simulation and climatic change scenarios for 2050 in order to identify areas of biodiversity's conservation in the French's Maritime Alps. In: 11th Urban Environment Symposium, September, 16–19, 2012, Karlsruhe, Germany, p. 44.

IGN, 2018. La forêt française: Etat des lieux et évolutions récentes. In: Panorama des résultats de l'inventaire forestier. Institution National de L'information Géographique et Forestière, Paris, p. 56 (in French).

IGN, 2017. Inventaire Forestier. Le Mémento, p. 30, édition 2017.

IPCC, 2014. Climate Change 2014: Impacts, Adaptation, and Vulnerability, Part B: Regional Aspects. Cambridge University Press, p. 696.

Iverson, L.R., Prasad, A.M., Hale, B.J., Sutherland, E.K., 1999. Atlas of Current and Potential Future distributions of Common Trees of the Eastern United States. United States Department of Agriculture, Forest Service, Northeastern Research Station, General Technical Report NE-265, p. 125.

Kamimura, K., Kuboyama, H., Yamamoto, K., 2012. Wood biomass supply costs and potential for biomass energy plants in Japan. Biomass Bioenergy 36, 107–115.

Kangas, A., Kurttila, M., Hujala, T., Eyvindson, K., Kangas, J., 2015.), Decision Support for Forest Management (Managing Forest Ecosystems, vol. 30. Springer, p. 307.

Kanzian, C., Holzleitner, F., Stampfer, K., Ashton, S., 2009. Regional energy wood logistics—optimizing local fuel supply. Sil. Fenn. 43 (1), 113–128.

Keenan, R.J., 2015. Climate change impacts and adaptation in forest management: a review. Ann. For. Sci. 72, 145–167.

Lindner, M., Fitzgerald, J.B., Zimmermann, N.E., Reyer, C., Delzon, S., van der Maaten, E., et al., 2014. Climate change and European forests: what do we know, what are the uncertainties, and what are the implications for forest management? J. Environ. Manag. 146, 69–83.

Lucier, A., Ayres, M., Karnosky, D., Thompson, I., Loehle, C., Percy, K., et al., 2009. Forest responses and vulnerabilities to recent climate change. In: Seppälä, R., Buck, A., Katila, P. (Eds.), Adaptation of Forests and People to Climate Change: A Global Assessment Report, Vol. 22. IUFRO, Helsinki, pp. 29–52, World Series.

Mendoza, G.A., Spouse, W., 1989. Forest planning and decision making under fuzzy environments: an overview and illustration. For. Sci. 35, 481–502.

Niua, S., Luob, Y., Li, D., Caod, S., Xiab, J., Li, J., et al., 2014. Plant growth and mortality under climatic extremes: an overview. Environ. Exp. Bot. 8, 13–19.

ONERC, 2015. L'arbre et la forêt à l'épreuve d'un climat qui change. Rapport au Premier ministre et au Parlement. La documentation Française, p. 181.

Reichstein, M., 2005. Severe Impact of the 2003 European Heat Wave on Ecosystems. Potsdam Institute for Climat Impat Research. <https://www.pik-potsdam.de/news/press-releases/archive/2005/severe-impact-of-the-2003-european-heat-wave-on-ecosystems>.

Ruiz, P., Sgobbi, A., Nijs, W., Thiel, C., 2015. The JRC-EU-TIMES Model. Bioenergy Potentials for EU and Neighbouring Countries. JRC Science for Policy Report, European Union, p. 172.

Talandier, M., Jousseaume, V., Nicot, B.-H., 2016. Two centuries of economic territorial dynamics: The case of France. Reg. Stud. Reg. Sci. 3, 67–87.

Thuiller, W., 2003. BIOMOD — optimizing predictions of species distributions and projecting potential future shifts under global change. Global. Change Biol. 9, 1353–1362.

Thuiller, W., Lavorel, S., Sykes, M.T., Araújo, M.B., 2006. Using niche based modelling to assess the impact of climate change on tree functional diversity in Europe. Divers. Distrib. 12, 49–60.

Tuck, G., Glendining, M.J., Smith, P., House, J.I., Wattenbach, M., 2006. The potential distribution of bioenergy crops in Europe under present and future climate. Biomass Bioenergy 30, 183–197.

Vennetier, M., Ripert, C., 2009. Forest flora turnover with climate change in the Mediterranean region: a case study in South-Eastern France. For. Ecol. Manag. 258S, 56–63.

Viana, H., Cohen, W.B., Lopes, D., Aranha, J., 2010. Assessment of forest biomass for use as energy. GIS-based analysis of geographical availability and locations of wood-fired power plants in Portugal. Appl. Energy 87, 2551–2560.

Appendix A List of the 25 trees species distributed into the main four woody forest types

Taxa and forest types
Mesophilous conifers
 Abies alba Mill.
 Larix decidua Mill.
 Picea excelsa (Lam.)
 Pinus cembra L.
 Pinus sylvestris L.
Mesophilous deciduous trees
 Acer campestre L.
 Acer monspessulanum L
 Acer opalus Miller
 Carpinus betulus L.
 Celtis australis L.
 Fagus silvatica L.
 Fraxinus excelsior L.
 Fraxinus ornus L.
 Ostrya carpinifolia Sco.
 Populus alba L.
 Populus nigra L.
 Populus tremula L.
 Quercus lanuginosa Lam.
 Quercus pedunculata Her.
 Quercus sessiliflora Sa.
Thermophilous conifers
 Pinus halepensis Mill.
 Pinus pinaster Soland.
Thermophilous sclerophyllous trees
 Arbutus unedo L.
 Quercus ilex L.
 Quercus suber L.

CHAPTER 9

Simulating together multiscale and multisectoral adaptations to global change and their impacts: A generic serious game and its implementation in coastal areas in France and South Africa

Bruno Bonté[1], Clara Therville[1,2], François Bousquet[3,4], Cédric Simi[1], Géraldine Abrami[1], Chloé Guerbois[5,6], Hervé Fritz[5,6], Olivier Barreteau[1], Sandrine Dhenain[1,7] and Raphaël Mathevet[2]

[1]G-EAU, Univ Montpellier, AgroParisTech, CIRAD, IRD, INRAE, Institut Agro, Montpellier, France, [2]CEFE, CNRS, Univ. Montpellier, EPHE, IRD, Univ. Paul Valéry Montpellier 3, Montpellier, France, [3]CIRAD, UPR GREEN, F-34398 Montpellier, France, [4]GREEN, Univ Montpellier, CIRAD, Montpellier, France, [5]Sustainability Research Unit, Nelson Mandela University-. Madiba drive, George, South Africa, [6]REHABS International Research Lab, CNRS-Université Lyon 1-NMU, Nelson Mandela University-. Madiba drive, George, South Africa, [7]TEC Conseil, Marseille, France

Chapter Outline

Ecosystem and Territorial Resilience.
DOI: https://doi.org/10.1016/B978-0-12-818215-4.00009-2
© 2021 Elsevier Inc. All rights reserved.

9.1 Introduction

In a changing world, ensuring territorial and ecosystem resilience requires perpetual adaptations. Polycentric governance have been acknowledged as robust and resilient (Ostrom, 2001, 2005) and, indeed, human activities can be split up in different sectors of activities interacting among themselves but having different stakes and dynamics. Regions are organized in administrative territories and subterritories, all equipped with governance structures that implement land-use zoning and development plans providing a scope for their developments but which need to comply with and adapt to the intrinsic dynamics of each sector of activity. In order to organize the consistency of multilevel and multisectoral adaptations, master plans are nowadays being built by governance structures composed of representatives of various sectors of activities and representatives of various levels of governance.

The MAGIC international research project focused on multiscale adaptations to global change and their impacts in coastal areas. The project included two case studies: a coastal area in Languedoc, France and another coastal area in the Garden Route, South Africa. In the two case studies, we could identify these master plans dedicated to large-scale and long-term territorial coordination. These plans are supposed to give guiding principles on important sectors for territorial development in the next decades. For the two case studies, the plans involve several local administratives. They provide diagnosis and recommendations for agricultural development, urban and demographical growth, and environment and biodiversity. However, the various sectors considered in these plans have their own dynamics and objectives; moreover, these dynamics and objectives within a sector may be different from place to place within the area of the plan. And still, these sectors and places impact each other in some inextricable ways typical of complex systems so that they need to be considered together in long-term plans. They share the same limited resources of the territory (water, space, etc.) and sometimes some human-made infrastructures such as roads. Some sectors may also directly depend on other ones such as natural tourism that depends on the environmental quality. The internal dynamics of a sector may also depend locally on the internal dynamics of another sector. For instance, extensive local high-quality farming production systems will not remain viable if the local urban sector does not provide enough consumers sensible to, and able to afford, these kinds of products. Hence, building these plans requires to identify the external risks, but also the internal trends and objectives of the major sectors, in the various places of the area. People responsible for these plans may feel helpless in their missions because of the multiplicity of points of view and dynamics to account to and because of the close interactions of these dynamics.

In order to improve the integration of multilevel and multisector adaptations, we imagined a role-playing game (RPG) framework based on the Social and Ecological System (SES) robustness analysis framework. The SES robustness analysis framework proposes to see an SES as users using resources through Infrastructures provided by infrastructure providers and we wanted to see if it could be used as a metamodel to build specific RPGs. We based our work on a first experience during which we validated and enhanced our framework when using it to specify a board-game RPG representing the French Case study. In this game, we took into account existing planning and collective decision instances at different scales (the whole area, subterritories) and for the different sectors. We were asked to organize a game session for the reviewing of the SCoT Sud Gard and showed that this session, understood as a geoprospective exercise, could in fact reproduce, and help to discuss, potential vulnerability transfers across sectors, across places, and across scales (Bonté et al., 2019).

In this chapter, we present the generic computerized device that was developed to implement our generic RPG Framework and that can be used to specify local instances of games based on this framework. It is called "Playners" (contraction of planners and players). We first detail the territorial stakes of the case studies of the project and explain what specific kind of geoprospective tool is needed in such situations. Then, we describe our platform "Playners" and explain in detail the set of conceptual and software system tools used so that the global methodology can be fully integrated by the reader. We then explain how we used this device to reproduce the French game in a computerized form and to create a South African version of the game based on the MAGIC project case study. We used five game sessions of the different versions which were played with stakeholders in various places to provide feedbacks about the use of such devices. We finally discuss the contributions of this work to geoprospective in general.

9.2 Multiple stakes of two coastal territories adapting to global change

9.2.1 Case study overviews

9.2.1.1 French case study: Languedoc

The study area is located in the Gard "département," belonging to the Occitanie/Languedoc Roussillon Region. The game was developed for the reviewing of the Scheme of Territorial

Consistency (SCoT) of a littoral territory of the Gard "département" located in the South of France. SCoT is a mandatory French urban planning document specifying 15 years objectives and reviewed every 5 years. Thus the gaming area fits to the territory covered by this plan and that we hereafter called its French name: the SCoT Sud Gard. The SCoT Sud Gard is formed of seven federations of municipalities (represented in Fig. 9.1) that represent 81 municipalities and around 400,000 inhabitants in an area of 1700 km². Half of the territory is made up of agricultural land. The other part is composed of natural land (36%) and urban land (14%). The extreme south of this territory is a coastal area that includes lagoons with serious flood risk.

9.2.1.2 South African case study: Garden Route

The area represented by the game is George Municipality, a coastal municipality typical of the Garden Route, characterized by sectorized urban and agricultural areas, protected areas, and a diversity of ecological infrastructures such as rivers, estuaries, and foredunes. The George Municipality Integrated Development Plan that we called hereafter George IDP specifies is a 5-year plan based on 10−15 years projections.

George Municipality extends on an area of 5191 km², divided into 23 wards (represented in Fig. 9.1) and host almost 200,000 inhabitants, with 80% of the population residing in the town of George, 12% of the population is in other towns, and only 8% in the rural area. Urbanization, demographic growth (6.9%/year), and high level of poverty are the main issues of this area (George Municipality Integrated Development Plan, 2017; Spatial Development Plan, 2013).

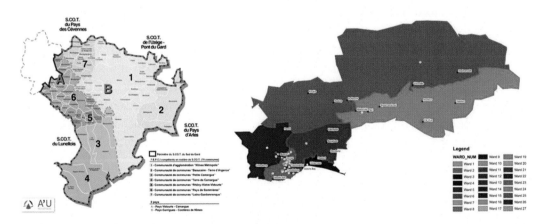

Figure 9.1
Administrative subdivisions of case studies. Intercommunalities (colors) of SCoT South of Gard perimeter (Syndicat mixte du SCoT Sud Gard, in prep) on the left. Wards of George Municipality perimeter (Draft Integrated Development plan).

The Garden Route is famous for its landscapes and attracts many tourists especially during the summer. The Wilderness National Park, attracted 101,938 tourists in 2015 (Wesgro Survey, 2016) and includes three lakes and a temporarily open and closed estuary classified as Ramsar sites. The area is also part of the Garden Route Biosphere Reserve proclaimed in 2017. Shortage of water, fire, and flood risks are important drivers of the future of the area.

9.2.2 Rich and fragile natural environments

The French study area is located along a major European axis, between Spain and northern Europe. It is characterized by the Mediterranean climate and is one of the world hotspot of biodiversity. The main landscape types which occur in this area are flat coastal landscapes with sandy coastal line, lagoons, salt pans, marshes and dunes, extensive "garrigues" and vast areas of vines in the lowland plains, mosaic landscapes of cultivated areas, and a limestone plateau of "garrigues" with green valleys of coastal rivers. These coastal wetlands, lagoons, and Mediterranean cultural landscapes, shaped by an age-old history of human presence, are characterized by high levels of complexity and diversity and by an important biological and esthetical interest. In the 1960s, the coastline was submitted to important land-planning measures ordered by the State, with the construction of seaside resorts to promote mass market tourism development. Today, the area has a fast demographic growth mainly in cities, along the main roads, and next to the shoreline, including in seaside resorts. These new inhabitants generate an important urban sprawl and artificialization process even if agriculture is still important in the area, shaping the landscapes and contributing to the local economy.

In the South African case study, we observe a high level of endemism in biomes (Fynbos, afromontane forests, estuaries...) and species (Outeniqua Yellowwood, Proteas, Knysna seahorse...). In terms of natural resources, this area has been overexploited in the 19th century to sustain the demand associated with urban development which resulted in the ban of forest exploitation in the mid-1900s (Mathevet et al., 2018). The introduction of exotic species for wood production such as gum trees (eucalyptus sp.), black wattle (acacia sp.) and the escape of commercial species (eucalyptus sp., pinus sp.) from forestry plantations is today considered as a major threat for biodiversity and landscapes but also for environmental risks such as water shortage and fires. Flood risk management and urbanization also had a great impact on local biodiversity affecting natural dynamics at the land—ocean interfaces. Estuaries in the area both the most threatened and the most important in terms of biodiversity. More than 30% of the Garden Biosphere Reserve is formally protected; hence, tourism constitutes an important economic sector in terms of land uses in the Garden Route. However, this sector remains extremely vulnerable to environmental changes and natural hazards such as drought, floods, and fires. Ecosystem-based adaptation is being advertised as the way forward to respond to global change but concrete action plans at local levels are still to be developed (Guerbois et al., 2019).

In both case studies, environmental risks are a major driver of land-use planning. Planners also face water management issues, with a double constraint: providing drinkable water for permanent residents, newcomers, and tourists, as well as freshwater for irrigation, rivers, and wetlands, especially in the context of climate change and sea level rise associated with the salinization of coastal lands and groundwater tables. Both areas are historically submitted to important risks such as seashore erosion and submersion along the coast, and flash flood and fires in the scrublands, and has an important background in terms of risk management policies (Therville et al., 2019; Guerbois et al., 2019; Quinn et al., 2018, 2019). In the context of urbanization, land abandonment versus intensification, wetland drainage or management specialization versus traditional multiuses, protected areas' establishment, critical water management issues, road and train infrastructure development, gentrification processes, demographic growth, and land market price change, all these changes lead to the implementation of several tools and policies aiming to identify and map ecological networks, ecosystem services within the context of national and (European for the French case study) biodiversity conservation policies.

9.2.3 A diversity of inhabitants, place meanings, and place attachments

With a particularly large population, the French study area is experiencing strong demographic and artificialization pressures. Urban areas have expanded, often located in areas exposed to flood risk, and the associated land-cover transformations have increased and amplified rates of surface water run-off. As a result, migration patterns have created new configurations of exposure and vulnerability to flood risk. The density per square kilometer varies and reveals an unequal distribution of the population mainly concentrated along the axis connecting the agglomerations of Montpellier and Nîmes and in the seaside resorts. There are different patterns of mobility. People from Italy, Spain, Morocco, and Eastern Europe have been immigrating for the last century, looking for jobs and they settled in the region. Wealthy retirees move from northern parts of France at the moment of retirement. They come for the amenities such as the sun, the sea, the landscape and are also demanding on services for health and care. The connection to Paris by fast train and European cities by plane also allowed the settlement of people who do not want to leave in big cities. They want to be connected and benefit from the landscape for activities in nature, healthy food, and high education for their children. Last but not least, the tourists. In the years 1960, the development of this region was planned to attract tourists and compete with the Spanish Costa Brava.

With a smaller population, the South African case study is also subjected to a high demographic growth rate of 2.9% according to the 2011 South African National Census. In a survey realized in 2016 in order to study sense of place of residents of the study area, researchers of the MAGIC projects identified a plurality of social and economical categories that they relate to different place meanings and place attachments (Quinn et al., 2019) with high spatial segregation showing, for instance, that 63% of the coastline properties are owned by retired people and 89% by white South Africans (Guerbois et al., 2019).

In both case studies, we find different profiles of residents with tourists and newcomers living together with the so-called local people who have been developing a local culture and lifestyle. They have different attachment to places and different wishes for the development of the region (Quinn et al., 2018, 2019).

9.2.4 A dynamic and challenged agricultural sector

The 2007 SCoT Sud Gard diagnosis distinguishes seven agricultural entities in the SCoT territory, highlighting both the spatial segregation and the diversity of the territory where some places are historically specialized in vineyard and extensive ranching with protected designation of origin for vineyard, olive oil, cheese, and bull and some places also dedicated to cereal large-scale farming (SCoT Sud Gard, 2007).

In the Garden Route, forestry has historically been the main agricultural sector in terms of employment, followed by dairy farming. However, these production systems are challenged by a decade of drought and are now exiting, with thousands of hectares being reallocated for conservation. Estimates are that the forestry industry's contribution to the George economy has declined from 80% in the 1980s to a mere 15% today. Small-scale farming (hop farms, orchid, vegetable, honey production) also contributes to the sector. Organic small-scale food production systems which are meant to be more waterwise are being developed but the sector is not yet self-sufficient at the moment.

In both case studies, agriculture is facing climate change, competing for land with urban sprawl and biodiversity conservation.

9.2.5 Adaptations to global change

In Languedoc and in the Garden Route, climate change is highly intricate with those other issues at stake. These coastal areas are facing multiple and combining challenges: demographic growth, economic difficulties, land transformation, evolution of biodiversity, and a multitude of coastal risks such as erosion, submersion, and flooding converge and challenge the capacity of local communities, organizations, and institutions to adapt. Since 2007, French local authorities legally have to elaborate climate change adaptation measures. All authorities larger than 20,000 inhabitants are now legally bound to develop local climate change adaptation strategies and action plans. In South Africa, the down-scaling climate change issues to local authorities are harder (Pasquini et al., 2015) but there is a political will to do so (Western Cape Province, 2017).

Therefore, in both case studies, local decision-makers and technicians are at the forefront of adaptation. They manage adaptation through specific actions mainstreamed into existing plans related to urban planning, floods and water management, or through specific climate change adaptation plans.

9.3 Specific kind of geoprospective tool for long-term adaptations

Von Korff et al. (2012) showed in the field of water management that quality decisions can be achieved by bringing together and integrating during discourse different types of knowledge. Participatory methods can at least build social capital by having participants interacting intensively and differently. As different situations and different steps in the decision process require different kinds of tools, Ferrand et al. (2017) advocate for "coupling for coping." They explain that all sorts of segmentation (sector, administrative, disciplinary...) are forms of protective specialization, driven by optimization and control issues, which restrain adaptive capacity and synergies. Consequently, they aim at providing support tools for coupling actors, scales, sectors, solution types, decision steps, etc. through an integrated set of transferable tools ranging from monitoring and evaluation to modeling and simulation and planning. In this line, we believe that middle-term to long-term adaptation to global change needs a dedicated tool. In order to have a place in the democratic processes, these tools need to be generic for many reasons: they can be explicitly mentioned in regulatory documents, they can be easily (understand non-expensively) transferred, learned, and appropriated in several places, feedbacks can be used from many places. However, these tools also need to be adaptable to specific situations so that we ensure that as little assumption as possible is brought from outside of the participatory process (Barreteau, 2003).

At the end of the MAGIC project, in regard to the stakes and local specificities within each case study (agricultural territories, spatially and culturally segregated residents, specific habitats of flora and fauna, local governance, etc. as presented in Section 9.2), we identified a need for a specific tool for a specific kind of RPG dedicated to the participatory analysis of long-term adaptation to global change that would integrate spatial, multisector, and multilevel dimensions of the problem. This tool should at least be able to represent the adaptations logics that we observed on the field.

Dhenain (2018) analyzed adaptation actions in Languedoc and discriminated four types of policies with different political logic have already been developed in face of global change (Smit and Wandel, 2006):

1. Engineering and technological solutions (dikes, hydraulic infrastructures, etc.) are built to protect territories and inhabitants against natural hazards. Local authorities want to protect economic zones and current development pathways linked to mass tourism on

the coastline, to secure current way of life, to "control and maintain" those pathways. But financial costs are high for those kinds of infrastructures, whereas public finances decrease and there is a risk that problems are displaced in time and space.

2. Vulnerability assessment and website interfaces dedicated to climate change constitute the second type of adaptive actions. The idea is to shed a light on the vulnerability of the territory, its exposure, its sensitivity. Those actions aim at highlighting relevant indicators, making visible the invisible changes so that municipalities or individuals take themselves the responsibility to implement actions (Epstein, 2012). Experts play a major role. Accountability is moved from the national or regional level to municipalities, local levels, or individuals.

3. The third type of adaptive actions consists in minor adjustments to "regulate" and absorb the major changes. Decision-makers elaborate soft measures to regulate land consumption in urban master plans, negotiate water resources reallocations between users in water management plans, or use ecological engineering against coastal erosion. The idea is to accept the risks and absorb the effects. The focus is to avoid the system to collapse, to support the resilience of the territory and its adaptability through new compromises, new equilibrium for resources, new trade-off. This can lead to localism with horizontal governance, shared accountabilities, but also to spatial inequities and transfer of vulnerabilities.

4. Relocations of economic activities and housing from the coastline to the hinterland, reconfiguration of economic pathways (from mass tourism to other types of economic activities) are the last type of adaptive actions. Losses of economic, ecological areas are accepted and lead to transformations of the territory. The focus is to find new structures of economic developments, with, for example, complementary activities between the coast, and the territories in the hinterland. Those types of actions are redistributive. Collaborations between many institutions and decision-makers at different levels are required. Political and financial costs are high and these processes are complex.

The tool we designed aims at providing words and experience to help discuss and understand the interactions between the various stakes and dynamics involved in these adaptations. We see that ex ante evaluation of adaptations action of types 2, 3, and 4 requires to take into account cross-sector and cross-scales interactions. Hence, the RPG designed must be abstract enough to consider long-term scenarios but also be contextualized enough to account for local socioeconomic (residents, agriculture, tourism), and environmental stakes in their heterogeneity (different types of people with different values and economic conditions, different types of agriculture) and complexity (mutual interactions between the stakes). Considering long time scales and corresponding large spatial scale, we also need to account for the multiscale aspect of decision-making and adaptation. As shown in Section 9.2.1, the SCoT Sud Gard territory and the George municipality territory are composed of several watersheds and subterritories that may have objectives or adaptation logics incompatible with other levels of scales (Daniell and Barreteau, 2014).

9.4 Tools and methods: How we used them and what did they bring to us?

9.4.1 Building up the generic Playners device

We justified above our ambition to provide a generic tool that can be specified to various places. To that we need dedicate tools. In our geoprospective approach, we consider the game device as a spatialized, dynamical model of the area under study. Furthermore, we consider the Playner device as a tool to integrate different issues: impacts of a place to another place, impact of a sector to another sector, path dependency, etc. Building such a generic tool, and making it available to the scientific and societal community, requires the use of existing frameworks and tools at various levels of abstraction. In this section, we try to present in the most unambiguous way, the tools and frameworks we used. We mainly rely on the RPGs that are powerful methods to engage stakeholders, the Multi-Agent System (MAS) paradigm that is commonly used in Companion modeling approaches, and the Coupled Infrastructure System framework that we used as a boundary object.

9.4.2 Using the robustness framework as a boundary object

Building such a geoprospective game to explore together multiscale and multisectoral adaptations to global change across several case studies requires the use of a common framework of design and analysis. Numerous frameworks exist to facilitate the communication among scholars who are all working on the sustainability of complex SESs, but who are also adopting different disciplinary perspectives and are interested by multiple territorial issues (Binder et al., 2013). These frameworks help to cumulate their knowledge by providing a common vocabulary and a set of potentially relevant variables to use in the design of data collection, the conduct of fieldwork, and the analysis of findings (Ostrom, 2009).

During the MAGIC project, we used the robustness framework (Anderies et al., 2004; Anderies, 2015; Anderies et al., 2016) as a common tool for the analysis and for the building of the Amenajeu device (Bonté et al., 2019). The robustness framework, which is an extension of the Institutional Analysis and Development framework (Ostrom, 2005), categorizes the entities of an SES in four main types of components (resources, resource users, infrastructures, and infrastructures providers) and focuses on the interactions between these components. It is used in the SES Library1 developed by the Center for Behavior, Institutions and the Environment (CBIE) to describe SES from around the world, facilitate their comparison, and develop theory on governance in SES. The three MAGIC case studies were analyzed using the robustness framework (see, e.g., Therville et al., 2019) and are included into this database.

The game was initially designed in the French case study on the basis of this analysis conducted by an interdisciplinary team of researchers mixing modelers, geographers, political scientists, and ecologists. During the 3 years of the project, each of them was in charge of analyzing a sector of activity represented in the game in terms of adaptation in the face of global change. In a second step, they were asked to synthesize their knowledge and to feed the framework in order to design their part of the game. Regular meetings were organized to integrate all sectors of activities, calibrate, and bring consistency to the game. The framework provided a common structure to identify the agents, their margin of maneuver and formalize the rules of the game. It also facilitated the adaptation of the game from one context to another, and more particularly from the French to the South African case study.

The researchers involved in the project, although coming from very different fields, acknowledged that the use of the robustness framework, as well as the process of cobuilding the game, was efficient boundary objects (Star and Griesemer, 1989; Barreteau et al., 2007; Hertz and Schlüter, 2015), improving the multidisciplinary understanding of vulnerability transfers on various scales and levels in coastal areas. When the framework provided a common identity to structure interdisciplinary knowledge, the game construction permitted the integration of this knowledge in a multiscale perspective, beyond the limits imposed by this same framework which is struggling to address multiple focal perspectives on an SES (Therville et al., 2019).

9.4.3 Using the multiagent system paradigm and unified modeling language to build a metamodel

A MAS is formed of several autonomous and intelligent agents that dynamically interact between each other and with objects in an environment (Ferber, 1999). They can act by moving, communicating, or doing actions on the objects of the environment (like collecting natural resources). They can make decisions and changes in behavior, memorize states and actions, and perceive the space including the other agents. It is a convenient paradigm to build models of natural resources management situations (Bousquet and Le Page, 2004). With appropriate methodology, MAS is also a helpful tool to perform participatory modeling because elements of the situations (Actors, Resources, etc.) are easily mapped to elements of the MAS paradigm (Agents, objects, etc.) (Etienne et al., 2011). The use of object-oriented programming and the associated graphical language UML (Unified Modeling Language) is suitable to implement MAS conceptual models into computer models (Le Page et al., 2015a; Le Page and Bommel, 2005). We mainly used the class diagram in order to specify the types of entities that compose the model. Each type of entity is implemented as a Class in object-oriented program. The diagram enables to represent the main characteristics (as a list of attributes) and the main processes (as a list of process names) of each type of entity. It also enables to represent two kinds of relations: first, the relations between entities of different types (for instance, an entity of the type territory is composed of entities of the type subterritory), second, a generalization/specialization relation between the types of entities (for instance, the type of entity User is a specialization of the type of entity Agent and thus an entity of type User is a User and also an Agent). The most general type is called the supertype and the most specialized is called the subtype. This last relation between types of entities (also called inheritance in object-oriented programming) is very useful to design and implement metamodels because of its polymorphism feature: the User type, as a subtype of Agent must implement all processes of the Agent type but not necessarily in the same way. For instance, Agent entities have a "move to a place" process that is called when the player takes an agent and puts it on another place. This process has been rewritten in our model so that the same process can be called ("move to a place") but it will have different effects according to the subtype of User that is moving. Consequently, when designing a specific version of the game, all code that is general can be unchanged, such as processes telling to all users to move or processes that enable a player to take a user and put it in another place, and only specific things need to be specified, such as the subtype of Hosting Infrastructures in which a subtype of User can go (see Section 9.5.2).

9.4.4 Implementing participatory simulations with RPG

The RPG concept characterizes a device in which the player takes a role that represents an actor that exists in the real world. The activity does not have to be "fun" or "playful" in the sense of serious game where the features associated with games, such as the need to be challenging and rewarding (you can lose or win) is important. Voinov et al. (2018) showed that a RPG is a useful method to exchange knowledge among stakeholders in a desired context. RPGs involve creation and use of a virtual world, with simplified real-world conditions, to collect information, explore and understand context and situation, and develop and explore collectively possible solutions. Compared to other participatory modeling approaches, RPGs can create more effective teams, help identify and address various stakeholders' common or conflicting interests, effectively build a supportive coalition, and increase the effectiveness of implementation. RPGs may also reveal implicit social rules and interactions between actors that might not have been evident during interviews and other interactions. As players are asked to write down their individual objectives at the beginning of the session and as they simulate the evolution of their territory by controlling its evolution during the game, we consider that the Playner game is both a RPG and a serious game. Results from game-based case studies indicate that the methods aid in demonstrating the dynamic nature of systems over time together with the dependency of system responses to decisions made by participants. Gaming environments serve to create shared knowledge spaces where interactive and iterative actions can be tested or "played out" by participants (Voinov et al., 2016).

9.4.5 Implementing computer models with the Cormas platform and take benefit from hybrid simulation features

Agent-based models should not be dependent on the platform in which they are to be implemented (Le Page et al., 2015b) and several agent-based modeling and simulation platform exist with which we can program implementations. However, some platform may be more adapted than others for some specific applications. We choose the Cormas (COmmon pool Resource Multi-Agents System) agent-based modeling and simulation platform which is the most adapted in this case because of its close relation with object-oriented programming (Le Page and Bommel, 2005), its transparent features of coupling human decisions with computerized dynamics (Bommel et al., 2015), and its users community centered on renewable resource management (Le Page et al., 2012).

9.4.6 Participatory design using crash-test sessions

To build the South African model, a pilot test session has been played as a crash test so that the stakeholders can test the model and suggest improvements. Coconstruction of a model allows to involve stakeholders and gather their value and perception. This kind of method (participative modeling and simulation) is used in ComMod approach as a powerful tool to share knowledge, point of views, and cobuild a model that gathers stakeholders.

The model has thus evolved following participants' feedback. For instance, players who played urban planners asked to have more infrastructure in order to help their users to access to better conditions. The concept of social evolution thanks to education and labor has been added. Indeed, in the first version, the model considered only the fact that a category of people, the wealthiest, can drive out the poorest (priority concept) and the fact that depending on the type of housing and services, the users will not be the same (attractiveness concept).

Another point raised during this pilot session is a lack of chart to track the users updating process. Players wanted to see what had changed during the step by seeing previous state and asked for charts so that they can follow the evolution and the impact of climatic event that occurred during some turn. For that reason, we artificially split each turn into two phases: the normal evolution first, and then the effect of climatic events.

9.5 The Playners generic model and its specialization for specific places

9.5.1 The Playners metamodel

In Fig. 9.2, we used UML class diagram to represent the Playners metamodel. The playners metamodel combines the concept of the SES robustness analysis framework (see Section 9.4.2)

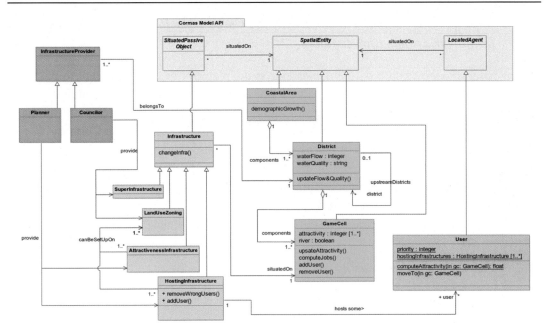

Figure 9.2
Class diagram of the Playners metamodel.

that we can find in the main superclasses User (in pink), Infrastructure (in blue) and Infrastructure providers (in orange), and resources (in green) with the MAS paradigm (see Section 9.4.3) that are the three superclasses of Cormas implementation of the MAS paradigm: SituatedPassiveObejct class for objects, SpatialEntity class for elements of the environment, and LocatedAgent class for agents. Note that Infrastructure Providers are human players.

The French game was the first Playners game so it has not been built in a systematic way. The design process was iterative and new kinds of elements were added on the fly. On the other hand, the South African game was built following a predefined design process based on the Playners metamodel and we can take benefit to this experience to describe how to specialize the Playners metamodel to a specific case study. To build a place-based model from the generic Playners metamodel, we rely on the "heritage" feature of object-oriented modeling as explained in Section 9.4.3. Specializing the Playners metamodel consists in three phases: defining the types of entities represented in the model, defining the initial state of the model, and defining the role of the players.

9.5.1.1 Types of entities

Specifying the types of entities in the specialized models consists in defining subclasses of the classes User, HostingInfrastrucutre, AttractiveInfrastructure, SuperInfrastructure and

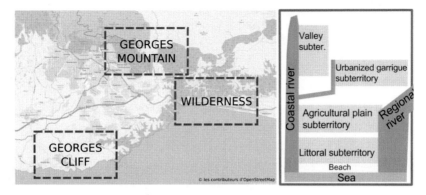

Figure 9.3

Splitting the territory into subterritories. On the left: subterritories of the South African game. On the right: subterritories of the French Game.

LandUsePlan of the Playner metamodel. On the conceptual side, interactions with thematic experts can be done by the use of tables specifying the name of the type, its sector of activity, and the effect it has on users. See in Appendix A such tables for the South African game given as an example. It is also important to find icons that will represent each entity on the game board. See in Appendix A some icons of the South African game given as an example.

9.5.1.2 Initial state

To define the initial state, we instantiate the types of entities described above in subterritories representing the real place spatial organization. Fig. 9.3 presents the spatial organization of the South African game (on the left) and the French game (on the right).

The spatial organization of Wilderness subterritory is given as an example in Fig. 9.4. Each colored cell is a GameCell entity. If we look at the top right-hand game cell, for instance, we observe that it has an agricultural zoning (yellow color), that it contains hosting infrastructures on the top white horizontal stripe (a drop icon and a camping icon), two attractiveness infrastructures on the right vertical white stripe (a sign icon, and a guinea fowl icon), and that it contains seven users (icons in the middle of the cell) that are hosted by HostingInfrastrucutres: two subsistence farmers are hosted by the dam represented by a drop, one ecotourist by the camping site, and four natural species users in the empty Infrastructure and the Nature infrastructure represented by a green dot (see Appendix A for an explanation of icons). Only concerned players can understand the meaning of the icons. For instance, the agricultural player has an explanation of agricultural users and infrastructures in the description of his role (see Section 9.5.1.3).

9.5.1.3 Players' roles

Players' roles are then defined by telling to which users each player must pay attention and which infrastructure he or she can implement. See Fig. 9.5 for the example of the

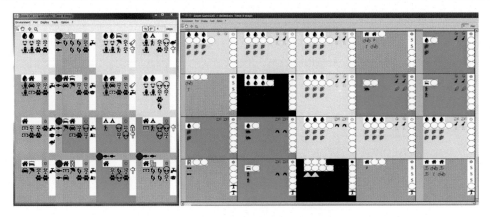

Figure 9.4
Spatial organization of a subterritory. On the left: Wilderness subterritory initial state (from South African game). On the right: Littoral subterritory initial state (from the French Game).

Agricultural planner

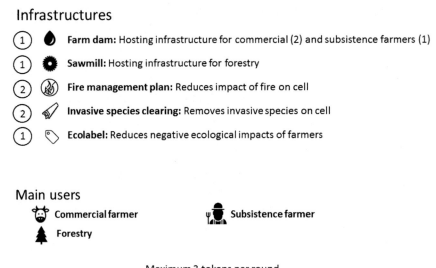

Infrastructures

- (1) 💧 **Farm dam:** Hosting infrastructure for commercial (2) and subsistence farmers (1)
- (1) ✳ **Sawmill:** Hosting infrastructure for forestry
- (2) 🔥 **Fire management plan:** Reduces impact of fire on cell
- (2) 🪚 **Invasive species clearing:** Removes invasive species on cell
- (1) 🏷 **Ecolabel:** Reduces negative ecological impacts of farmers

Main users

🐄 Commercial farmer   👨‍🌾 Subsistence farmer

🌲 Forestry

Maximum 3 tokens per round

Figure 9.5
Agricultural player of South African game.

agriculture player in the South African game. After having their roles explained to them, we assign each player to a subterritory or to a part of a subterritory (a set of cells) according to the number of players and a facilitator describes the territory for all players. Only then we ask each player to write down his or her objective for the game, namely, how they want to develop their sector in their subterritory for the 20 next years. For instance, during a game session, the agricultural player of Wilderness subterritory chooses the objective: "grow commercial agriculture & forestry in the area."

9.5.1.4 A typical game session

A typical game session is similar to a board-game session as presented in Bonté et al. (2019) apart that instead of having facilitators updating users tokens, the facilitators enter the new infrastructures set up by players, and the computer computes the evolution of all subterritories by integrating the effects of events and the description of users dynamics. Thus, for the players, a typical Playner game session is as follows:

- Tables have been initialized (existing infrastructures and users).
- Distribution and presentation of Roles (Players) and distribution among tables.
- Write-up of individual objectives.
- Simulation of four rounds of 5 years each and split each into five phases:
    (1) Announcements of the events and their effects;
    (2) Players put their infrastructures and councilors meet to discuss actions taken in terms of superinfrastructures;
    (3) Board councils with councilor, modification of land uses, and addition of superinfrastructures;
    (4) Update of users (1 minute): Players note their strategies;
    (5) Players can observe the evolution of their territory on each board.
- Debriefing on achievement of the objectives by table.
- Debriefing about the evolution of the whole territory all together.

Fig. 9.6 illustrates phase 2 with the players discussing the attribution of super infrastructures at the Councilor table (on the left) and sector players putting their infrastructures on the Valley subterritory table (on the right). Game boards of each territory, as illustrated in Fig. 9.4, were projected by a video-projector on the wall. Players could stick their infrastructures with fixing paste and user's icons were automatically updated each turn.

9.6 Game sessions and analysis of results

9.6.1 Game sessions

Table 9.1 presents the game sessions played in France and South Africa on which we base our experience of the use of Playners game. We distinguish two kinds of session: Action-Research sessions in which an observation protocol was set up, and Action sessions in which no observation protocol could be set up by lack of means of the research team. We consider that all sessions have an "Action" part because participants of the territory under study were involved and all the sessions constituted a debate arena of them. This table is per se a first result because it shows the need for this kind of tools. Sessions history in France is quite different from session history in South Africa.

Figure 9.6
Pictures of session 5 in Vauvert. On the left: Councilors' table. On the right: Valley subterritory table. Photo credits: C. Therville.

Table 9.1: Game sessions (*NCO: Nature Conservation Organization).

Game session	Date and place	Participants	Model
1. Review of SCoT Sud Gard (Action-Research)	23rd of May 2016, Center of Scamandre, Vauvert, Gard, FR	50 elected people and experts and territorial private managers members of the SCoT Sud Gard	Initial French Game (Board game)
2. Pilot session of SA Game (Action-Research)	1st of June 2017, Nelson Mandela University, George, Garden Route, RSA	Students, a Councilor from George, Researchers	First version of SA Game (Playner game)
3. Workshop at NMU (Action-Research)	14th of June 2017, Nelson Mandela University, George, Garden Route, RSA	5 Students, 3 South African researchers, 1 San Park ranger and 1 NCO* leader	Second version of the SA Game (Playner game)
4. Garden route Interface meeting (Action-Research)	18th of September 2017, SANParks scientific services offices, Rondevlei, Garden route, RSA	12 participants including 3 researchers, 6 local NCO, 1 National NCO, 1 local gov., 1 environmental police	Second version of the SA Game (Playner game)
5. CAET of "Petite Camargue" (Action)	14th of October 2017, administrative headquarters of Petite Camargue inter communality, Vauvert, FR	28 participants including residents, NCO* members, elected people, private experts, territorial project managers	Computerized French game (Playner game)
6. CAET of "Pays de l'Or" (Action)	23rd of May 2018, administrative headquarters of Pays de l'Or inter communality, Mauguio, FR	8 participants including students from Montpellier, 3 private experts in renewable energy and energy transition, 2 territorial project managers	Computerized French game (Playner game)

In South Africa, we, as researchers of the MAGIC project (both South African and French), were at the initiative of all three sessions. The advantages were that we could set up an observation protocol for the three sessions but the drawback was that these sessions were not included in a decision-making process. In France, we suffered a snowball effect where participants of previous sessions contacted us to implement new sessions. Indeed, after the first game session for the review of the SCoT Sud Gard (session 1), we were contacted by the territorial project manager of the Climate Air and Energy Territorial (CAET) plan of "Petite Camargue" who participated in session 1 and wanted to use the game in the setting up of the CAET plan in her territory (session 5). Then, after session 5, we were contacted by a private consulting agency specialized in renewable energy and energy transition. They participated in session 5 because they were contracted for the setting up of the "Petite Camargue" CAET plan and when they were contracted to set up the CAET plan of "Pays de l'Or," they contacted us to implement a game session on this close-by territory (session 6). Hence, all French sessions were related to a collective decision process.

9.6.2 Different kinds of adaptation actions observed during the game sessions

As our aim was to coanalyze the effects of adaptations, we needed to make sure that the kind of adaptations identified in the field (see Section 9.3) could also be observed during the game. In game session 1, we found three of the four adaptation types identified in the field.

We found the first type "control and maintain" in the littoral subregion. Players decided to use one superinfrastructure against sea surge. The littoral has been submitted to significant damages during the game rounds. Surprisingly, players did not choose to protect mass tourism but decided to maintain urbanization pathway. The superinfrastructure was negotiated during the meeting between the four councilors of the four tables. We could observe that the councilor of the littoral subregion could coordinate with other councilors for the attribution of an infrastructure.

The players of the table "Garrigues" used some soft regulations to regulate changes due to high demographic pressure and to absorb potential or effective chocks (droughts or floods). To stabilize the demographic growth, they tried to make limited adjustment in urbanism plans, by allocating small pieces of lands for urbanization whereas keeping agricultural and biodiversity zones. Therefore they try to attract middle-class families and try to keep social homogeneity on the territory.

In the table "Valley" and "Plain," some stakeholders, at the individual level, tried to develop alternative economic pathway, based on green tourism or combining green tourism, agriculture, and biodiversity. But high demographic pressure in the short term was an obstacle to change completely the economic pathway based mainly on residential economy and mass tourism in summer.

The second type of adaptation (Vulnerability assessment and website interfaces) has not been observed nor much discussed in the game. It was not surprising due to the fact that this type of adaptive actions are developed at larger scales (regional, large watershed scales) and that it affects the finer level of individual decision-making of users. In the way we model users in the Playner games, we are representing populations of users and not individual users with their psychology and social networks. Such models exist (Erdlenbruch and Bonté, 2018) but they require a discussion of their validity with stakeholders hardly compatible with their availability if we want to also discuss decisions at the scale of the territory.

9.6.3 Effect of political will: Comparing sessions

Even though the game is complex, we could show that different political will had an impact on the territory built by players. We compared, for instance, general figures of sessions 3 and 4 that were played on the same version of the South African model but with different players. A significant number of participants of session 4 were residents involved in local NCOs while in session 3 we had mostly local students and researchers. The first observation is that they had different objectives. For instance in the Wilderness subterritory which is the most natural territory, all five players of session 4 had an objective related to nature preservation (including the councilor, the Urban Planner, and the Agricultural planner) while in session 3 only the biodiversity and Tourism players related their objectives to nature. Simulation result shows that policies implemented during the game sessions had different results and we can observe that ecotourism, which is a good indication of nature integration, was much higher at the end of session 4 and that commercial farming on the contrary was much higher in session 3 (see figures in Appendix B). We did not do it but we think that comparing such session results could be a powerful tool in order to organize discussion and debate among various groups of stakeholder that are hardly able to communicate, clearly showing that even at the large-scale and long-term horizon, the territory built is a matter of choice.

9.6.4 Experiencing long-term cross-scales processes

In Bonté et al. (2019), we describe extensively through narratives two experiences of integrated adaptations that occurred during session 1. In the first narrative, we could trace how through multilevel and multisector negotiation processes a protected area has been set up in the upstream urbanized territory in order to compensate urban sprawl planned in this upstream territory, each self-motivated to make room for displaced persons from the coastlines cities of the downstream territory. Setting up a protected area is a difficult and long process requiring high level of integration and consultation. Having this kind of action implemented during a game session involving the stakeholders responsible for it in the real life enables those stakeholders to have a first experience on how to justify it precisely from the territorial context by passing through negotiations at the coastal area scale and how to integrate it in the local context of the urban subterritory in which other sectors, agricultural, urban, tourism, also face global change.

9.6.5 Adaptation versus transformation

We initially designed the Playner game to analyze how the adaptation actions that some sectors or scales undertook to reduce their vulnerability to global change could in fact increase the vulnerability in other sectors, other places, or other scales (Bonté et al., 2019). As we asked the players to write down their objectives when the game starts, we could evaluate vulnerability in their ability to fulfill their objectives or not. However, some players decided to modify their objectives during the game. Doing so, they become actors of a transformation of their sector. The agricultural planner of the wilderness table in session 3 is a good example. At the beginning, her objective was "to grow commercial agriculture & forestry in the area." Unfortunately, during a drought occurring at the second turn, several farm dams had been abandoned (see red circles). Moreover, the sawmill that she had just built on the last turn was not efficient due to the rule of forestry: the water flow was too low and thus no forestry users came (see the black circle) (Fig. 9.7).

The agricultural planner had therefore decided to change her objective to: "water-friendly farming development." Consequently, her strategy was now, from turn 3, to promote water-friendly farms, that is to say subsistence farms. Indeed, in the rules, commercial farm users

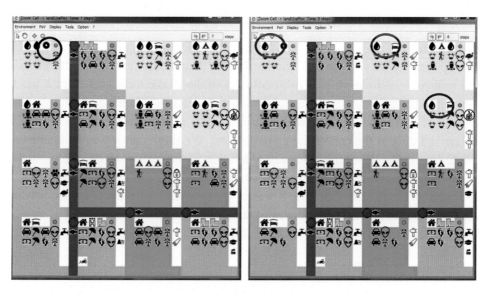

Figure 9.7
Screenshot before (left) and after (right) the drought event.

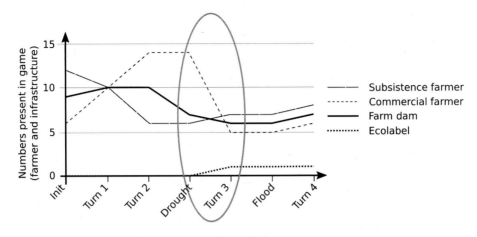

Figure 9.8
Farmers evolution graphic of session 3.

can set up in cells only if it contains at least two farm dams, while only one is needed for subsistence farms, representing the fact that commercial farming in the area is based on large livestock herds requiring large irrigated grazing land. We can see in Fig. 9.8 the moment when the strategy is changed, resulting from the drought (in grey the circle). The player decided not to rebuild farm dams. Instead, she puts ecolabel to reduce the water use of farms. Even though the player did not succeed in increasing significantly subsistence farming, as she wanted, we can see in Fig. 9.8 that she reversed the trend from initial situation. At turn 4, there are more subsistence farmers than commercial farmers when it was the opposite in turn 1, and the number of subsistence farmers is increasing while it was decreasing at turn 1.

In the debriefing, the player related that the ecolabel had no effect and she thought there was a computer malfunction. In fact, the reason was that we designed the rule related to ecolabel to their need to be at least two ecolabels in the whole territory before it has an effect on subsistence farmers in order to represent the fact that their need create a market for this kind of tool to be useful (threshold effect). The player was frustrated by this answer but that arrived only at the end of the game session. We think that it was an interesting point to discuss in the debriefing and we argued that in the reality we do not know the rules of the world and that there is always a part of experimentation where we can assume things but never predict them with accuracy. It was also interesting for players to discuss during the debriefing the reasons and challenges of such a transformation.

9.6.6 Participants learning

An online questionnaire was sent out to assess participants learning experience following the workshop in South Africa. The questionnaire was meant to gather feedback on representation, learning experience, and knowledge transfer and application. It should be noted that the workshop themselves were affected by environmental issues which prevented some participants to partake. For the first South African workshop (session 2, 1st June 2018), some George Municipality agents sent their apologies because they had to attend an extraordinary meeting to implement water restriction in the municipality, and the second workshop (session 3 held on the 14th of June) coincided with dramatic fires in Knysna, the neighboring municipality. Some participants raised the points that it would be more fruitful if the workshop timing had coincided with the Spatial Development Plan revision as more people would have attended. Only 12 participants completed the survey and most of them agreed that the workshop had changed the way they think about adaptation (75%), that they developed a better understanding of their role for building adaptive capacity (83%), and believe that this tool can assist their institution to address wicked problems (83%). Most of them also acknowledged that playing this game helped them to be more sensitive to other's perspectives and other ways of seeing (67%). Almost half of them (58%) had shared their experience of the workshop with others who could not attend. In terms of impact, 50% of the participants said that what they learnt will change what they do. However, one should consider the impact of the participants represented as only 42% of the participants felt that all knowledge holders were present.

9.7 Discussion

9.7.1 Building up an adaptive territory

Overall these participatory workshops on multidimensional and multiscale interactions seem very useful for exploring collectively expected and unforeseen outcomes of any adaptation policy to change. This approach is effective and efficient to make a hidden social−ecological interdependence or transfer of vulnerability visible to decision-makers, land planners, and land users. The RPG and its debriefing led the participants to discuss how to adapt various sectoral regional planning policies to global change diversity of systemic impacts. Protecting farm lands and biodiversity remains a key challenge in the face of any demographic pressure and its linked urban development (Bonté et al., 2019). Although if financial means are of course key to coping with changes such as floods or drought, it is also important to have free space in terms of regulation to be able to innovate and adapt provided to have any political will (Therville et al., 2019). Such kind of experience is about learning individually and collectively to draw what we can call the "adaptive territory" (Mathevet et al., 2015), a region that is able to reconcile human activities and conservation of landscapes, identities, ecological functions, and biodiversity in the context of changing ecology. A territory whose users and stakeholders are able to take into account the social and ecological interdependencies but also the issues of social and environmental justice in their decision-making processes. An adaptive territory whose users and inhabitants are able to think collectively and implement a social, ecological, economic, and energy transition policy despite the globalization forcing. Integrating transfers of vulnerability and thinking social and ecological solidarity between regions and subterritories require actions but also such kind of experiment to name and make sense of the changes and actions undertaken (Mathevet and Godet, 2015). Such kind of RPG session shows that since a social group or community is part of a process of knowledge of interdependencies, it builds a stewardship approach of its landscapes and consolidates both its general and specific resilience to future social, economic, and ecological changes.

9.7.2 Integrating thematics, methods, and tools: How to deal with heterogeneity?

As R. Duboz states in his PhD (Duboz, 2004): "In computer science, heterogeneity concerns every level of abstractions from the material level up to the conceptual level, including applications or software systems." Hence, many initiatives and development exist in computer science to integrate heterogeneity at different levels of abstractions in application to the use and building of simulation models to help collaboration or decision. In his effort of building a conceptual framework for Modeling and Simulation, Zeigler (2000) proposes a conceptual hierarchy composed of six interdependent layers. Duboz (2004) summarizes these layers as follows: The "network" layer contains all physical computational elements and the software systems linked to networks. The "simulation" layer is a software system that executes the models. The "modeling" layer includes the model developments in unambiguous formalisms that should be independent of the "simulation" layer implementation. The "research" layer is where the models are built from the analysis of the real world. The "decision" layer is where model exploration is performed and analyzed (which set of parameters are used, how simulation results are analyzed, how debriefing is organized in the case of participatory simulation). The "collaboration" layer describes the phases of concertation among the participants to the model building.

As we consider the Playner model as a modeling tool in its ability to represent the territory under study, we can situate the conceptual and operational tools developed (white boxes) and used (gray boxes) in this hierarchical conceptual framework in Fig. 9.9. We see that we developed tools at different operational levels and different levels of specificity. The Playners CorMAS Metamodel (presented in Fig. 9.2) based on the SES Framework is at the heart of the Playner tool. It is generic since it worked to conceptualize the French board game and both South African and computerized French game. It is not operational enough to reach the simulation layer (it helps design the conceptual models but not the simulation

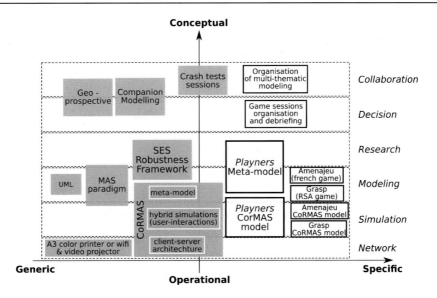

Figure 9.9
Integrating heterogeneity.

computer model). The Playners CorMAS model is a Cormas Model that implements the Playners metamodel in the CorMAS agent-based modeling and simulation platform. It is specified at the simulation level since it implements the main functions to simulate a planner model. It is quite generic since we used it to implement the Amenajeu and Grasp CorMAS models that have been simulated during the game sessions presented in this paper. Collaboration for multithematic modeling was organized with the use of tables as presented in Appendix A and rely of course a lot on the SES Roubstness framework. Game sessions organization and debriefing, explained in detail in Bonté et al. (2019), describes how the Playner model is participatory simulated and how simulation results are discussed during the debriefing part of the session.

On the other hand, we also consider the Playner model as a tool to cope with a heterogeneity of points of views and of stakes of the stakeholders that has to shape together a common and viable future for their territory. In that sense, its use should also be integrated specifically in collective decisions processes and associated arenas that correspond to the scale and issues represented in the Playner models. Hence, we believe that all the tools developed should be integrated to sets of tools dedicated to the organization of participation such as the CoOPLAaGE toolkit (Ferrand et al., 2017). We clearly saw the lack of integration of our game sessions in the political agenda when very few people attended session 6 because the territorial project manager of the "Pays de l'Or" CAET plan did not have a clear idea on how and why she would mobilize the session in her public consultation plan.

9.8 Conclusion

We designed a generic tool and associated methodology to implement long-term, multiscale, and multisector geoprospective workshops during which participants simulate and then discuss together the potential evolution of their territories over the 20 next years. The implementation of the tool on two case studies through six game sessions in France and South Africa showed that the tool is useful to discuss adaptation of territories to global change because it enables them to experience together: different logics of adaptations observed in the field, the dilemma between adaptation and transformation, long-term cross-scales processes and more generally the effects of political will at this time scale.

Building on this experience, during which the device was used in various arenas such as interface meetings between South African National parks and South African local politicians, or the setting up of master plans and climate and energy plans of two French intercommunalities, we must now work to integrate the geoprospective workshops based on this tool into formal decision-making processes.

Appendix A South African conceptual model

Table 9.A.1

Table 9.A.1: Infrastructure table.

Sector	Name	Effect
Town planner	Social housing	Attract poor and middle resident
Town planner	Private housing	Attract all except poor
Town planner	Service	Attract richer resident and seasonal
Town planner	Drinking water provision?	Protection against droughts?
Town planner	Gabion	Reduce flood impact/soil erosion on the cell but rise neighborhood cells
Agricultural	River dam	Open space for new commercial farmer—impact on water level downstream
Agricultural	Hillside dam	Protection against droughts—no impact on water level?
Agricultural	Problem animal control	Remove x baboon users
Agricultural	Invasive species control	Destroy x invasive species users
Agricultural	Commercial label	Open space organic farm? or allow subsistence farmer to stay
Agricultural	Fertilisant and animal feed manufactures	Open space for new commercial farmer
Biodiversity	Trail	Attract TB + minor—impact on biodiv if too much tourist
Biodiversity	Protected area	Councilor cannot land use's cell. Number of tourists is limited
Biodiversity	Invasive species control	Destroy x invasive species users

(Continued)

Table 9.A.1: (Continued)

Sector	Name	Effect
Biodiversity	Problem animal control	Remove x baboon users
Seaside tourism	Resort	Open user place on urban cell
Seaside tourism	Golf	Attract seaside tourists
Seaside tourism	BnB	Open user place on urban cell
Ecotourism	Lodge	Open user place on nature cell
Ecotourism	Trail	Attract backpacker tourists
Ecotourism	Camping site	Open user place on nature cell
Councilor	N2 Road	Repair the road
Councilor	Retreat Plan	Move all the users of selected coastal cells
Councilor	River Dam + +	Reduce flood impact on cells
Councilor	Mouth estuary	Repair in case of breaching?
Councilor	Seawall	Reduce soil erosion and sea surge impact

Table 9.A.2

Table 9.A.2: User table.

Sector	Name	Where	Effect
Town planner	Rich class	U	
Town planner	Middle class	U	
Town planner	Poor class	U	
Town planner	Seasonal	U	
Agricultural	Commercial farmer	A	
Agricultural	Subsistence farmer	A	
Agricultural	Fishing ladies	A	
Agricultural	Organic fair farming	A	
Biodiversity	Knysna Loerie	A/N	Attract "green" tourist and residents middle/seasonal + good for biodiversity
Biodiversity	Milkwood tree	A/N	Endemic species = > biodiversity interest
Biodiversity	Pine tree	A/N	Invasive species
Biodiversity	Seahorse	A/N	Water quality indicator (can be related to fish ladies)
Biodiversity	Baboon	A/N	Attract tourist but can be annoying for residents/farmers
Seaside tourism	Tourist seaside	U Coastal	
Ecotourism	Tourist backpacker	N	

Appendix B Comparing figures of different sessions

Figures 9.A.1—9.A.3

Figure 9.A.1
Icons of users (left column) and infrastructures (right column, non-exhaustive).

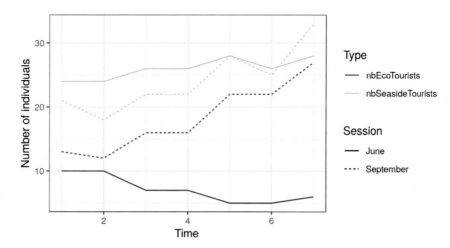

Figure 9.A.2
Evolution of Tourism sector in sessions 3 (June) and 4 (September).

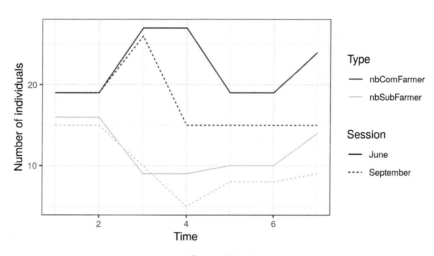

Figure 9.A.3
Evolution of Farming sector in sessions 3 (June) and 4 (September).

References

Anderies, J.M., 2015. Understanding the dynamics of sustainable social-ecological systems: human behavior, institutions, and regulatory feedback networks. Bull. Math. Biol. 77, 1−22. Available from: https://doi.org/10.1007/s11538-014-0030-z.

Anderies, J.M., Janssen, M.A., Ostrom, E., 2004. A framework to analyze the robustness of social-ecological systems from an institutional perspective. Ecol. Soc. 9 (1), 18. Available from: https://doi.org/10.5751/ES-00610-090118.

Anderies, J.M., Janssen, M.A., Schlager, E., 2016. Institutions and the performance of coupled infrastructure systems. Int. J. Commons 10 (2), 495−516. Available from: https://doi.org/10.18352/ijc.651.

Barreteau, O., 2003. Our companion modelling approach. JASSS, 6(2):8. < http://jasss.soc.surrey.ac.uk/6/2/1.html > .

Barreteau, O., Le Page, C., Perez, P., 2007. Contribution of simulation and gaming to natural resource management issues: an introduction. Simul. Gaming 38, 185−194. Available from: https://doi.org/10.1177/1046878107300660.

Binder, C.R., Hinkel, J., Bots, P.W.G., Pahl-Wostl, C., 2013. Comparison of frameworks for analyzing social-ecological systems. Ecol. Soc. 18. Available from: https://doi.org/10.5751/ES-05551-180426.

Bommel, P., Bécu, N., Le Page, C., Bousquet, F., 2015. Cormas, an agent-based simulation platform for coupling human decisions with computerized dynamics. In: Kaneda, T., Kanegae, H., Toyoda, Y., Rizzi, P. (Eds.), Simulation and Gaming in the Network Society. Translational Systems Sciences, vol. 9. Springer, Singapore, pp. 387−410. Available from: https://doi.org/10.1007/978-981-10-0575-6_27.

Bonté, B., Therville, C., Bousquet, F., Abrami, G., Dhenain, S., Mathevet, R., 2019. Analyzing coastal coupled infrastructure systems through multi-scale serious games in Languedoc, France. Reg. Env. Change 19, 1879−1889. 10.1007/s10113-019-01523-6.

Bousquet, F., Le Page, C., 2004. Multi-agent simulations and ecosystem management: a review. Ecol. Model. 176, 313−332. Available from: https://doi.org/10.1016/j.ecolmodel.2004.01.011.

Daniell, K.A., Barreteau, O., 2014. Water governance across competing scales: coupling land and water management. J. Hydrol. 519 (Part C), 2367−2380. Available from: https://doi.org/10.1016/j.jhydrol.2014.10.055. ISSN 0022-1694.

Dhénain, S., Barreteau, O., 2018. Policy instruments and political dynamics of adaptations to global change in coastal areasNat. Sci. Soc. 26 (4) 407−417, Available from: https://doi.org/10.1051/nss/2019005.

Duboz, R., 2004. Heterogeneous models integration for the modelling and simulation of complex systems : application to multi-scale modelling in marine ecology. PhD thesis University of Calais, France, 230 pp.

Epstein R., 2012. De la différenciation territoriale à la libre conformation, in Douillet A.-C., Faure A., Halpern C. (Eds), L'action publique locale dans tous ses états. Différenciation et standardisation, Paris, L'Harmattan, 127-138

Erdlenbruch, K., Bonté, B., 2018. Simulating the dynamics of individual adaptation to floods. Environ. Sci. Policy 84, 134−148. Available from: https://doi.org/10.1016/j.envsci.2018.03.005.

Etienne, M., Du Toit, D.R., Pollard, S., 2011. ARDI: a co-construction method for participatory modeling in natural resources management. Ecol. Soc. 16 (1), 44. <http://www.ecologyandsociety.org/vol16/iss1/art44>.

Ferber, J., 1999. Multi-Agent Systems. An Introduction to Distributed Artificial Intelligence. Addison Wesley, London.

Ferrand, N., Abrami, G., Hassenforder, E., Noury, B., Ducrot, R., Farolfi, S., et al., 2017. Coupling for Coping, CoOPLAaGE: an integrative strategy and toolbox fostering multi-level hydrosocial adaptation. In: Conference_item. Proceedings of the ACEWATER2 Scientific Workshop, Accra, Ghana, 31 October−3 November 2016. <http://agritrop.cirad.fr/585578/>.

Guerbois, C., Brady, U., de Swardt, A.G., Fabricius, C., 2019. Nurturing ecosystem-based adaptations in South Africa's Garden Route: a common pool resource governance perspective. Reg. Environ. Change 19, 1849−1863. Available from: https://doi.org/10.1007/s10113-019-01508-5.

Hertz, T., Schlüter, M., 2015. The SES-Framework as boundary object to address theory orientation in social−ecological system research: the SES-TheOr approach. Ecol. Econ. 116, 12−24. Available from: https://doi.org/10.1016/j.ecolecon.2015.03.022.

Le Page, C., Abrami, G., Barreteau, O., Becu, N., Bommel, P., Bonté, B., et al., 2015a. Concevoir et développer un modèle informatique. In: Michel, E. (Ed.), La modélisation d'accompagnement: partager des représentations, simuler des dynamiques. FormaSciences, FPN, INRA, Nantes, pp. 95−112. (FormaSciences, 4) ISBN: 978-273-801-3828.

Le Page, C., Abrami, G., Becu, N., Bommel, P., Bonté, B., Bousquet, F., et al., 2015b. Multi-platform training sessions to teach agent-based simulation. In: Conference on Complex Systems, Tempe, États-Unis, 28 September 2015−2 October 2015. Complex Systems Society, Résumé, Tempe, 1 p.

Le Page, C., Becu, N., Bommel, P., Bousquet, F., 2012. Participatory agent-based simulation for renewable resource management: the role of the Cormas simulation platform to nurture a community of practice. J. Artif. Soc. Soc. Simul. 15 (1), 10. Available from: https://doi.org/10.18564/jasss.1928.

Le Page, C., Bommel, P., 2005. A methodology for building agent-based simulations of common-pool resources management: from a conceptual model designed with UML to its implementation in CORMAS. In: Bousquet, F., Trébuil, G., Hardy, B. (Eds.), Companion Modelling and Multi-Agent System for Integrated Natural Resource Management in Asia. IRRI Publications, Los Banos, Philippines, pp. 1−17.

Mathevet, R., Godet, L., (Eds.), 2015. Pour une géographie de la conservation. Biodiversités, Natures & Sociétés. L'Harmattan, Paris. p. 400.

Mathevet, R., Peluso, N., Couespel, A., Robbins, P., 2015. Using historical political ecology to understand the present: water, reeds and biodiversity in the camargue biosphere reserve (Southern France). Ecology and Society 20(4), 17.

Ostrom, E., 2001. Vulnerability and polycentric governance systems. Newsletter of the International Human Dimension Programme on Global Environmental Change 3, 3−4.

Ostrom, E., 2005. Understanding Institutional Diversity. Princeton University Press, Princeton.

Ostrom, E., 2009. A general framework for analyzing sustainability of social-ecological systems. Science 325, 419−422. Available from: https://doi.org/10.1126/science.1172133.

Pasquini, L., Ziervogel, G., Cowling, R.M., Shearing, C., 2015. What enables local governments to mainstream climate change adaptation? Lessons learned from two municipal case studies in the Western Cape, South Africa. Clim. Dev. 7, 60−70. Available from: https://doi.org/10.1080/17565529.2014.886994.

Quinn, T., Bousquet, F., Guerbois, C., Heider, L., Brown, K., 2019. How local water and waterbody meanings shape flood risk perception and risk management preferences (2019). Sustain. Sci. in press. Available from: https://doi.org/10.1007/s11625-019-00665-0.

Quinn, T., Bousquet, F., Guerbois, C., Sougrati, E., Tabutaud, M., 2018. The dynamic relationship between sense of place and risk perception in landscapes of mobility. Ecol. Soc. 23 (2), 39. Available from: https://doi.org/10.5751/ES-10004-230239.

Smit B., Wandel J., 2006. Adaptation, adaptive capacity and vulnerability, Global Environmental Change, 16, 3, 282−292, doi: 10.1016/j.gloenvcha.2006.03.008

Star, S.L., Griesemer, J., 1989. Institutional ecology, translations, and boundary objects: amateurs and professionals on Berkeley's museum of vertebrate zoology. Soc. Stud. Sci. 19, 387−420.

Therville, C., Brady, U., Barreteau, O., 2019. Challenges for local adaptation when governance scales overlap. Evidence from Languedoc, France. Reg. Environ. Change 19, 1865−1877. Available from: https://doi.org/10.1007/s10113-018-1427-2.

Voinov, A., Jenni, K., Gray, S., Kolagani, N., Glynn, P.D., Bommel, P., et al., 2018. Tools and methods in participatory modeling: selecting the right tool for the job. Environ. Model. Softw 109, 232−255. Available from: https://doi.org/10.1016/j.envsoft.2018.08.028.

Voinov, A., Kolagani, N., McCall, M.K., Glynn, P.D., Kragt, M.E., Ostermann, F.O., et al., 2016. Modelling with stakeholders—next generation. Environ. Model. Softw 77, 196−220. Available from: https://doi.org/10.1016/j.envsoft.2015.11.016.

Von Korff, Y., Daniell, K.A., Moellenkamp, S., Bots, P., Bijlsma, R.M., 2012. Implementing participatory water management: recent advances in theory, practice, and evaluation. Ecol. Soc. 17 (1), 30. Available from: https://doi.org/10.5751/ES-04733-170130.

Western Cape Province, 2017. State of environment outlook report for the Western Cape Province 2014-2017. Cape Town, South Africa.

Zeigler, B.P., Kim, T.G., Praehofer, H., 2000. Theory of modeling and simulation: Integrating Discrete Event andContinuous Complex Dynamic Systems. Academic Press.

CHAPTER 10

Geoprospective as a support to marine spatial planning: some French experience-based assumptions and findings

Françoise Gourmelon[1], Brice Trouillet[2], Romain Légé[2], Laurie Tissière[2] and Stéphanie Mahévas[3]

[1]*CNRS, UMR LETG, Institut Universitaire Européen de la Mer, Plouzané, France,*
[2]*Université de Nantes, UMR LETG, Nantes, France,* [3]*Ifremer, Nantes, France*

Chapter Outline

10.1 Introduction

With around 60 initiatives worldwide (IOC-UNESCO, EC-DGMARE, 2017), marine spatial planning (MSP) is asserting itself on a global scale in a context of rapid change due to a combination of two phenomena which are cited in the preamble of European Directive

2014/89 on establishing a framework for MSP. First, the need for ocean space is intensifying because of the development of "new" uses, such as marine renewable energies to meet the challenge of climate change. Second, tools for protecting marine ecosystems and biodiversity—marine protected areas (MPAs) currently being the most developed of these—are multiplying with the aim of incorporating 10% of the world's ocean surface area by 2020 (i.e., about 36 million km^2) compared with just under 4% currently.[1]

The expressions "marine spatial planning," "maritime spatial planning," and "maritime space planning" are often used analogously to signify a *planned* approach to developing maritime space, in other words, an interventionist approach. However, the first two terms put an emphasis on *spatial planning* (Smith et al., 2011), an approach that has actually evolved over time and changed in essence by being applied to the maritime domain (Kidd and Ellis, 2012), whereas the third term (used by the French authorities and translated as such in the French version of European Directive 2014/89) comes from wanting to highlight the strategic character of the planning process at the expense of the spatial aspect. This position adopted by the French authorities at the end of the 2000s and until the start of the 2010s can be explained by the experience of planning initiatives mainly based on *zoning* (Trouillet et al., 2011), notably the *Schémas de Mise en Valeur de la Mer* initiatives. In a certain way, this reluctance shown by the French authorities echoed Mintzberg's (1994) view: "The goal of those who promote planning is to reduce managers' power over strategy making." Besides, European institutions also prefer the adjective "maritime" as opposed to "marine" when employing this expression, the aim being to encompass everything to do with the ocean and not just refer to what is produced by the sea.

However, implementing MSP is a complex and even controversial process. In the European Union, for example, at least two opposing points of view show themselves to be in favor of either planning based on an ecosystemic approach (hard sustainability), in keeping with its original spirit (Douvere, 2008; Degnbol and Wilson, 2008), or planning as a support to "blue growth" (soft sustainability) (Frazão Santos et al., 2014; Qiu and Jones, 2013). This opposition gives rise to an ambiguity, particularly when it comes to coordinating MSP and MPAs. It also brings the "spatial" aspect into question, which is viewed differently depending on the dominant logic and on the "temporal" aspect, as conservation takes place over the long term whereas economic rationales require flexibility over the short term. Another problem arising when implementing MSP is the uncertainty surrounding possible changes in maritime uses and how these evolving uses can cohabit. Uncertainty may well be inherent to planning and development; however, the problem becomes more acute with regard to sea uses for which knowledge gaps often exist, especially in terms of their spatiotemporal dimension and the speed at which changes take place (St. Martin and Hall-Arber, 2008). Fishing is a prime example of this problem, both due to the highly variable spatiotemporal nature of the activity

---

[1] http://www.mpatlas.org/map/mpas/ (viewed in January 2018).

(Le Tixerant et al., 2010; Le Guyader et al., 2017; Tissière et al., 2016) and its strong dependency on external factors that often lead to conflicts (Trouillet, 2015), falling in line with the observation made by the UK government in one of its first plans: "The lack of uniformity and stakeholder consensus regarding fisheries data combined with the difficulties in predicting the future of fisheries, makes formulating prescriptive marine plan policies for this sector a challenge" (HM Government, 2014). Lastly, the ocean remains "unfinished territory" as, without mentioning the number of maritime borders that have not yet been agreed upon as to their demarcation, these nascent sea territories still lack a component (Trouillet, 2004) as they hover somewhere between the institutional territory of planning [e.g., *façades maritimes* (coastal regions) in France] and the territories of activity and use. In short, speaking of MSP here raises the question of what a "territory" is and consequently of what "makes" a territory (actors, projects, etc.).

These three main challenges faced when implementing MSP, especially in a context of rapid spatial changes and the need to adapt to long-term climate changes, bring three geoprospective elements to mind: "space" as a support and agent, "participation" as a performance-related device and, situated at interface of these two aspects, "models" as complements or supports for scenario setting (Emsellem et al., 2012; Houet and Gourmelon, 2014; Voiron-Canicio, 2012). These three elements are inseparable because geoprospective implies that actors participate in discussions about assumptions based on knowledge and models used to describe and represent current trends. These models and knowledge are in turn based on data, notably spatial data, which refers to questions of availability and quality that condition the use of other types of model (GIS, simulators, etc.) (Le Tixerant et al., 2018; Pinarbasi et al., 2017).

We posit that MSP is a pertinent framework for testing the heuristic potential of geoprospective, which we view as an aid to social learning, or in this particular case, to "socio-spatial learning." However, beyond the interest of social learning as a means of generating new knowledge, of acquiring technical and social competencies, or of developing relations that may contribute to a common understanding of the system and thereby lead to agreements and collective action (Muro and Jeffrey, 2012), in this chapter, we focus more specifically on the characteristics of geoprospective processes themselves, as well as the contextual factors that contribute or are an obstacle to their implementation and the results obtained. In fact, although participation has gradually become a key principle for MSP, it can be confronted not only with differently appropriated types of model used in geoprospective—from the more traditional (map) to the more sophisticated (numerical simulation model), and from the more abstract to the more realistic (Le Page, 2017)[2]—but also with various ways of using space as a support or agent (data, types of space, relations

---

[2] According to Le Page (2017), an empirical model is a means of representing observations with the aim of sharing them. He distinguishes between realistic empirical models built by integrating georeferenced data (e.g., from a Geographic Information System) and stylized empirical models that represent space schematically.

to space, etc.). The variety of possible links between the questions of space, participation, and models suggests that there is a potential for thinking about and using geoprospective in different ways, each way liable to offer advantages and challenges depending on the implementation context and methods.

In order to test this general hypothesis (i.e., the diversity of viable geoprospective approaches and tools vs the diversity of publics along with the variety of implementation contexts and methods), we developed an analysis based on experiments performed over several scales of time and space, with a variety of publics and using different objects (Fig. 10.1).

We think that the implementation of MSP in France provides a pertinent application framework as it was rolled out as a national strategy first, before becoming a basis for producing regional MSP documents with a strong strategic quality and ultimately, enabling the coordination of spatial planning tools and efforts at a local level (Trouillet et al., 2011). Based on this logic, we chose three experiments which enabled us to approach the space-participation-modeling trio from different angles: a first case combines one use and regional scale (Section 10.2), a second links multiple uses and regional scale (Section 10.3), and a third focuses on multiple uses and local scale (Section 10.4). Following on from this, we analyzed the benefits and limits of geoprospective using a discussion grid based on this trio (Section 10.5).

10.2 A geoprospective experiment concerning one activity on a regional scale

10.2.1 Implementation context and objectives

The aim of this first experiment was to test a geoprospective approach on a case focusing on fishing activities, which are often at the heart of MSP stakes (variability of production factors, large spatial distribution, resource conservation challenges, activities confronted with the development of new uses, and the implementation of MPAs). Although this in itself is an interesting observation, at least three other reasons inherent to the fishing sector bolster this interest: the question of data, and thus of models, is particularly sensitive in this domain; participation is considered a key element in the management of fisheries; and the area at stake rarely matches the management area.

For the main fisheries off the French Atlantic coast, the Bay of Biscay (Fig. 10.1) is the most appropriate management area with regard to fishing resources. The conservation of resources, the good environmental status, the distribution of access to the area, or the renewal of fleets are all important goals for fishery management policies and more broadly, policies on the development of maritime space in the Bay of Biscay. Whereas public

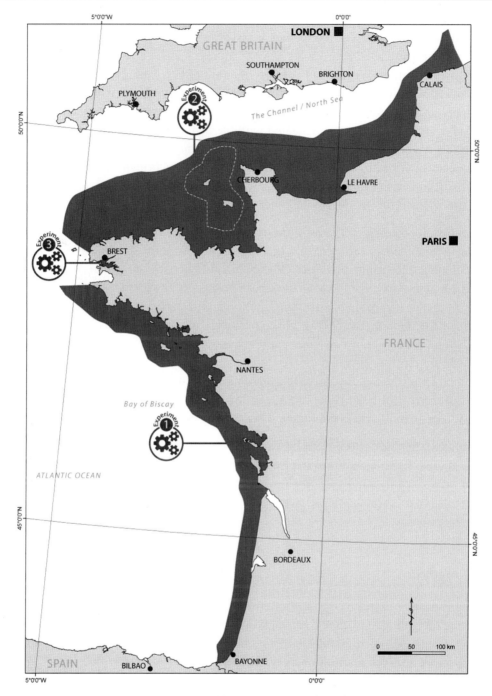

Figure 10.1
Location of the three case studies.

administrative and scientific institutions have always been involved in management policies, the regionalization of the Common Fisheries Policy now justifies the increasing participation of new public and private actors, namely, fishers, local, or regional authorities and nongovernmental organizations (Van Hoof et al., 2012). The involvement of stakeholders—with fishers being at the forefront—in the decision-making process is increasingly encouraged to guarantee the success of the measures adopted. Given that these management measures are evaluated with the help of simulation models, the issue of adapting this type of tool to the actors becomes primordial. The decentralization of fishery management as defined by Pomeroy and Berkes (1997) thus brings into question the tools used to assist in decision-making, particularly models, which have gradually become more complex to account for the different elements of fishery systems (Thébaud et al., 2014).

The ISIS-Fish model, which spatially and temporally charts fishery dynamics, was designed to measure the effects of different management scenarios on the dynamics of both exploited fish populations and fishing activities (Mahévas and Pelletier, 2004). This feature makes it usable for geoprospective in terms of fishery management and support for MSP. More globally, it makes it capable of shedding light on model-participation interactions (Kieken, 2005; Houet and Gourmelon, 2014). From a technical point of view, as its precision is determined by the level of knowledge and the availability of input data, ISIS-Fish simulations are mostly generated on the statistical scale of the International Council for the Exploration of the Sea (i.e., a surface area of between 4169 and 4474 km^2 around the latitudes of the Bay of Biscay) and on a monthly basis. It produces a cartographic interface, making it user-friendly and encouraging discussion between actors.

In this context, this first experiment had three targets: (1) to contribute to reflection on the future of marine fisheries in the Bay of Biscay, given the changes taking place there in terms of occupancy dynamics and the implementation of MSP tools, (2) to accompany the participation of several types of actor, and (3) to analyze the use of tools in a participatory context.

10.2.2 Methods

The geoprospective exercise consisted of cowriting several scenarios for 2050 by spatializing the narrative to enable interaction with several scientific methods, including modeling. Using ISIS-Fish in this context required us to establish a procedure involving about 20 actors from the maritime fishery sector in the Bay of Biscay, including government department representatives, fishers, community activists, other users of maritime space, and researchers from different disciplines. Three rounds of surveys were led by a foresighter and a geographer, with modelers sometimes brought in to work with them (Fig. 10.2).

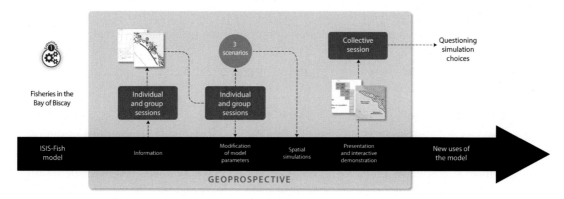

Figure 10.2
Fisheries in the Bay of Biscay: methodology.

From the outset, the participants were informed that the ISIS-Fish model was going to be used. The model was then presented in detail during the third round of surveys via a short film explaining parameters in layman terms[3] and an interactive demonstration of the different forms of simulation, such as the spatial distribution of populations and of fleet fishing efforts, or the evolution of biomasses, catches, and fishing mortality. The model was also accompanied by different qualitative support materials to enable discussion about its parameters and to cobuild hypotheses to be tested.

10.2.3 Results

The first round of surveys consisted of individual and group sessions with the aim of identifying actors' concerns for the future. Three topics came to the fore, around which the rest of the geoprospective experiment was organized: (1) management of fisheries in terms of the organization of decision-making processes and the measures adopted; (2) cohabiting with other uses from the point of view of dialogue between activity sectors and the distribution of spatial resource; and (3) the state of the environment, which was a secondary topic perceived through global changes. These concerns were represented on mental maps[4] (Fig. 10.2). During the second round of surveys, also consisting of individual and group sessions, these three concerns were used to establish change hypotheses (Fig. 10.2). The latter were then brought together into three contrasting scenarios whose respective narratives were progressively adjusted through to the end of the experiment. At first, the scenarios took on a narrative, written, and illustrated form before being transposed numerically into the ISIS-Fish model. They continued to be developed during the third

---

[3] The short film and the results generated by the model can be viewed at this address: www.isis-fish.org.
[4] "Mental maps are understood here as physical, hand-drawn maps on a blank sheet of paper where the subject expresses, graphically, a subjective reality of space using individual and social memory" (Dernat et al., 2016).

round of surveys. This round drew on different types of support material (Fig. 10.2), which were valued by the actors present to varying degrees. Some of the support materials were rejected whereas others were created by the participants in the same vein as an influence graph, which enabled the chain of causal links for one of the scenarios to be highlighted (Tissière et al., 2018).

The quantification of the effects of the scenarios on the Bay of Biscay fishery shows the real pertinence of using the ISIS-Fish model, which was at the heart of the third round of surveys. The modeling demonstration was based on one of the scenarios called "Jaws in the Bay," or more specifically, on two of its hypotheses—(1) the standardization of the fishing fleet and (2) the appropriation of the coastal band by new uses—which required further interviews to be quantitatively transposed. The discussions that ensued once the simulation results had been presented in a plenary meeting showed that all the participants had a good understanding of how the model worked. The criticisms expressed were not aimed at the parameters but at the choices made in transposing the hypotheses (faithfulness to the original narrative, probability of orders of magnitude selected, etc.) (Provot et al., 2018). Similar comments were made regarding the representation of scenarios in the shape of maps. In the end, these criticisms equated to a questioning of the decision taken in transposing the narratives of the actors interviewed, given that the more strategic aspects of their participation had been formalized.

10.3 A geoprospective experiment concerning multiple uses on a regional scale

The goal of this second experiment was to test a geoprospective approach in the context of a French regional scale initiative prefiguring the production of the strategic document for the Channel/North Sea French coastal region.

10.3.1 Implementation context and objectives

MSP needs to respond to an increasing number of challenges at sea, especially with regard to the Channel/North Sea coastal region due to the creation of an MPA, the development of marine renewable energies, an increase in maritime traffic, and a high level of fishing activity. Given the context, the general aim of this second experiment was to prepare and provide support for the new decision-making processes concerning the sea for this coastal region (Fig. 10.1).

Throughout the different stages, it brought together about 20 actors representing public government departments and agencies, as well as professionals and users involved in a variety of activities. Our main hypothesis was first that participation in a geoprospective experiment would encourage the convergence of perceptions held by actors with different points of view, leading to the implementation of shared strategic visions, and second that these visions could not be constructed based on spatial representations that were worlds apart.

In practice, the approach primarily aimed to use participatory mapping methods (Moore et al., 2017) as suitable tools for understanding and clarifying the perceptions of actors and monitoring the way these developed throughout the geoprospective process. This would then confirm whether these approaches encourage or not both the emergence of converging perceptions of maritime space among the actors and the drafting of strategic visions.

10.3.2 Methods

To meet these objectives, the project was split into two main stages (Fig. 10.3). The first stage consisted of gathering the perceptions of participants during semistructured interviews using participatory mapping methods. We thus interviewed 28 actors from the Channel/ North Sea French coastal region to identify their perceptions of the current state-of-play and their visions for the future. The method relied on using mental maps and stake maps. This is understood as being a free representation of the stakes on a map, based exclusively on individual perceptions which are meant to embody social perceptions. The second stage involved organizing two 2-day seminars to create the conditions for cobuilding contrasting and spatialized scenarios based on a combination of hypotheses and to observe the process itself. These seminars unfolded following the usual three-step approach: phase-in, scenario building, and discussion (Mermet, 2005). The first seminar drew 11 participants out of the 26 actors met beforehand. This made it easier to gather qualitative data while keeping a broad range of the profiles identified during the individual interviews. Three months later, the second seminar counted 10 participants, all of which had been present at the first seminar,

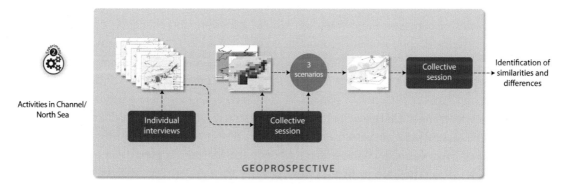

Figure 10.3
Activities in the Channel/North Sea: methodology.

ensuring continuity in our experiment. The aim was not so much representativeness as to check our main hypothesis.

During these seminars, we favored the use of qualitative tools given that planners generally only have access to data which is scattered across several institutions and organisms has information gaps (St. Martin and Hall-Arber, 2008) and is nonstandardized (Abramic et al., 2018). Therefore gathering qualitative data from actors was quicker than collecting quantitative data, which involves fieldwork over a relatively long period. The qualitative approach, which is more adapted to the frequently very tight public policy-making time schedule, was our preferred approach. It was based on mapping aids that were deliberately schematic and adapted to research on the unpredictable future of vast entities.

10.3.3 Results

The preparatory phase of the geoprospective experiment provided us with two types of maps for each actor (Fig. 10.3). A mental map enabled us to gather information concerning activity locations, whereas a stake map encouraged a first step toward identifying perceptions by asking actors to spatialize and qualify these stakes (i.e., activities and actors concerned, major changes to be anticipated, trends, etc.).

During the first seminar, the phasing-in stage enabled us to establish a shared perception based on a collectively produced regional scale map. This map enabled us to group together activities, uses, and zoning (Fig. 10.3). To build spatialized scenarios (second part of the seminar), we started by identifying possible future changes, distinguishing between strong trends, whose evolution can be anticipated over time, and critical uncertainties, whose pathways are less predictable. Different hypotheses arose on how the main critical uncertainties could evolve. The group spontaneously chose three hypothesis combinations as a basis for three contrasting scenarios characterized by a narrative and a graphical spatial modeling for the year 2050 (Fig. 10.3). Scenario 1, "channeling together," was founded on strong dynamics of cooperation between the United Kingdom and France to centrally manage ecosystems despite serious economic crises. Scenario 2, "the company as guarantor and fostering blue growth in the Channel," described a weakening of public policies and the rise to power of an ultraliberal model leading to the depletion of resources following a period of strong growth. Scenario 3 presented the Channel as the "blue gate to Europe," boosted by ever stronger European policies. The general leanings of the scenarios show that environmental stakes were all but absent in all three, despite the participation of representatives from this sector (the French Agency for Marine Protected Areas, the French Water Agency, etc.). They are in fact a lower priority than governance stakes, probably due to the political context in which the exercise took place, that is, the reorganization of government departments coupled with the rise of MSP. The link between current events and the "game" proposed by the workshop seemed obvious, both in the opinions held by the actors and the scenarios themselves.

These three spatialized scenarios were discussed in a second seminar (Fig. 10.3), enabling their similarities and differences to be identified. Although the diagnosis and analysis of stakes led to a consensus, the discussions on the solutions to be adopted to meet these challenges did not result in a shared strategic vision common to all the participants.

10.4 A geoprospective experiment concerning multiple uses on a local scale

This third experiment consisted of testing a geoprospective approach from the point of view of challenges linked to the multiple uses of maritime space, and on a local scale potentially corresponding to that of the spatial planning tools (e.g., marine natural parks) which need to be coordinated in the local strategic documents.

10.4.1 Implementation context and objectives

The geoprospective workshop dealt with the maritime activities which take place in the Bay of Brest, located on the western-most tip of Brittany (France) (Fig. 10.1). This maritime basin faces (1) potential space conflicts between commercial fishing, maritime transportation, and supervised nautical activities (windsurfing, sailing, kayaking, rowing, and scuba-diving) and (2) ever-growing regulations and coastal conservation policies. This new situation is handled through an Integrated Coastal Zone Management (ICZM) process, which aims to promote cooperation among stakeholders. The geoprospective workshop exploited a GIS-driven Spatio-Temporal Database (STDB), which had gathered heterogeneous data describing the spatial and temporal scope of supervised activities on a daily basis over a year (Le Guyader, 2012). This STDB was developed to (1) model the spatial and temporal dynamics of the activities and (2) produce different kinds of artifacts (spatial or not: maps, geovisualization, statistics, networks, etc.) for local stakeholders (Gourmelon et al., 2014). In short, the geoprospective workshop aimed to evaluate the usability of the STDB in a professional context as a support for debates as well as a way to optimize the simulation based on the collective scenario process, while providing different types of representations.

10.4.2 Methods

The geoprospective workshop was organized by three researchers and took place within an applied research process (Fig. 10.4). During two successive sessions, one representative from local commercial fisheries and five representatives from local agencies involved in the management of the Bay of Brest were brought together. Government departments and

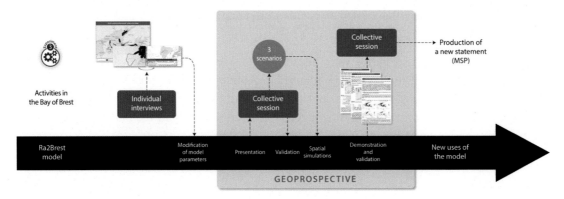

Figure 10.4
Activities in the Bay of Brest: methodology.

representatives from nautical activity centers and from maritime transportation declined the invitation. During the workshop, a moderator and an observer from the scientific team were appointed to manage and record the reactions and debates between the participants.

During the first 3-hour session, one of the researchers presented the different phases of knowledge production concerning the supervised human activities that simultaneously occur in the Bay of Brest: data gathering, notably from what actors have said, the creation and analysis of a spatiotemporal database, and the generation of results in different graphical formats highlighting potential conflicts between activities (Fig. 10.4). A back drop was then created for applying the geoprospective tool, using a scenario proposed by the researcher, which simulated the siting of a wind farm in the Bay of Brest and the consequences it would have on the current activities. The participants were then invited to propose future scenarios for maritime activities in the Bay of Brest.

The second workshop session was programmed to take place a month later (Fig. 10.4). It lasted 3 hours and featured the same participants. A researcher presented the simulations resulting from the scenarios proposed by the stakeholders during the first session. The debriefing was structured by an evaluation form concerning the approach and the output, filled in individually each by each actor. It gave rise to a collective discussion on the advantages of the geoprospective approach in this kind of application context and notably, the benefits of the spatiotemporal aspect of representations.

10.4.3 Results

During the first workshop session, the spatiotemporal modeling of human activity in the Bay of Brest was interactive. Participants reacted positively to the approach in general and to the value added by the spatiotemporal aspect. Spatiotemporal interactions between activities were presented in different formats: summary maps (e.g., density of entities over a year), dynamic geovisualization (e.g., the spatial footprint of activities as it unfolds on a daily basis over a year), interaction matrix and network (e.g., the number of times two activities share the same space at the same time over a year) (Fig. 10.4). Several questions were raised by participants during the course of the presentation. They concerned among other things the inclusion of physical parameters in the model (bathymetry, currents, tide, weather forecast), the possibility of obtaining an even more precise temporal granularity (a half-day) and of ranking the spatiotemporal interactions highlighted, and the inclusion of benthic habitats to evaluate the pressure put by some activities on the resource. With regard to the scenario-setting phase, once the participants had overcome their initial surprise at the scenario proposed by the researcher, they all took part seemingly unreservedly, except for the fisher representative who took a back seat. After an hour of spontaneous discussions often held in pairs between representatives from local authorities, without any research intervention, three simulation leads came to the fore: the creation of a new passenger maritime transport route (scenario 1), ulva harvesting via trawler offshore from the urban beach (scenario 2), and reintroduction of marine aquaculture (scenario 3). These three scenarios were essentially proposed by three local authority representatives.

During the second workshop session, the context, data used, and results for the three simulations made using the STDB were presented for each scenario, despite the difficulties encountered in collecting complementary data and in remotivating the local actors between sessions, even though they had been very motivated during the first session. These presentations were made using statistics and maps and highlighted the activities impacted (scenarios 1 and 3) and the days with minimum restrictions (scenario 2). The debriefing provided an opportunity for coming back to the advantages of the approach and usefulness of the output for managers in the ICZM process. All the participants thought that the interaction matrices for the activities were very useful for the diagnosis phase. Equally, for the ICZM prospective and planning phase, they all generally appreciated the dynamic geovisualization depicting the occurrence of activities and the prospective simulations. During the presentation, the issues surrounding MSP were evoked by two participants as a future challenge for the Bay of Brest. With regard to this, they asserted that the approach could be of interest for two stakes: (1) showing the spatiotemporal interactions between activities to anticipate potential conflicts and (2) the collective development of planning scenarios to help with decision-making. The other participants agreed with this and a discussion ensued between participants and researchers on the transferability of the method to another maritime space.

10.5 Discussion—Conclusion

The common objective of these three geoprospective experiments was to test the advantages and limits of this approach in terms of social learning, with a focus on the role of the different types of spatial representation and whether they aided or hindered the participatory process. As with all empirical approaches, these three exercises only serve as illustrations and are intrinsically characterized by the conditions in which they were conducted (type of participants, current issues, etc.). In addition, the aim was not necessarily to compare them but rather to multiply the analysis angles with the aim of adding to the discussion. In this context, it seems logical that these three cases involved publics that were more or less homogeneous (managers of territories, user representatives, etc.) and maritime territories of varying size (regional or local scale) confronted with MSP stakes in different ways. Although not a perfect approach, due to the numerous gaps between the three terms of the space-participation-modeling trio, it can still serve as a guide for transversal analysis, aimed at gaining knowledge on applying geoprospective to the maritime area in terms of planning.

First, in terms of participation, the assiduity and involvement of the participants throughout the three experiments show both their interest in the exploratory geoprospective approach and their willingness to participate, on this occasion, in an unprecedented discussion proposed by the researchers. Geoprospective thus comes to the fore as a social learning aid for better understanding interdependency. Although meetings to initiate dialogue on fishery management are often a place for expressing power relations between different groups of actors, with one group predominantly taking the floor (Tissière et al., 2018), the first case shows that as long as dialogue between participants is constant and does not lead to a deadlock, debates in the context of a geoprospective exercise can be more constructive and balanced than in the more traditional participatory exercises. This can clearly be explained by the fact that projecting oneself into the relatively long-term future (50 years in this case) makes it easier to cooperate and even adhere to the principle of participation. The second case shows us that if a diversity of actors is involved, geoprospective improves the understanding of each other's goals and perspectives and sheds light on the underlying values and visions of the future. Furthermore, using spatial scenario setting on the scale of a coastal region encourages consideration for the different system compartments (geographical, socioeconomic, ecological) and the links between them. However, such a vast territory can also actually hinder collective actions, which rely on mutual agreement between actors who do not share the same areas of interest and/or issues. At a local level, the third case shows that with a relatively homogeneous public made up of local stakeholders, the creation of collective scenarios raises questions of legitimacy, of active involvement, and of access to additional information which is not always readily provided by the actors, as has already been shown in other participatory experiments (Becu et al., 2008; Dupont et al., 2016). Some participants tend to remain on the sidelines and at a group level, the geoprospective workshop raises questions about the risk this could pose to policies and actors who are not represented (Steyaert et al., 2007). Nevertheless, in this local experiment, the geoprospective approach taken as a whole helped not only to encourage actors to cooperate at certain times, but also to produce a new statement to which they all adhered. This substantiates what Ritschard et al. (2018) have already shown: spatiotemporal representations and other forms derived from their analysis serve not only to spark the interest of and enroll stakeholders in a participatory process, but also to smooth the way for the emergence of a collective statement of new stakes for the territory.

Second, where spatial and modeling issues are concerned, although the various categories of public involved may have appropriated the multifarious geographical information in different ways, they all showed an interest in the spatial representations based on qualitative data, that is, data gathered from what actors have said. This supports the current trend and the interest in geographical content and knowledge as produced by the general public (Haklay, 2013). For users of maritime territory, it would appear that when their knowledge is recognized by the other stakeholders, all the actors are brought onto the same footing, which is essential for a collaborative approach based not only on an exchange of knowledge and proposals but also on the individual power to influence the process (Barreteau et al., 2010). For the researchers, by using the information gathered from actors interviewed prior to the scenario-setting phase, they were able to acquire new data and better understand the perceptions and visions of the stakeholders, which is often restricted at first to the space in which they practice their activity and their field of activity. In fact, collecting data from interviews probably makes the notion of cobuilding more concrete for managers and policy-makers and additionally contributes somewhat to stakeholder enrollment in the management process (Ritschard et al., 2018). The first case supports this idea by showing that the participants' understanding of the model was improved by cobuilding input variables and combining participation support material prior to the simulations. The second case also illustrates that using mental maps based on interviews with the actors and schematic graphical models enables participants to move on from the technical questions linked to data and representations and concentrate on the strategic questions involved in policy decisions, especially where access to space and marine resources is concerned. On a local scale, the third case shows us that realistic or hyperrealistic representations using data from geographical databases can be successful and therefore prove to be interesting in certain contexts. The same applies for the temporal dimension of geographical data on a daily scale, which played a major role in the diagnosis, planning, and environmental foresight phases according to the participants taking part in the third experiment. This result, which confirms the relevance of realist models as a support for management approaches, is probably linked to the homogeneity of the group in terms of spatial competencies (Noucher, 2009). With other types of actor, using spatial representations—particularly those based on realist models—in whatever form they may come in, can prove to be more complex or even controversial, depending on the contexts in which and the subjects for which they are used (Ritschard, 2017). The experiment undertaken with fishers from the Bay of Biscay in case 1 illustrates this situation. The spatial dimension became a mediator when it gave rise to stakes shared by the different categories of participants, both in terms of the fishing components (e.g., identifying key zones) and in terms of the representation of these components (e.g., choosing the appropriate scale). The spatial dimension also played a mediation role when it enabled the narratives to be viewed in all the graphical presentations as well as in the ISIS-Fish model display interface. However, conversely, it not only created a deadlock when it embodied much debated management modes (especially MSP and MPAs) but also when obstacles relating to methods (e.g., the difficulty of mapping a piece of information) or strategy (e.g., not wanting to map a piece of information) were met.

Lastly, from a more global point of view, among the numerous issues to be resolved in geoprospective projects, the question of an individual's appropriation of the reality of a given territory, including a reality expressed by others, is clearly raised. In our three cases, we have considered that all types of spatial representation that a priori enable individual understanding and the expression of different viewpoints are worth using. Therefore we have drawn on a broad range of spatial representations based on qualitative and quantitative data. They are used both as intermediary objects (Vinck, 2009) to provide local actors with a dynamic and integrated vision of the territory, facilitating cooperation between them, and as inscriptions used by certain actors to argue and convince their colleagues of a certain statement (Latour, 1985). The results drawn from our three case studies illustrate the nuanced appropriation of the different forms of spatial representations and somewhat evoke the MORE reference framework (Modeling for resilience thinking and ecosystem stewardship) proposed by Schlüter et al. (2013) in socioecological system models (Fig. 10.5).

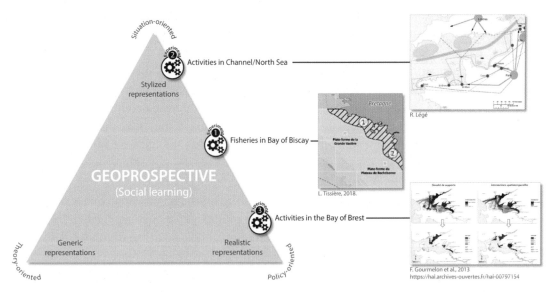

Figure 10.5

The different forms of spatial representations produced by the three geoprospective experiments versus MORE reference framework (Schlüter et al., 2013).

According to the authors, schematic models support inter- and transdisciplinary communication and integration for the development of societal strategies (adaptation and transformation) as shown by the experiment conducted in the Channel/North Sea coastal region (case 1), whereas approaches based on structurally realistic models and case-specific models are policy-oriented as shown by the two other experiments (cases 2 and 3), with varying degrees of possible actions and appropriation. The question of implicit choices that guide the debate is particularly meaningful for fishery models (case 1) as management measures are prerequisites and debates can only set certain parameterizable elements such as thresholds (e.g., quotas) and limits (e.g., fishing areas). In case 3, the participants adopted the realist model without any difficulty and "played along" during the scenario-setting phase. Their level of innovation with regard to possible futures was nevertheless limited by their feeling of illegitimacy in taking decisions.

To conclude, the challenges and some of the different stages of the MSP process can benefit from a geoprospective approach as shown by Pinarbasi et al. (2017). Our study highlights the interest in having a plurality of model types and tools adapted to different application contexts and to different stages of the MSP process. The various forms of spatial representation, based on qualitative and/or quantitative data, and the different modeling techniques used must be inscribed in these contexts and carefully introduced to ensure that they facilitate rather than hinder an often-fragile collaborative process (Barreteau et al., 2010). In any case, further evidence of the beneficial outcomes is now required, which justifies the increasing number of new experiments.

Acknowledgments

These experiments were conducted in the context of research programs: the COSELMAR project (Pays de la Loire region), the Liteau IV "*Dessine-moi... un système mer - terre!*" project (The French Ministry of Environment, Sustainability and Energy in partnership with the French Agency for Marine Protected Areas and the *Direction interrégionale Manche est − mer du Nord*), and the LITEAU III "*Rad2Brest*" project (The French Ministry of Environment, Sustainability and Energy). This paper also benefited from support for its publication from the CNRS (LETG UMR 6554) and from the contribution of cartographer Laurence David to the design and production of the illustrations. The authors would also like to thank all those who participated in the experiments, along with the anonymous reviewers for their precious comments which helped improve the first draft of this paper.

References

Abramic, A., Bigagli, E., Barale, V., Assouline, M., Lorenzo-Alonso, A., Norton, C., 2018. Maritime spatial planning supported by infrastructure for spatial information in Europe (INSPIRE). Ocean Coast. Manag. 152, 23−36. Available from: https://doi.org/10.1016/j.ocecoaman.2017.11.007.

Barreteau, O., Bots, P., Daniell, K., 2010. A framework for clarifying participation in participatory research to prevent its rejection for the wrong reasons. Ecol. Soc. 15 (2), 1−22. <http://www.ecologyandsociety.org/vol15/iss2/art1/>.

Becu, N., Neef, A., Schreinemachers, P., Sangkapitux, C., 2008. Participatory computer simulation to support collective decision making: potential and limits of stakeholder involvement. Land Use Policy 25, 418−509. Available from: https://doi.org/10.1016/j.landusepol.2007.11.002.

Degnbol, D., Wilson, D.C., 2008. Spatial planning on the North Sea: a case of cross-scale linkages. Mar. Policy 32 (2), 189−200. Available from: https://doi.org/10.1016/j.marpol.2007.09.006.

Dernat, S., Johany, F., Lardon, S., 2016. Identify choremes in mental maps to better understand socio-spatial representations. Cybergeo Eur. J. Geogr. 800. Available from: https://doi.org/10.4000/cybergeo.27867.

Douvere, F., 2008. The importance of marine spatial planning in advancing ecosystem-based, sea use management. Mar. Policy 32 (5), 762−771. Available from: https://doi.org/10.1016/j.marpol.2008.03.021.

Dupont, H., Gourmelon, F., Rouan, M., Le Viol, I., Kerbiriou, C., 2016. The contribution of agent-based simulations to conservation management on a Natura 2000 site. J. Environ. Manag. 168, 27−35. Available from: https://doi.org/10.1016/j.jenvman.2015.11.056.

Emsellem, K., Liziard, S., Scarella, F., 2012. La géoprospective: l'émergence d'un nouveau champ de recherche. L'Espace Géogr. 41 (2), 154−168. Available from: https://doi.org/10.3917/eg.412.0154.

Frazão Santos, C., Domingos, T., Ferreira, M.A., Orbach, M., Andrade, F., 2014. How sustainable is sustainable marine spatial planning? Part I-Linking the concepts. Mar. Policy 49, 59−65. Available from: https://doi.org/10.1016/j.marpol.2014.04.004.

Gourmelon, F., Le Guyader, D., Fontenelle, G., 2014. A dynamic GIS as an efficient tool for Integrated Coastal Zone Management. ISPRS Int. J. Geo Inf. 3 (2), 391−407. Available from: https://doi.org/10.3390/ijgi3020391. <http://www.mdpi.com/2220-9964/3/2/391>.

Haklay, M., 2013. Citizen science and volunteered geographic information: overview and typology of participation. In: Sui, D., Elwood, S., Goodchild, M. (Eds.), Crowdsourcing Geographic Knowledge. Springer, Dordrecht, pp. 105−122, <http://link.springer.com/chapter/10.1007/978-94-007-4587-2_7>.

HM Government, 2014. East Inshore and East Offshore Marine Plans. Department for Environment, Food and Rural Affairs, London, p. 193.

Houet, T., Gourmelon, F., 2014. La géoprospective. Apport de la dimension spatiale aux démarches prospectives. Cybergeo Eur. J. Geogr. 667. Available from: https://doi.org/10.4000/cybergeo.26194.

IOC-UNESCO, EC-DGMARE, 2017. The 2nd International Conference on Marine/Maritime Spatial Planning, March 15−17, 2017, UNESCO, Paris. IOC Workshop Reports Series, 279.

Kidd, S., Ellis, G., 2012. From the land to sea and back again? Using terrestrial planning to understand the process of marine spatial planning. J. Environ. Policy Plan. 14 (1), 49−66. Available from: https://doi.org/10.1080/1523908X.2012.662382.

Kieken, H., 2005. Les prospectives environnementales fondées sur des modèles. Quelle dialectique entre modélisation et forum de débat? In: Mermet, L. (Ed.), Etudier des écologies futures. Un chantier ouvert pour les études prospectives environnementales. EcoPolis. P.I.E. Peter Lang, Bruxelles.

Latour, M., 1985. Les "vues" de l'esprit. Cult. Tech. 14, 4−29.

Le Guyader, D., 2012. Modélisation des activités humaines en mer côtière, Thèse de Géographie, Université de Bretagne Occidentale. <https://tel.archives-ouvertes.fr/tel-00717420v2>.

Le Guyader, D., Ray, C., Gourmelon, F., Brosset, D., 2017. Defining high resolution dredges fishing grounds with automatic identification system (AIS) data. Aquat. Living Resour. 30 (39), 1−10. Available from: https://doi.org/10.1051/alr/2017038.

Le Page, C., 2017. Simulation multi-agent interactive: engager les populations locales dans la modélisation des socio-écosystèmes pour stimuler l'apprentissage social. Mémoire d'HDR, UPMC, <https://collaboratif.cirad.fr/alfresco/s/d/workspace/SpacesStore/45d837d8-da99-46a2-90b0-17487d15e94d/LePage_2017_DossierHDR.pdf>.

Le Tixerant, M., Gourmelon, F., Tissot, C., Brosset, D., 2010. Modelling of human activity development in coastal sea areas. J. Coast. Conserv. Plan. Manag. 15. Available from: https://doi.org/10.1007/s11852-010-0093-4.

Le Tixerant, M., Le Guyader, D., Gourmelon, F., Quéfellec, B., 2018. How can Automatic Identification System (AIS) data be used for maritime spatial planning? Ocean Coast. Manag. 166. Available from: https://doi.org/10.1016/j.ocecoaman.2018.05.005.

Mahévas, S., Pelletier, D., 2004. ISIS-Fish, a generic and spatially explicit simulation tool for evaluating the impact of management measures on fishery dynamics. Ecol. Model. 171, 65−84. Available from: https://doi.org/10.1016/j.ecolmodel.2003.04.001.

Mermet (dir.), L., 2005. Étudier des écologies futures, Un. chantier Ouvert. pour les. Rech. prospectives environnementales, 5. P.I.E.-Peter Lang, EcoPolis.

Mintzberg, H., 1994. The Rise and Fall of Strategic Planning. Harvad Business Review. January−February, pp. 107−114.

Moore, S., Brown, G., Kobryn, H., Strickland-Munro, J., 2017. Identify conflict potential in a coastal and marine environment using participatory mapping. J. Environ. Manag. 197, 706−718. Available from: https://doi.org/10.1016/j.jenvman.2016.12.026.

Muro, M., Jeffrey, P., 2012. A critical review of the theory and application of social learning in participatory natural resource management processes. J. Environ. Plan. Manag. 51 (3), 325−344. Available from: https://doi.org/10.1080/09640560801977190.

Noucher, M., 2009. La donnée géographique aux frontières des organisations: approche socio-cognitive et systémique de son appropriation, Thèse de Géographie, Ecole Polytechnique Fédérale de Lausanne (EPFL). <https://halshs.archives-ouvertes.fr/tel-00654203/document>.

Pinarbasi, K., Galparsoro, I., Borja, A., Stelzenmükller, V., Ehler, C.N., Gimpel, A., 2017. Decision support tools in marine spatial planning: present applications, gaps and future perspectives. Mar. Policy 83, 83−91. Available from: https://doi.org/10.1016/j.marpol.2017.05.031.

Pomeroy, R.S., Berkes, F., 1997. Two to tango: the role of government in fisheries co-management. Mar. Policy 21 (5), 465−480. Available from: https://doi.org/10.1016/S0308-597X(97)00017-1.

Provot, Z., Mahévas, S., Tissière, L., Michel, C., Lehuta, S., Trouillet, B., 2018. Using a quantitativ model for a participatory geo-foresight: ISIS-Fish and fishing governance in the Bay of Biscay. Mar. Policy. Available from: https://doi.org/10.1016/j.marpol.2018.08.015.

Qiu, W., Jones, P.J.S., 2013. The emerging policy landscape for marine spatial planning in Europe. Mar. Policy 39, 182−190. Available from: https://doi.org/10.1016/j.marpol.2012.10.010.

Ritschard, L., 2017. Représentations spatiales et processus de Gestion Intégrée des Zones Côtières (GIZC): application à deux territoires côtiers, Thèse de Géographie, Université de Bretagne Occidentale. <https://tel.archives-ouvertes.fr/tel-01512946>.

Ritschard, L., Gourmelon, F., Chlous, F., 2018. Différencier les représentations spatiales selon leurs statuts: expérimentation en Gestion Intégrée des Zones Côtières. Rev. Int. Géomat. 28 (1), 39−67. Available from: https://doi.org/10.3166/rig.2017.00037.

Schlüter, M., Müller, B., Frank, K., 2013. How to use models to improve analysis and governance of social-ecological systems. The reference frame MORE. Working paper. <http://ssrn.com/abstract = 2037723>.

Smith, H.D., Maes, F., Stojanovic, T.A., Ballinger, R.C., 2011. The integration of land and marine spatial planning. J. Coast. Conserv. 15, 291−303. Available from: https://doi.org/10.1007/s11852-010-0098-z.

St. Martin, K.S., Hall-Arber, M., 2008. The missing layer: geo-technologies, communities, and implications for marine spatial planning. Mar. Policy 32, 779−786. Available from: https://doi.org/10.1016/j.marpol.2008.03.015.

Steyaert, P., Barzman, M., Billaud, J.P., Brives, H., Hubert, B., Ollivier, G., et al., 2007. The role of knowledge and research in facilitating social learning among stakeholders in natural resources management in the French Atlantic coastal wetlands. Environ. Sci. Policy 10, 537−550. Available from: https://doi.org/10.1016/j.envsci.2007.01.012.

Thébaud, O., Doyen, L., Lample, M., Mahévas, S., 2014. Building ecological-economic models and scenarios of marine resource systems: workshop report. Mar. Policy 43, 382−386. Available from: https://doi.org/10.1016/j.marpol.2013.05.010.

Tissière, L., Mahévas, S., Michel, C., Trouillet, B., 2016. Les pêches maritimes, un terrain d'expérimentation de la géoprospective. Cah. de. Géogr. Québec 60 (170), 287−301. Available from: https://doi.org/10.7202/1040536ar.

Tissière, L., Mahévas, S., Trouillet, B., 2018. Findings from an exploratory study on the governance of a French fishery. Mar. Policy. Available from: https://doi.org/10.1016/j.marpol.2018.01.028.

Trouillet, B., 2004. La mer côtière d'Iroise à Finisterre. Etude géographique d'ensembles territoriaux en construction, Thèse de doctorat en géographie, Université de Nantes, 293p.

Trouillet, B., 2015. Les enjeux spatiaux: la reconfiguration des espaces halieutiques. In: Guillaume, J. (Ed.), Espaces Maritimes et Territoires Marins. Ellipses, Paris, pp. 53−88.

Trouillet, B., Guineberteau, T., de Cacqueray, M., Rochette, J., 2011. Planning the sea: The French experience. Contribution to marine spatial planning perspectives. Mar. Policy 35 (3), 324−334. Available from: https://doi.org/10.1016/j.marpol.2010.10.012.

Van Hoof, L., Van Leeuwen, J., Van Tatenhove, J., 2012. All at sea; regionalisation and integration of marine policy in Europe. Marit. Stud. 11 (9), <https://maritimestudiesjournal.springeropen.com/articles/10.1186/2212-9790-11-9>.

Vinck, D., 2009. De l'objet intermédiaire à l'objet frontière. Vers la prise en compte du travail d'équipement. Rev. D'anthropol. Connaiss. 3 (1), 51−72. Available from: https://doi.org/10.3917/rac.006.0051.

Voiron-Canicio, C., 2012. L'anticipation du changement en prospective et des changements spatiaux en géoprospective. L'Espace Géogr. 41 (2), 99−110. Available from: https://doi.org/10.3917/eg.412.0099.

CHAPTER 11

Simulating the interactions of environmental and socioeconomic dynamics at the scale of an ecodistrict: urban modeling of Gerland (Lyon, France)

Christine Malé[1] and Thomas Lagier[2]

[1]MUG, Metropole of Lyon, France, [2]SoLyft, Lyon, France

Chapter Outline

11.1 Introduction

Cities face many challenges. Different issues have to be simultaneously addressed: demography issues (68% of the world inhabitant will live in a city in 2050); financial issues: cities have to constantly arbitrate between new infrastructure and maintenance of the existing one; environmental issues: for example, air quality; policy issues: waste

Ecosystem and Territorial Resilience.
DOI: https://doi.org/10.1016/B978-0-12-818215-4.00011-0

© 2021 Elsevier Inc. All rights reserved.

management, building norms, etc., are more and more strict and numerous requirements have to be taken into account.

Since 1990s, globalization and economic competition led to the emergence of city marketing and led cities to be more competitive and attractive, to benchmark and grade in the international survey and referential.

More and more cities have to take into account stakeholders, which are operating the city (energy provider, waste management company, real estate), which are funded the city (bank, state, private project, etc.), which are living in the city (associations, citizens, etc.). Cities decision is less and less in a "top down" model and become more horizontal. Considering the stakeholder became crucial. Decision must be share, project must be cobuild, and mayor and his team are accountable for their choice.

Among issues, environmental awareness leads to a systemic approach of the cities. It is no longer possible to separate the investment phase (construction of building, infra, etc.) and the operational phase (energy, transport, water supplying, etc.). City makers have to take decision with consequences on the short term but also on the middle and long terms, the vaguer it is, but to reach sustainability, middle and long-term issues have to be taken into account in day-to-day decision to check that the global target will be reached.

11.1.1 Toward a new urban paradigm

The emerging needs and issues described above led to the idea that "city makers" must be able to address globally, systemically, consistency the way to think and the way to build the city. The main evolution in the last decades is the massive use of Geographic Information System and the sharing of data through open data movement (EU Directive Inspire for Europe) but surprisingly, despite an active academic research on this topic, the stakeholders use very few operational tool and go on mainly with document, excel sheets, and PowerPoint.

We can ask the question of the level of maturity of a new urban paradigm: are the conditions met? This question can be answered on different axes:

- On the city stakeholder point of view, the diagnosis is well shared, and the global need is expressed. Some stakeholders can have conservative and "silo" vision to protect interests but the vast majority of them are convinced and find out a more collaborative and integrative way to "make" cities;
- On a technical point of view, the digital offers tangible progress: (1) massive access to data; (2) important computational capacity; (3) intuitive interface (UI/UX design) allowing the nonexpert person to access powerful tools; (4) SaaS business model allowing fast development and deployment of operational solutions; (5) ability to collaborate (sharing partially or total access, information);

- On an economic model point of view: the standardization of data and the generalization of SaaS allow the cities to access to powerful digital services for acceptable prices.

In conclusion, there is an emerging market for powerful decision software allowing a systemic and a collaborative way to think and to build cities. This opens the way to a new urban paradigm.

The Metropole of Lyon understood these stakes and decides to develop a project to address a new way to think and build cities.

The point is that cities are complex systems that need to be grasped comprehensively in order to preserve and improve the quality of life of its inhabitants. Cross-sectoral activities and services, such as the provision of transport infrastructure and the supply infrastructure of energy and water, as well as access to education and health care, have to be considered simultaneously.

The four main objectives of this project are

- Accompanying the process of urban and social transition in the urban district Gerland;
- Merging visionary urban planning with computerized modeling;
- Linking different domains in urban planning in a holistic and integrated way;
- Creating a decision support for the long-term development of the district Gerland.

The project intends to provide the Métropole de Lyon with an interactive planning platform as tool for decision support. This is enabled by the diverse competences of the project consortium allowing to integrate comprehensively urban challenges into computerized models, creating a tool that makes interrelations and interactions visible. Furthermore, the planning platform enables the planning authority of Lyon to develop different scenarios for the futures, to engage in interdisciplinary and transdisciplinary exchange and helps breaking the traditional silo approach to urban planning. This has been achieved via simulating and visualization of interdependencies through a software SaaS tool named MUG.

11.1.2 Experimenting the new urban paradigm: the MUG project

The MUG project has been an opportunity to develop and experiment this new urban paradigm. The goal of this 4-year long project consists in (1) the codevelopment of a powerful decision tool based on urban modeling and (2) the test of the MUG software in real condition by the Métropole of Lyon and its stakeholders.

The whole city has been modeled but a specific focus has been given to the district of Gerland where the MUG software has been extensively tested.

Gerland is a very dynamic district of the city under strong development. In the context, there is a necessity to make a lot of choices: because of an intensive development of the district, choices will have effects need to be assessed (directly, indirectly, short, middle, and long term the consequence), and cross-sector activities and services, such as transport infrastructure and supply of energy, as well as an access to education and health care, have to be considered.

An urban area cannot be anymore described as an addition of many weakly dependent systems from each other, it is necessary to favor the competitive city or the including city.

A problem is to decide in the uncertainty with consideration to

• Interactions between various components;
• Temporal and spatial scales according to the phenomena;
• Feedback loops;
• The functioning of the global system does not deduct of the study independent from every component;
• The need to anticipate and measure the impacts;
• The new capacities: the contribution of digital technology, ability to represent, capacity to be coupled.

A consortium proposes a decision-making tool dedicated to the urban projects:

• To integrate and estimate any urban project with regard to the existing;
• To integrate the interactions between the various components of urban ecosystem (transport, use of the ground, services in the energy, the water, the waste, buildings, etc.).

11.2 Context, challenges, stakes, and issues of the experimentation

11.2.1 District of Gerland: specificity and singularity

The Lyon Métropole has elected the Gerland territory, being the southern entrance to the center of Lyon, for the experimentation of this R&D project, a vast 700 ha area. This territory represents a strong actual diversity constantly reinventing itself: an agricultural area in the 19th century, then a popular district with its metallurgical, chemical, and food industries; in the 1980s, Gerland became the biotechnology stronghold; and since 2014, it has been an attracting laboratory focusing on life sciences in its southern part, known as the "Biodistrict."

This territory is in the grip of a powerful active real estate for an urban renewal integrating the environmental, economic, and social dimensions. 400 housing units are built each year along with this leading economic hub in progress throughout the metropole. Gerland's population will have doubled from 20,000 in 2005 to 40,000 by 2025.

This real estate dynamism in Gerland is partly due to the many public transport systems that link this territory to the rest of the urban area.

11.2.2 The ambitions of the Métropole of Lyon for the district of Gerland

The strategic plan designed by the landscape workshop Alain Margherit and the Obras firm is organized around three identity sectors, in front on the Rhône with large facilities and great residential area, commercial polarities and residential areas are mixed in the center of district and in the east a transitional zone between residential areas and economic/industrial activities.

Besides being an academic and scientific hub of excellence, Gerland is also becoming a place to live where quality and diversity are at the forefront. The different land transfers have transformed this district in a real town center, with living spaces, sports, and cultural spaces (playground, Gerland Stadium, Tony Garnier Hall), commercial and tertiary activities, local services, an efficient public transport network, a place for sustainable transport means, quality housing, and of course the ever-present nature in this urban environment.

Nature in the city is an important part of Gerland's identity. Gerland Park, the 2nd largest park in the City of Lyon, and Parc des Berges are two key factors that offer Gerland real spaces for breathing and relaxing. Gerland has also benefited from the Berges du Rhône project and all its spaces have been redeveloped.

11.3 The MUG project methodology

The project intends to provide the Métropole de Lyon with an interactive planning platform as tool for decision support. This is enabled by the diverse competences of the project consortium which allow to integrate comprehensively urban challenges into computerized models, creating a tool that makes interrelations and interactions visible. Furthermore, the planning platform enables the planning authority of Lyon to develop different futures, to engage in interdisciplinary and transdisciplinary exchange and helps breaking the traditional silo approach to urban planning. This shall be achieved via simulating and visualization of interdependencies. The approach developed throughout the project was based on the consultation of the experts of each of the consortium members, backed by the Metropolitan experts, to provide the tool best suited to the community's needs. This coconstruction was an opportunity to develop everyone's skills and to better understand the new challenges and systemic aspects at stake within the territories. In this sense, the tool is appropriate to the imperatives of integrated planning encouraged in the current transition of the territories.

11.3.1 The birth of the project and the network of stakeholders

In 2014, the Métropole of Lyon joined the EcoCité network gathering 19 cities in France following a call for projects launched in 2011, entitled "Ville de Demain" (City of Tomorrow), working alongside the State, the Caisse des Dépôts et Consignation (CDC) and the Commissariat Général à l'Investissement.

Thus the Métropole of Lyon was contacted by a consortium composed of two major companies Veolia group and EDF and two start-up ForCity and The CoSMo Companies. The initial assumption is that one can transpose to the urban world the approach to complex integrated systems developed in other fields (like biology).

As part of a French National Research Call "Ecocité," the project was granted 950k€ from Métropole de Lyon and 510k€ from "CDC," for a total cost of the project of 4.8 million euros. This funding is made possible thanks to the French program "Ville de Demain," which aims at building a network of actors developing innovation for cities. The project began in November 2014 for 33 months, three phases: project targeting (9 months), development of models and tools (18 months), and test (6 months).

A great progress has been made with enriching and learning understanding for all the members of the group and the Metropole. Urban modeling in multiple sectors is complex and demanding, requiring the support of different stakeholders, including experts from the Métropole and the Urban Planning Agency, industry, operators, and research. In the case of MUG, the exchange provided an interesting overview of the city's organization and processes, its challenges/constraints and needs, as well as the experience already existing in the field of modeling.

An interactive 3D platform allows the decision-makers of Métropole de Lyon to access the tool, set a scenario, and visualize the results either in the form of the 3D visualization or charts and tables. Gerland is considered here as a test case. The approach and models are reproducible for other urban areas.

ForCity is project representative for the consortium design of the decision, support tool, and urban modeling platform (Fig. 11.1)—trustee of the project, integrator of models into the 3D platform, land use and transport models—with Veolia Research and Innovation—air quality and waste model, research and innovation division of Veolia.

11.3.2 Project expectations and objectives

The main objectives of MUG are accompanying the process of urban and social transition of the urban district Gerland, merging visionary urban planning with computerized modeling, connecting urban topics in a holistic way (original journey into the genesis of

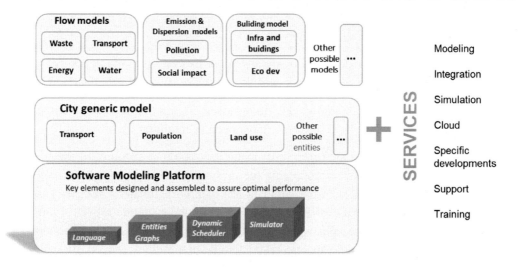

Figure 11.1
The urban modeling platform.

BOX 11.1 Specific objectives of the Métropole of Lyon

1. Development of territorial models feed by data to describe the different field of a territory. Each model can be seen as a component and can be coupled to the other one.
2. Professional "application": design of a political and technical decision-making tool, which allows, on the basis of the coupling of territory modeling, to simulate several scenarios for the evolution of the territory (based on the variation of criteria).

---

these project workshops, interviews and round tables, conferences, etc.), and creating a decision support for the long-term development of the district Gerland (Box 11.1).

How to achieve this? By embracing diversity of understanding to increase our knowledge, by engaging in an iterative inter- and transdisciplinary exchange to overcome silos in urban planning culture, by vocalizing the differences between infrastructure development (fixed hardware of city with long development and maintenance periods), planning, and social processes (software of the city).

Modeling can only provide a simplified model of reality. It could provide additional information and support traditional processes, the challenge would be to include social aspects in the rational approach; this has been initiated with the watchers of the Foresight and Public Debate Department needs to be improved.

To foster the integration of the different stakeholders, the modeling process must be an open exchange process, which was the case in the MUG. Interesting discussions with city representatives led to a better understanding of the needs, which led in some cases, for example, to a gradual change in models.

The challenges on the data side did not always allow us to go into the depth that would sometimes have been necessary to fully understand the dynamics of the Gerland territory.

In order to acquire and create knowledge on multisectoral modeling, a more complete integration of the models would of course have been preferable.

The systemic and integrated approach must make it possible to account for the complexity of the phenomena without overcomplexing the messages and losing the user/analyst. Such a tool must help people understand the major issues facing the territory in order to make a documented decision without drowning the user. While the systemic aspect of transport policies was highlighted, the intertwining of the city's built elements, their economic function, and associated social aspects should also be highlighted. Indeed, for example, the renovation of buildings has an impact on the number of households likely to fall into fuel poverty, but it can also increase the attractiveness of a block/district and as a result it can become more attractive.

In addition, this approach must be complementary to the one usually practiced in the community. Indeed, the latter must be able to continue to use the models and tools at its disposal in a complementary manner. The chaining of tools allows for validating the process, the compatibility of the measurements, and possibly highpointing the counter-intuitive effects produced by the whole system. MUG is a tool that makes it possible to identify trends and measure impacts. It is not intended to replace primary and complementary sectoral expertise but to strengthen it.

In the context of urban planning or urban project development, such a digital tool would provide stakeholders with a more comprehensive expertise than that currently available on many points, particularly in terms of anticipating impacts on mobility and the environment, by extension on the prospects for modeling the negative and positive effects of development on health (air quality, green spaces, cycle paths, noise, so on) or energy, nature in cities, which are and will be central elements of an urban project, themes that have been lately integrated in the projects.

These future models would give the opportunity to design urban planning at the scale of the life area, beyond the Metropole, where the residential choices of the inhabitants, the CT service strategies, and the development of economic land are really at stake (Box 11.2).

11.3.3 Organization and governance

A specific point is dedicated to the project governance. This appears to be a major issue to address and a condition of success for the project.

BOX 11.2 Operational goals

- To test a powerful modeling approach on a territory under real estate pressure;
- To work transversally on new digital tools;
- To overcome silo logic, iterative operational, and strategic approach;
- To think about the construction of the city of tomorrow in partnership with major companies and local start-ups;
- To create a positive dynamic by involving the scientific community and all the stakeholders in Gerland's territory future;
- To be precursor in thinking of interfaces on various urban themes by using emerging and disruptive technology;
- To bet on this experimentation to enrich the conduct of the action, decision support;
- To consider new forms of governance: this project disrupts the conventional project management acquired within the Métropole of Lyon and requires inventiveness in terms of the geometry of the profiles to be mobilized during the construction phase of the models;
- To be prepared for tomorrow's urban planning challenges, integrating financial scarcity and multicriteria Kpi's.

An efficient governance of the project involves different committee and working groups:

- A Technical Committee (monthly).
- A Steering Committee (once a quarter).
- Operational meeting (every 15 days) between the project manager and the ForCity representative.
- Organization of in-depth thematic workshops with the professional experts of the Métropole of Lyon, the City of Lyon, and the consortium: setting up a monthly working group on data, between OVD (Observatory and Data Valuation) within the STPU and ForCity; regular briefings with the Director of Mission Gerland and the Director of STPU; annual examination of EcoCity projects in the presence of the CDC and the Vice Presidents of the Métropole of Lyon.

A specific action was implemented to ensure a good communication between the different partners of the project.

- Two days of launching workshop in 2014 to explain the goal, the issues, and the stakes of MUG project (70 + experts).
- Six newsletters describing the main stages of the project, constituting a basis for capitalization.
- Two days of scientific activities welcoming 80 people with all the entities of the Metropole and combining the world of Research, urban planning agency, and CEREMA. The success of these days is due to the themes tackled: exchanges on complex systems and urban planning, examples of approaches to support urban

planning in a European context, definition of integrated urban planning, social and economic issues in urban modeling, data and the role of simulation platforms in urban projects.

- One day of Feedback to the project partners and 60 contributors.

11.3.4 Data collection and management

The data collection is a key point of the project. The process was carried out using an original, long, iterative methodology. The time dedicated to data collection and treatment has been underestimated. It was a question of specifying the Group's requests regarding the models to be developed but also of studying the availability and existence of the various necessary datasets. The data transmitted by the Métropole, a commitment in accordance with the R&D agreement, were then audited by the members of the consortium.

Various departments of the Métropole as well as the Urban Planning Agency have been mobilized: OVD, the Geomatics and Metropolitan Data Unit that manages the Open Data site of the Grand Lyon, 135 datasets have been collected, 72 of which come directly from the Métropole (28 different and valued producers in the OAD MUG).

Data constitute essential components for the current representation of the territory and its possible futures through systemic modeling.

In the MUG project, the data have three main uses:

- they are used to describe the territory (air cover, ground elevation, building footprint, height, transportation networks, so on.);
- they supply the models with input data (population and employment data, building uses, urban planning, construction rules, demand matrix and transport time, etc.);
- they allow the models to be set (section loads and observed speed, counting points for transport modeling).

11.3.4.1 Assessment

Data within the Lyon Métropole were not sufficiently structured and not exhaustively referenced and cataloged. Based on this situation, it has been difficult to have an overview of data availability. A more efficient qualification of the data would have been useful and efficient for MUG Project. We faced different situations: (1) Some data are sensitive, reserved for local authorities (heating, electricity, and gas networks, VP transport, UTVE) and it has been long and difficult to get the right to use those data (more than 12 months in certain cases); (2) some data are under licenses reserved for local authorities (e.g., MAJIC, RIL) and the Métropole de Lyon does not have the right to share this type of data; (3) recent, sensitive, and no fully validated data has also been difficult to access. For example, the Movement Survey Large Territory—2015 (producer SYTRAL) and the Origin

Destination public transport matrix which are key data and the MUG software; (4) in most of the cases, the historic of the data were not available.

On the Métropole de Lyon point of view, the main interest of this work for services administering territorial knowledge data is the creation of a territorial reference frame within the MUG tool; this result calls for the creation of a "catalog" of available data, which could then be enriched for other approaches. The decision to historize data is also a positive consequence of MUG project.

The MUG project is taking place in a context of exponential growth in the creation of individual and collective data, their openness (open data), their administration (big data); If the opening of public data is constant, that is not the same for the so-called commercially sensitive data (e.g., energy). This retention presents difficulties when one wishes to carry out cross-operating operations, like for MUG, or to update the information used by models.

The long-term data administration after the experimental phase:

To run models over a long period of time, it is necessary to have datasets that are robust over time, which means that the quality of the metadata must be checked on a regular basis, annually, for example, for statistical data. As a result, consolidating a modeling and visualization tool/data "catalog" is essential; one could even talk about dataset maintenance, in the same way as the maintenance of the application itself.

11.3.5 Decision tool's operation

The development of the decision tool is a key milestone of MUG project. The decision tool is composed of four main functions: territory knowledge, action, scenario, and results.

11.3.5.1 Territory knowledge

It provides stakeholders with an extensive description of the urban territory through crossing information from data collected. The description is available at different spatial scales and under three media: map, table, and graphs.

11.3.5.2 Actions

Actions are the levers that city stakeholders are likely to play with. The 17 actions come from the workshops carried out at the beginning of the project. Actions are the following:

- Building in programmed real estate production;
- Configuring a public space;
- Testing the construction rules;
- Creating a lane;

- Building a bridge;
- Extending a tramway;
- Transforming highway into an urban boulevard;
- Building the West City Ring;
- Optimizing the park car offer;
- Strengthening the real estate offer of the Biodistrict;
- Changing the organization of waste collection;
- Modifying the amount of waste produced by incentives;
- Installing decentralized and local means of production: photovoltaic on roof and façade;
- Connecting or disconnecting buildings to the heating network;
- Rehabilitating housing according to recent energy performance regulations;
- Implementing infrastructure for charging electric vehicles;
- Identifying energy poverty.

Most of these actions are already planned by Métropole de Lyon and can describe precisely (capacity, building phase, etc.) but the main interest of MUG software is the possibility to change information attached to each action. It is possible to drag and drop the commissioning date of the new bridge because in the real life, this kind of things has a strong occurrence. Instantaneously, consequences are assessed.

11.3.5.3 Scenarios

A scenario is the combination of city tendencies (e.g., evolution of demography) and actions. The creation of a scenario needs to consider the time. Each action must be activated within the time (e.g., Building the West City Ring is activated in 2025) and some key information—traffic capacity, for instance—must be filled by the user.

Contrary to conventional approach of the prospective, there is no need to have an initial "intention" to build a scenario, to draw tendential 0 carbon BAU (Business As Usual) scenario. The observation of the user of MUG software indicates that experts are "playing" to scenarize their city development based on the operational feasibility of the actions. Once results, there is a loop of trial-and-error to modify parameters of actions and scenario.

On this point of view, MUG software brings a difference with prospective scenarios mainly due to: (1) the fact that the users are day-to-day experts of the city and know precisely what is feasible. This allows a "bottom up" construction of the scenarios; (2) the fact that MUG software is easy to manipulate that allows for building as many scenarios as the user wants. In this case, it becomes interesting to understand in detail which action is contributed to what more than having a "one block," global, and consolidated trajectory difficult to modify easily (Fig. 11.2).

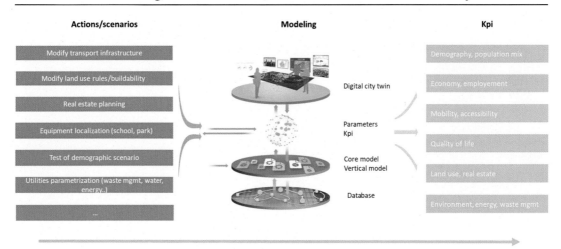

Evolution of the city and geospatial dynamics

Figure 11.2
The interactive planning platform.

11.3.5.4 Representation of the results of the scenarios

Results of the scenarios can be explored in a specific space. More than 60 indicators are calculated and can be represented by graphs, tables, and maps (2D and 3D). Informations are available at different spatial scales: from building to the whole city, and at different time scales: with a timeline, making it possible to move through time (year per year, from now to 2040). It allows for comparing scenarios together easily (Fig. 11.3).

11.4 Use cases of the MUG decision tool

A decision tool has been developed within the MUG project based on 140 + dataset, 20 + models coupled together, and 17 actions. This tool is available on SaaS and allows for

- simulating scenario for the City of Lyon and particularly for the Gerland district as a function of the time from now to 2040;
- helping the decision-making with more than 60 indicators available at different spatial and time scale;
- comparing scenarios each other.

Numerous scenarios can be generated and simulated through the Decision Tool. For the sake of brevity and illustration, two use cases are developed:

- The use case of air quality,
- The use case of urban development.

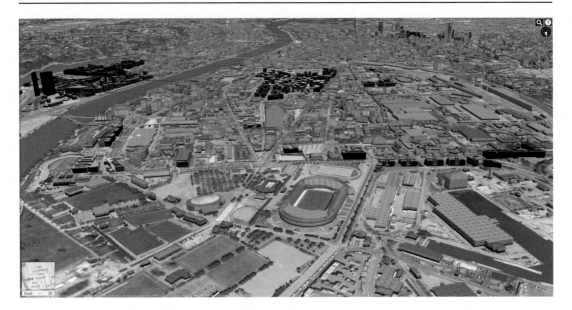

Figure 11.3
Example of 3D view.

11.4.1 Use case 1: Air quality

Air quality is a good example chosen to illustrate the use of the tool developed because deeply systemic. Air quality is an issue for the Métropole of Lyon. The air quality is out of the limits more than 30 days per year and the strong economic development of the city led to increase the emission and the exposition of the population. The development of the Gerland district will contribute to the air pollution.

MUG gives the opportunity to estimate year after year for 10 years the emission of air pollutants and the exposure of population on each point of the city and for each category of population.

The model considers the main factors of contribution of air emission:

- The emission of the transport (car, bus, trucks, etc.). The model takes into account the traffic and the speed of the vehicles. For each person, as a function of its sociodemo profile, the model gives a probability to take a car or to take alternative ways (public transport, walk, bike, etc.);
- The mix of vehicles year after year with the possibility to change the proportion to simulate more or less proactive scenarios;
- The emission of urban development: once a building is constructed, the model fills it with people. This people need to move and will generate emission of pollution;

- The shape of the building has also an influence, in particular, the canyon effect. This effect is taking into account in the 2,5D dispersion model;
- The pollutant emission is in a strong with infrastructure's choices: for example, the building of a bridge will have an impact on the traffic in all the city of Lyon. It leads to smooth traffic flows but will also increase the mean speed of vehicles with a consequence of air emission;
- The emission of energy heating system.

The other sources of air pollution (e.g., agriculture, industry, and diffuse emission) are not modeled.

The tool allows for modeling the population repartition at a building level. The sociodemographic characteristics of the population are also described. Various analyses can be carried out including the impact of air quality on the prevalence of diseases and health problems.

The threshold of the year of reference of each source is taken into consideration and a general tendency for the future is proposed (based on the projection of the past). The user of MUG can also modify it manually with a yearly step (Fig. 11.4).

Some actions will increase or decrease the impact on the population of air quality. MUG takes into account actions that can be taken to reduced emission:

- National regulation: car classification as a function of pollutants atmospheric emissions is compulsory since 2016 in France.

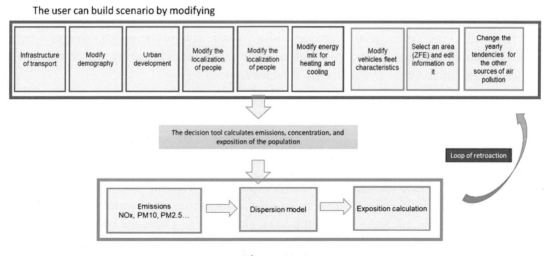

Figure 11.4
The modeling stages.

- National grants: to help people to switch from conventional engines to electric cars. The model allows to modify the car mix to simulate a strong proactive decision for electric vehicles.
- Local regulation: the model allows for forbidding the city to low-level Crit'air classes car. MUG allows the user to select on a map an area such as a low-carbon emission area, and to define in this area the right.
- Shape and high of the buildings: strong effect on dispersion and diffusion of atmospheric pollution (the canyon effect).

Two scenarios of city development were modeled with the aim of detecting the consequences of the projections of population on the car-traffic, the mean speed of vehicles, and the atmospheric air emission of pollutants, directly linked to traffic flows (Fig. 11.5A). Intermediary outputs of MUG provide information about the increase of car-traffic and the repartition of the concentration of air pollutants (NO_2) within the Gerland district (Fig. 11.5B).

NO_2 concentration rates range from dark purple (low rates) to lighter yellow (high rates). The pollution is stronger nearby the streets. New buildings are in green on the maps. We can see that in the left scenario, there is a more intensive urban development that generates the highest concentration of pollutants due to an increase of transport demand and reduction of mean speed of circulation.

Changing the scale enables the user to measure the impact of the decision related to the Gerland district on the air pollution (NO_2) of the metropolitan area. Strong urban development of Gerland will generate transport demand and will increase air pollution emission of the whole city, as shown in Fig. 11.5C.

The user can test the impact of an action, such as increasing the share of electric vehicles in the fleet. MUG enables the user to change easily the objective of the action—mix of vehicles and regulation, for instance—and to define the trajectory year by year to simulate different rates of adoption of the electric vehicles (Fig. 11.6).

Maps of Fig. 11.7 illustrate the impact of the evolution of fleet of vehicles on the air pollution. All the other parameters have been frozen to evaluate the impact of car mix. We can see that the electric scenario (figure right) reduces the global air pollution of 9% within 10 years compared to the pessimistic scenario (figure left).

If the aim is to examine the impact of local policy on air pollution, the user determines a perimeter of a low emission area (ZFE in French) and specifies its characteristics (year of implementation, vehicles authorized in this area, etc.). He can assess the impact of this measure in a scenario where all the other parameters are frozen. The outputs on Fig. 11.8 concern a low emission area surrounding Perrache railway station where the air pollutant concentration is thus reduced to 10%. The user can also evaluate the effect of this measure

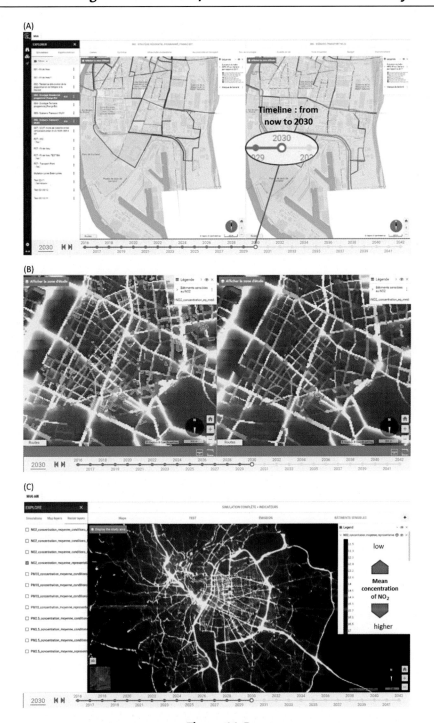

Figure 11.5
Maps of car-traffic (A); pollution levels in the Gerland district (B) and in the city of Lyon (C).

Scenario Electric – Objective 30% of the fleet in 2030

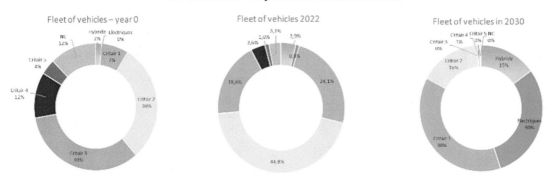

Figure 11.6
Defining the variation of values within the framework of a scenario.

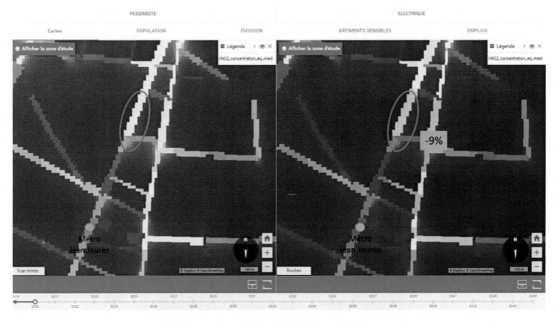

Figure 11.7
Evaluating the impacts of two different scenarios on the air pollution.

at the level of the whole city and therefore become aware of the transfer of air pollution emission to other parts of the city.

11.4.2 Use case 2: Urban development of the district of Gerland

In this use case, MUG has been used to build, run, and compare three scenarios of development of Gerland.

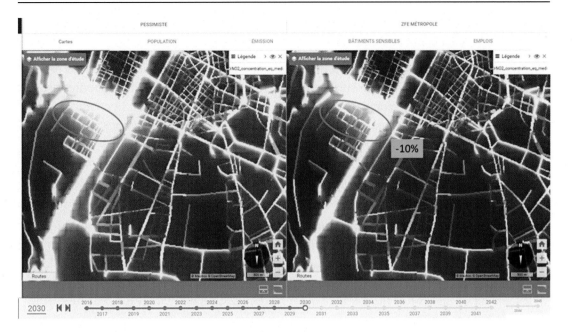

Figure 11.8
Detecting the impact of a low emission area on the concentration of NO_2 pollutants.

Scenario 1 consists in strengthening transportation connection between Gerland and the rest of the Métropole de Lyon. The main action is the building of a new bridge on the Rhone river. In this scenario 1, the bridge is inaugurated in 2025. The question is what the consequences of this new infrastructure on different questions are (environmental, demographic, traffic, quality of live, accessibility, etc.) as a function of year (from 2019 to 2040) and for each entity of the territory of the Metropole? On a macroscopic point of view, it is obvious that a new infrastructure contributes significatively to improve the accessibility of Gerland but such an approach presents two main interests:

1. The capability to asses quantitatively the scenario;
2. The capability to assess systemic and indirect effects.

On this scenario, the outputs highlight the saturation of the traffic of the Gerland district year after year. The street of the district remains the same capacity before and after the bridge arrival. Their capacity is drastically saturated after 2028 (+100% on a certain part of the road system). There is also an indirect consequence out of the Gerland district where the increase of the traffic in Gerland generates traffic around the Gerland district and particularly in the area at the opposite sides of the new bridge. In consequence, the bridge generates traffic indirectly in other districts. On an environmental point of view (e.g., CO_2), the total emission doubled taking into account the indirect effect of the bridge initially calculated without taking account these indirect effects.

On an economic point of view, the increase of the accessibility is improving the district's attractiveness with some consequences.

The population is increasing (in 2030, $+12\%$ compared to the scenario without the bridge) due to the reduction of the mean surface per inhabitant. The prices of the properties are also going up (in 2030, $+15\%$ compared to the scenario without the bridge). Population categories and the average age are changing ($+5$ years after 2025) with consequences on the purchasing power and on the number and age of children and consequently the need of kindergarten and school places. The urban transformation from industrial sites to office building and housing is accelerating after 2025 with a strong decrease of industrial sector and industrial employment.

The variation around this scenario 1 consists in assessing the impact of different options of development: one is clearly oriented on the housing development and the other one on the office and commercial building. Scenario 2 considers that all the programming of building is oriented after 2022 on the housing development while scenario 3 is based on a strong development of the east of the district where there is a strong potential for office building, between 2022 and 2027.

The interest of MUG is to propose a variation of each scenario as a function of assumption or information that stakeholders have. For example, if the user is informed that a developer considers building 10,000 more spare meter than initially plan and achieving the project 1 year later, it is easily possible to build a "daughter scenario" 3.1 and to modify this information on the scenario as well as the general assumption and actions to obtain variation on the root scenario 3, in this example.

Otherwise, MUG fosters the analysis of the systemic and multicriteria impacts of scenarios 2 and 3. The algorithm fills the housing and office building as a function of attractiveness and prices.

In scenario 2, the needs in waste collection point, school, water supply, etc. are increasing proportionally to the number of inhabitants, while in scenario 3 the increase of the demand of water, energy, and waste collection system but also school places are lower due to the land use chosen in the scenario. Indeed, for the community, there is less investment to plan in the next decade. But, as there is no many housing in the district, employees are coming from other districts and generate extra traffic compared to scenario 2. In this case, it is particularly interesting to measure the ratio inhabitant /job and to observe that the actual ratio (1:1) is changing with a consequence of the infrastructure's evolution of the needs of the district. The other systemic effect of scenario 3 has been the rate of inoccupation of offices in other districts. Indeed, scenario 3 generates many office buildings in the Gerland district but there are also many projects in the rest of the Métropole of Lyon. The relative attractiveness of Gerland compared to other districts (Carré de Soie) leads to an increase in the unoccupied rate in other districts.

11.5 Discussion and conclusion

As we can see with the two uses cases developed previously, the computational capacity of the tools is important and numerous of systemic issues of the cities can be addressed with such a tool but the question of MUG project was also: how does professional will adopt it?

The second goal of the MUG projects was to test in real condition the value of the decision tool. A representative pool of 15 beta testers has been selected. The main criteria were the diversity of the beta testers (urbanism, engineering, data, IT, etc.). The goal of the part of the project was to assess the value of the decision tool.

The protocol was (1) validation of the operational capabilities; (2) identification of relevant use cases; (3) effective simulation of scenario in the decision tool; and (4) interpretation and discussion of the results. Tests were conducted for 6 months.

More generally, a great progress has been made with enriching and learning understanding for all the members of the project. One of the main interests has been to extensive discussion between modelers, scientists, and urban world. Important differences in the way to address conceptual issues have been observed and reveal a strong need to redefine a common language. A lot of misunderstanding found their origin in the conceptual mean behind a word.

Urban modeling in multiple sectors is complex and demanding, requiring the support of different stakeholders, including experts from the Métropole and the Urban Planning Agency, industry, operators, and research. In the case of MUG, the exchange provided an interesting overview of the city's organization and processes, its challenges/constraints and needs, as well as the experience already existing in the field of modeling. Before modeling them, the different interactions between the different systems and stakeholders have been extensively described. By definition, modeling is a simplification of reality and each model aims to address a question. The context, the limits, the strength, and the weaknesses of each model have to be described and give relevant information and support for the decision-maker. One of the challenges of the project was to describe through the modeling social aspects in the rational approach; this has been initiated with the "watchers of the Foresight" and "Public Debate Department" but only very simple rules were implemented. Methodology able to capture "social issues" in order to model it needs to be improved and is probably a research field for the next decades.

To enable the appropriation of the tool by the different stakeholders, the modeling process must be an open exchange process. Interesting discussions with city representatives led to a better understanding of the needs, which led in some cases, for example, to a gradual change in models. The agile method applied to the whole process of elaboration of the decision tool has been crucial for the success of the project.

The challenges on the data side did not always allow us to go into the depth that would sometimes have been necessary to fully understand the dynamics of the Gerland territory. In order to acquire and create knowledge on multisector modeling, a more complete integration of the models would have been better.

The approach developed throughout the project was based on the consultation of experts to provide the tool best suited to the community's needs. This coconstruction was an opportunity to develop everyone's skills and to better understand the new challenges and systemic aspects at stake within the city. In this sense, the tool is appropriate to the imperatives of integrated planning encouraged in the current transition of the territories.

However, it seems that training on the use and handling of the tool have been underestimated. A powerful decision tool is necessary deeply changing the way to practise a job. Only a few services within the Métropole of Lyon have taken up the tool. The main reason is that each service is organized vertically with specific and focus tools. The mission of the service is focused on a specific topic and in most of the case does not need external systemic inputs. How can this be explained and how can the dissemination of such a tool be improved?

The systemic and integrated approach must make it possible to account for the complexity of the phenomena without overcomplexing the messages and losing the user/analyst. Such a tool must help people understand the major issues facing the territory in order to make a documented decision without drowning the user. While the systemic aspect of transport policies was highlighted, the intertwining of the city's built elements, their economic function, and associated social aspects should also be highlighted. Indeed, for example, the renovation of buildings has an impact on the number of households likely to fall into energy poverty, but it can also increase the attractiveness of a block/district and as a result it can become more attractive. Such antagonist effects need to be understood by the end user easily. The MUG project underlines that more than a technical challenge there is a change management challenge to move from conventional ways to think and build cities to new systemic, global, multicriteria, and multi-Kpi tools. The market is emerging and today, authors recommend targeting pioneers and early adopters in cities or cities facing to massive constraints. In addition, this approach must be complementary to the one usually practiced in the community. Indeed, the latter must be able to continue to use the models and tools at its disposal in a complementary manner. The chaining of tools makes it possible to validate the process, the compatibility of the measurements, and possibly to highpoint the counter-intuitive effects produced by the whole system.

The integration objective, that is, to "de-pilot" the sectors, was not fully achieved as several experts mentioned the difficulty of being able to validate results that did not concern their sector. Therefore it is worth considering how to overcome this difficulty. Coconstruction was mainly carried out on the themes of air, waste, and energy. There is probably a lack of more cross-cutting workshops to strengthen and acculturate oneself to a systemic approach.

The tests of the tool show that the agglomeration constitutes the right scale of deployment; at the scale of a district, some samples of territorial information are so small that they generate a projection error when the thematic models are run. This risk decreases significantly over larger areas, which are more relevant for a "generalist" modeling tool such as MUG. From this point of view, it can be seen more as a tool to support project owners than as a decision-making tool. It therefore does not compete with the "business" modeling tools already available in the community. MUG is a tool that makes it possible to identify trends and measure impacts. It is not intended to replace primary and complementary sectoral expertise but to strengthen it.

In the context of urban planning or urban project development, such a digital tool would allow more comprehensive expertise than that currently available on many points, particularly in terms of anticipating impacts on mobility and the environment, by extension on the prospects for modeling, the negative and positive effects of development on health (air quality, green spaces, cycle paths, noise, for instance) or energy, nature in cities, which are and will be central elements of an urban project, themes that have been lately integrated in the projects. These future models would make it possible to design urban planning at the scale of the life area, beyond the metropolitan area *stricto sensu*.

To conclude, the main interest of such a tool is the ability to have a systemic approach of the city decision by combining projects, actions (transport, infrastructures, building, etc.), and tendencies of different nature, at different geographic and time scales. MUG is based on simulations and comparisons of scenarios. The goal is not to offer a precise forecast of the future but to assess possible trajectories, to propose tendencies, and to challenge decisions. The decision-maker can have a more global, more systemic view of the consequences of its choices and decide fully informed.

CHAPTER 12

Geodesign for collaborative spatial planning: three case studies at different scales

Matteo Caglioni[1] and Michele Campagna[2]

[1]*University Côte d'Azur, CNRS, Laboratory ESPACE, Nice, France,* [2]*University of Cagliari, DICAAR, Cagliari, Italy*

Chapter Outline

12.1 Introduction

In the coming three decades, the world population will increase by 2 billion people, 68% of which will leave in a city (UN, 2018). Even though the United Nation world population projections show that growth rates are slowing down, in 2050 we will reach 9.8 billion people and a doubling of the urban environment. If we do not shape the cities in the right way—warns the American architect Peter Calthorpe (2012)—we will not be able to face climate change, and probably no technical solution in the world will save mankind.

How we build a city it is intrinsically interconnected to how we live, not only because of our environmental impacts, but also as a consequence of our social well-being, our economic vitality, and our sense of community and connectedness (Calthorpe, 2012). People behavior can shape better cities, overcome the most common urban issues (including climate change and overpopulation) and improve the quality of urban life.

Ecosystem and Territorial Resilience.
DOI: https://doi.org/10.1016/B978-0-12-818215-4.00012-2
© 2021 Elsevier Inc. All rights reserved.

Technological solutions from new Information and Communication Technologies (ICT) have been rapidly developed and proposed to face problems like urban traffic, air and noise pollution, energy efficiency, and insecurity: the Smart City model, riding the rise of digital, combines pervasive data, collected by sensor devices or by "citizens as sensors" (Foody et al., 2017), with the great computing power and the high performance in training algorithms of the artificial intelligence, in order to propose new services to city-dwellers. Sadly, a downward slide and a misleading interpretation of people wishes have been observed in all the smart cities: the citizen's desire of brightness and transparency leads to construction of glass buildings (with no more privacy) and the desire of nature leads to construction of green roof tops and façades (which are not really sustainable). Nevertheless, it would be extremely limited to think the Smart City just as a technological answer to all urban problems. A smart city goes with smart citizens (Hill, 2013), they need to be supported in the accomplishment of their place of life.

Still today, smart cities are often represented as an urban monitoring center, which can manage the city with efficacy, but where the citizens are invisible or absent. People are not just customers of the city, but they want to be active actors (Hill, 2013), understand how and why decisions are taken, have access to decision-makers, get informed and participate to decision process, and get involved in the formulation of laws which govern their daily lives. On closer observation, smart citizens seem to appear as fast as technology in the world: Citizens actions multiply, people get organized, they cooperate, share the resources, take care of the youngest and the oldest, give advices on social medias; they assume their responsibilities, and do not want to be passive citizens anymore.

In this panorama, actions like collaborative Geodesign appear to be extremely interesting and it meets people desires for participation, transparency in the decision-making process, and coconstruction of the place where they want to live.

12.2 Geodesign

Geodesign is a new word, intended both as a noun and as a verb, to describe a collaborative interdisciplinary activity which is prerogative of design professions, geographic sciences, and information technologies at once. Each participant, in a geodesign initiative, should be able to contribute according to his or her competences, yet during the process, no one needs to lose his or her professional, scientific, or personal identity (Steinitz, 2012). Geodesign is not a science or profession, rather it is a set of concepts and methods derived from both geography and design professions. It applies *systems thinking* to the creation of proposals for change and to the simulation of impacts in its geographic contexts, usually supported by digital technology (Dangermond, 2010; Flaxman, 2010; Ervin, 2011). Geodesign can be defined as an integrated spatial planning and design process informed by contextual environmental impact assessment, which includes project conceptualization, analysis, projection and forecasting, diagnosis, alternative design, impact simulation and assessment, and which involves a number of technical, political, and social actors in collaborative decision-making (Campagna, 2014).

Even though design has always been at the center of Geographical Information Sciences, nowadays we can witness a revitalization of the relation between design and geosciences. Michael Batty (2013) explains this sudden occurrence with the concomitance of four specific factors: (1) a sufficiently large arsenal of tools to support the design process, (2) the emerging of sharing technologies and participation platform for design, (3) spreading out of computing and related technologies, and (4) increasing of bottom-up activities. Moreover, the innovation in Geodesign, compared to older approaches in environmental planning and landscape architecture, falls rather on the extensive use of digital spatial data, processing, and communication resources (Campagna, 2014).

12.2.1 Geodesign framework

Geodesign is not a solitary activity, especially when handling serious societal and environmental issues, but rather a collaboration of different people making one final design (Steinitz, 2012). In order to work together people needs two things: a framework for collaboration, a set of rules for information exchange, and a basis for communication that means shared knowledge of the subject, shared assumptions, and shared language (Wiener, 1954). Workshops can be a useful format for collaboration in Geodesign, and diagrams are an essential common language of system thinking. Moreover, it is possible to perform Geodesign really fast (most of the workshops last 3 days), using small data (no need to acquire big data from urban sensors).

Carl Steinitz's framework (2012)—which currently can be considered the main methodology reference for Geodesign—is color coded: Design Professions are in yellow color, Geographic Sciences are blue, Information technologies are green, and People of the Place are violet. People are direct participants and not clients of Geodesign and they can belong to one, two, three, or even four groups at a time. It is not unusual, indeed, that professionals or scientists live in the area concerned by the project and participle to workshops.

According to Steinitz (2012), societal and environmental problems, which cut across size, scale, and theme, share six questions that are asked explicitly or implicitly, and the answers to these questions are models in the geodesign process. The six questions are the following:

1. How should the study area be described in content, space, and time?
2. How does the study area operate and evolve with no intervention?
3. Is the current study area working well?
4. How might the study area be altered?
5. What differences might the changes cause?
6. How should the study area be changed?

The first question is answered by representation models, the data upon which the study relies, which describe the study area evolution from the past up to the present (Fig. 12.1); the second question is answered by process models, which provide information about the study area and its likely evolution with no action; the third question is answered by evaluation models, which are dependent upon the values of the stakeholders involved in decision-making, they put information in context generating knowledge useful for decision-making; the fourth question is answered by change models, which generate data that will be used to represent future conditions; the fifth question is answered by impact models, which are forecast information produced by the process models under changed conditions; and the last question is answered by decision models, which, like the evaluation models, are dependent upon the values of the responsible decision-makers.

Questions 1−3 focus on *assessment*, they refer to the past and to the present condition of the study area, the geographic context. Questions 4−6 focus on *intervention*, they concern the future of the study area, the changes of geographic context. These questions are asked at least three times during the Geodesign process. The iterative process starts with stakeholders, the people of the place, rising specific issues, and asking professionals (designers, geographers, information technologists) to help them with those problems:

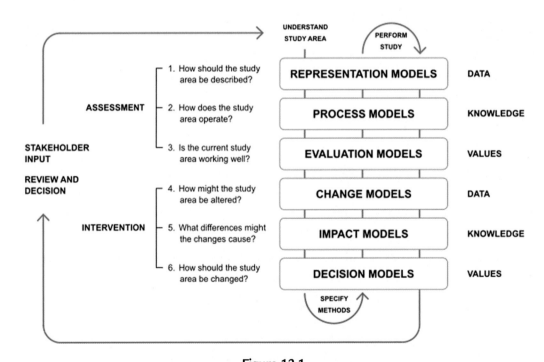

Figure 12.1

Geodesign framework. *Source: Adapted from Steinitz, C., 2012. A Framework for Geodesign: Changing Geography by Design. ESRI Press, Redlands.*

stakeholders need support in decision-making, in order to change things. The first phase is about understanding the study area (values, knowledge, and data of the place), trying to treat 1−6 questions as *Why?* questions, why people want to change the place context, basically why Geodesign is needed. The second phase is about to choose and specify methods of the study, answering the question *How?* in a reverse way, from questions 6 to 1. This phase, unlike the first one, is decision-driven, rather than data-driven. The third phase performs the study using methods emerging from the previous phase, answering the *What?*, *Where?*, *When?* questions. If the geodesign team reaches a consensus or agreement around one design, the study will be sent to stakeholders for revision and decision. If consensus is not reached, feedbacks will be sent to modify models or to change the project scale. Changing scale implies to answer at exactly the same questions, but the answers and the models will be different (Steinitz, 2012).

However, Steinitz (2012) warns that, in practice, Geodesign is neither a so ordered nor a so linear process, but rather participants jump from one question to another, from one model to another, from one phase to another, from one scale to another, and they come back to previous questions, models, phases, and scales with new feedbacks.

12.2.2 Change models

"The best way to predict the future is to create it," a quote credited to both Abraham Lincoln and Peter Drucker, perfectly resumes the core of Geodesign: the change models. How is it possible to move from the present state of a geographical area to its best possible future? The answer to this question is given by Carl Steinitz (2012) though the change models, the ways of achieving the goal in a geodesign study.

In the Steinitz's framework, any change model has four fundamental components. The first one is the *history* of the geographical area, all the past designs, decisions, or interventions, everything occurred in the past of the study area. The second ones are the *facts*, everything that is not supposed to change over the geodesign process, such as geology, rivers, cultural heritage, lifestyle, and place economy. The third component is the *constants*, or everything is certain to occur in the future during the realization of the project. For example, a new highway, which is already proposed, approved, designed, funded, but not yet constructed. Finally, the fourth component of a change model is the *requirements and their options* that is everything that should or could happen during the time, all the intermediary states through which we have to pass to reach the final desired future state.

Steinitz (2012) pointed out nine ways (anticipatory, participatory, sequential, constraining, combinatorial, rule-based, optimized, agent-based, and mixed) to approach change models in Geodesign, and he classified them in three groups. The first group is composed of all approaches that are more suitable to deal with *certainty*, that is, the geodesign team is confident in its ability to directly design for the future. In general, these approaches are based on the personal experience of people who are involved in the project (experts, designers, scientists). Anticipatory, participatory, and sequential change models make part of this group.

In an *anticipatory change model*, the designer is able to foresight the future state of the study area and uses deductive logic in order to arrive this future vision from the present conditions. In this approach, the designer is able to explicit several assumptions and to consider a set of requirements to reach the final solution. Anticipatory approach is successful for small and short-term projects.

A *participatory change model* is quite similar to the previous one, even though it involves more than one designer (often the people of the place) with different visions and aims, who can produce different future designs, which need to be aggregated in a consensus one, resolving conflicts. This is a more democratic and collaborative design approach, and it implies that designers have sufficient experience and sense of place to provide a future-oriented design.

In a *sequential change model*, the future design is developed step by step starting from the present state. The designer makes a series of confident hypothesis and choices that are systematically applied to produce the final design. Sequential decisions are taken by the designer based on his experience and preferences, considering the people's wishes. The main difference with the previous approaches is that the future of the place is not known *a priori* (designers cannot make a leap into the future), but still it is possible to build it during the design process.

The second group is composed of two approaches more suitable to work under *uncertainty*. Different kinds of uncertainty exist about data, about models and theories, about requirements and options, and about assumptions and choices. These uncertainties should be made explicit and they will affect the choice of a change model and how to use it. Constraining and combinatorial models make part of this group.

The *constraining change model* is appropriate when the geodesign team is not sure about decision models or about criteria to select among different options, that is, when there is no dominant solution from Pareto's point of view. To proceed using this approach, the geodesign team discusses every single option for each requirement and tries to rank them or reach a consensus around one. This process ends at the final state of the study area when all requirements have been discussed.

The *combinatory change model* is useful when there are very few requirements or when few requirements have options of similar importance. This approach allows the geodesign team to explore all possible alternatives for few requirements, using different combinations of the project options. The combinatory approach generates a large number of propositions, which are then systematically evaluated to get one of them for the future development.

The third and last group contains four approaches useful when the geodesign team cannot directly foresee the future of the place, but still it is not under fully uncertainty conditions. Indeed, these methods are *rule-based*, where the geodesign team is confident enough to specify a set of formal rules to build the final state of study area in the future. Most of the time these formal rules are represented as a set of computer algorithms for design development. Rule-based, optimized, agent-based, and mixed change models belong to this group.

The *rule-based change model* is organized as a set of computer algorithms or mental steps which can be followed manually. These algorithms are a precise sequence of rules for each requirement. The rules are combined together during a sequential approach to build step-by-step the future scenario. In a complex system, the rules can be depending on evaluations, taking into account feedbacks from the project. This approach needs a calibration on the historical data and a sensitivity analysis to understand the variability on assumptions, requirements, and options underlying the sets of rules.

The *optimized change model* is useful to understand automatically the relative importance of the requirements and their options. In order to do this, the requirements should be comparable through a single metric such as costs, energy, and votes. To use an optimized change model, geodesign team should be able to understand and integrate the decision process into the rules.

The *agent-based change model* is adapted to simulate behavior and interactions of individual agents, which can represent people, vehicles, households, or land parcels. Belong to this class Cellular Automata (Benenson and Torrens, 2004) and Multi-Agent Based (Heppenstall et al., 2012) simulation models, well known and already applied to geographical contexts. Cellular Automata are used to simulate future land use changes and agent-based models are interesting to simulate people interactions. This approach assumes that the future of the study area is the results of behavior and interactions of different independent agents, actors, and stakeholders.

Finally, the *mixed change model* is an approach that consists in combining together different types of change models. The number of possible combinations of change model is almost infinite. Each model can be combined in whole or in part. The outcome could be one or several designs, which should be analyzed and geodesign team should reach a consensus on one design.

The next steps in the geodesign process will be assessed and compare the impacts of each change model and reach a consensus on the final project to present to stakeholders for their decision and possible implementation. Geodesign has been developed and used for an *ex-ante* phase of strategic planning and helps stakeholders to work together and resolve project conflicts during a negotiation phase. Other tools are necessary to evaluate the project *in-itinere* or in the *ex-post* phase, that is when the project is definitely realized.

12.2.3 *Geodesign versus Geoprospective*

Between Geodesign and Geoprospective (Voiron-Canicio, 2006; Gourmelon et al., 2012), there are several analogies and differences alike we can observe between Design and Geography (Steinitz, 2012). If we look at them under the lens of project features, it becomes possible to define the most suitable application context for the two. Both of them do not work at any possible scales: from building room (1:10) to planet (1:100,000,000). Geodesign and Geoprospective share the will for place changing to make it better; however, the former will be more suitable for local or regional scales, and the latter for regional or global scales. Obviously, their projects will overlap over spatial scale, exactly like over any other scale (Fig. 12.2). Both Geodesign and Geoprospective take part in the foresight

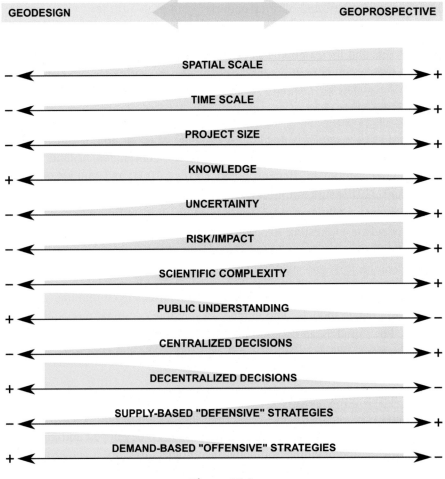

Figure 12.2

Comparison between Geodesign and Geoprospective as project function.

process to evaluate the spatial changes related to different planning options or to outgoing dynamics; nevertheless, Geodesign will better fit a smaller time scale, matching with project and planning activities, rather than a larger time scale like Geoprospective, studying long-term changes. The size of project and the size of study objects are different. Geodesign will take into account individuals, ideas, singularities, and will work on projects based on experiences, while Geoprospective will consider communities, hypotheses, typologies, and will work the most on models and scenarios, even based on policies.

Since scale does matter (Steinitz, 2008), when the project size becomes smaller, people have a better understanding of the context and the knowledge about their own city-block or neighborhood is logically greater. When we move to a bigger size or to a global scale, we deal with diversity, and our understanding and knowledge about every single element of the project is much more limited, increasing uncertainty. Nevertheless, Geoprospective can face a greater scientific complexity because the number of elements and their interactions increase in an exponential way with the project size, but at the same time the impacts and risks of stakeholder decisions in this case will reach a greater population and they will be heavier than the ones of projects at the local scale.

On the opposite of knowledge, Geoprospective have to treat a greater and multiple degree of uncertainty (Fusco et al., 2017). It should be made explicit and not hidden, because it could be an enrichment for the process. This pervasive uncertainty is not only related to data and information that we can acquire about the geographic context, but also about definition of project elements, theories, and models that we use in the project, representation of geographic objects, without to enumerate all subjective decisions that affect the study or the planning project.

Geodesign and geoprospective activities are naturally involved in decision support. Depending on the project size and scale, decisions can be centralized or decentralized (Steinitz, 2012). At the local scale, people are able to decide by themselves, due to their greater knowledge about the place and to the minimum risk related to their decision. When geodesign projects or geoprospective studies face a larger scale (in space and in time), the complexity and the risks increase, and the understanding of local people will be quite limited; therefore, decisions will be taken by a small group of experts, who will consider people desires.

Finally, geoprospective strategies are relatively more focused on conservation and organization than geodesign ones: they are meant to preserve landscapes, ecological systems, and cultural values, and they are supply-based because existing resources are their priority. This does not mean that nothing should change, but on the contrary it is necessary to definitely change the way to live, the way to occupy the space, and the way to plan the cities. Geodesign strategies are more related to development and changes, more offensive, focused on details and feelings, and driven by people demands.

The following paragraphs will introduce three case studies on Geodesign applied on participative spatial planning considering three different scales.

12.3 Geodesign applications at the regional and metropolitan scale

Although the geodesign approach is deeply rooted in the tradition of landscape and environmental planning and architecture (Steinitz, 2012), its current success since the last decade is more strictly related to advances in digital data availability and in planning support systems (PSSs). Among others recent technology innovations, the web-based Geodesign Hub PSS enabled the diffusion of intensive workshops as a way to implement collaborative Geodesign. In just few years after Geodesign Hub creation, many collaborative geodesign workshops were tested in research, education, and practice worldwide, generating wide interest in the planning and design community.

Geodesign Hub (*geodesignhub.com*) is a web-based collaborative application for Geodesign, created by Hrishikesh Ballal, PhD in Geodesign from the Centre for Advanced Spatial Analysis at University College London and with experience in leading and organizing more than 120 geodesign projects. Geodesign Hub is a project management platform used in the context of public policy, infrastructure investments, tourism, environmental management, citizen engagement, etc. It helps in analyzing, streamlining, organizing, and efficiently coordinating project activities. Geodesign Hub manages and coordinates activities across many organizational stakeholders in different geographical contexts: communities, cities, regions, urban, rural, marine, and littoral. It supports the creation of real-time plans including project schedule and sequencing. It is able to interrogate spatial data and social interactions even though it does not produce high-quality rendered final scenarios. It can use open data and customer's data. It offers management and analysis tools to understand financial implications and organizational stresses, systematic methods and technology to coordinate different stakeholders, and digital negotiations technology to develop consensus and agreements.

This section presents the objectives, the process, and the outcomes of this approach based on the literature review and on the direct experience of the researchers of the UrbanGIS Lab at the University of Cagliari in applying the approach to several case studies in Sardinia, Italy. A number of collaborative intensive geodesign workshops with Geodesign Hub within several geodesign studies aiming at testing its reliability in research, education, and real-work practice were conducted.

The intensive workshop is a collaborative geodesign format implementing the intervention phase of the geodesign framework, that is, the change, impact, and decision models of the Steinitz's framework. While the preparation of the workshop may require several weeks in order to implement the representation, the process, and the evaluation models in the study area, the workshop itself requires usually 2 full working days for design teams to create design alternatives based on their interest and values and to negotiate final syntheses based on consensus.

Since one of the core features of Geodesign is applying system thinking, the input of geodesign workshops is the output of evaluation models, that is, the evaluation maps relating to the systems of interest in the study area. Evaluation maps can be defined as the spatial pattern of values and preferences of the local community (however, it may be represented depending on the local context) with regard to the ongoing territorial dynamics. As such, they represent the common territorial knowledge on the basis of which change proposals (i.e., design alternatives) are created by the actors involved in the planning process. Being fast and intensive, geodesign workshops are not intended to necessarily produce detailed spatial plans, but rather to achieve consensus in spatial development strategies, and in rising the understanding and the awareness of the involved parties on the complex challenges of territorial development in the study area. They are collaborative in essence, as they usually involve up to 30 people in the design (but cases were developed with up to 70 participants) having a role of the planning actors (e.g., decision-makers, professional experts, community stakeholders, and citizens in some cases).

While the detailed workflow of each workshop is customized on the basis of the study objectives and of the local context, the process usually involves three main steps: (1) the collection of change action proposals (i.e., project or policies represented in form of georeferenced diagrams) in a sort of collaborative design crowdsourcing; (2) the selection of actions in order to develop comprehensive integrated planning alternatives, or syntheses; and (3) the negotiation to find agreement on a final spatial development scheme based on consensus.

The first phase of the workshop originates a matrix of possible change actions (Fig. 12.3) proposed by the participants and arranged by systems (i.e., columns). In Geodesign Hub, each element of the matrix is an interactive geobrowser where geographic diagrams are collected.

In the second phase of the workshop, diagrams can be selected by geodesign teams depending on their preferences and priorities, and once integrated comprehensive design syntheses are created, they can be shared and compared among teams (Fig. 12.4). It should be noted that syntheses are created along several design rounds in the light of their interactive performance assessment, which is based on the application of the impact models encoded in the platform to the current version of the change model.

By default in Geodesign Hub, the impacts of change actions on the actual system are calculated by multiplying the site evaluation/suitability maps (Fig. 12.3) of each system by the overlapping areas of project diagrams (Fig. 12.3) of the same system: low-density housing (LDH) to accommodate demography growth limiting fragmentation and sprawl; mixed high-density housing (MIX) to accommodate demography growth with densification

SITE EVALUATIONS

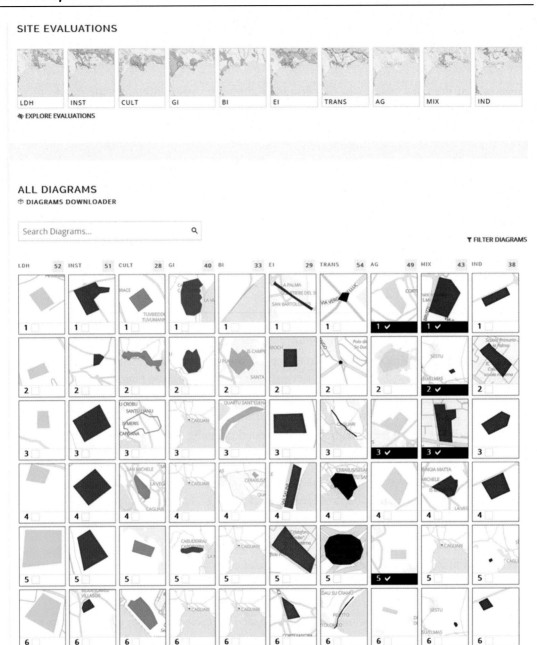

Figure 12.3
Geodesign Hub: projects and policies diagrams interactive matrix.

SYNTHESIS COMPARISONS

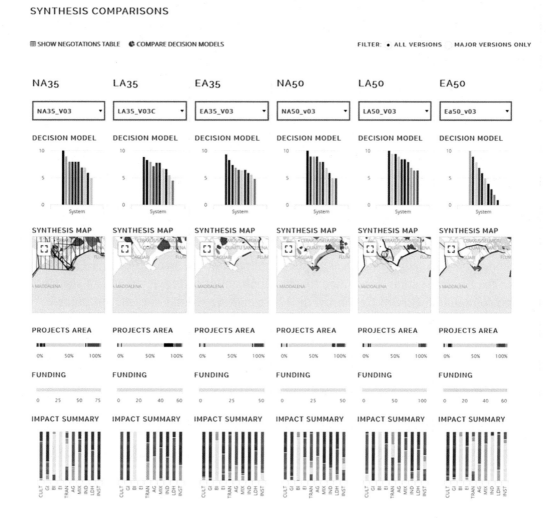

Figure 12.4
Geodesign Hub: integrated comprehensive design syntheses comparison.

and enhance accessibility to commerce and services; institutional uses (INST) to promote locational accessibility and balanced spatial distribution; industry and commerce (IND) to promote technology innovation in production and balance spatial distribution within the metropolitan city of cagliari to reduce workers commuting; cultural heritage (CULT) to preserve historic centers and protect archaeological sites while enhancing fruition; green infrastructure (GI) to enhance connectivity and expand protection to natural or seminatural areas; blue infrastructure (BI) to reduce hydrological risk and limit pollution of resources; energy infrastructure (EI) to increase green production through technology innovation, promote local production; transportation (TRANS) to improve accessibility to the most populated areas and the level of service of current road infrastructure, improve the light-rail network, foster active traveling in recreation, leisure, and cultural heritage accessibility; agriculture (AG) to protect prime soils, promote bioproducts, and foster innovation in production to address desertification processes, climate change, and possible future shortage of water. The impact of each project is synthetized in a score between -2 and $+2$. Quite simple, but calculated *on-the-fly*, which is really important during the negotiation phase, since it increases the interaction speed. This kind of approach is suitable during strategic phases where the accuracy is expendable for the benefits of interaction in dealing with complex problems with a systemic approach. By the way, the Geodesign Hub platform allows customers to modify interactions and impacts calculations.

In the third phase of the workshop, negotiation interactive support tools available in Geodesign Hub (Fig. 12.5) facilitate reaching consensus among the participants.

The geodesign workshop format was applied several times to two case studies at different scales. The first case study concerned the development of future development scenarios for the Metropolitan City of Cagliari (Fig. 12.6), and the second one, the sustainable tourism development in the Oristano Gulf (Fig. 12.7). While in both studies land-uses and infrastructures, as well as development and conservation policies were the main system under consideration to inform the design, the size of the areas was varying, involving 17 and 5 municipalities, respectively. In both cases, the approach was proven useful and effective. In addition, in the Cagliari case study (Fig. 12.6), a change of scale was also experimented as a first workshop was carried on in the study area (approximately 80 by 80 km), while a second one focused on a change on scale on a smaller nested area (approximately 20 by 20 km) in order to test an increase of design detail in the spatial development scheme.

Both of the case studies were carried on several times in education and research conditions, each time testing new settings both in the systems selection, in the workflow, and in the actors' roles. In the first sets of workshops, the selection of systems was adapted to a local baseline scenario, while in the second set of experiments the International Geodesign Collaboration standard settings were tested. Altogether, the various experiments contributed each to earn further insights into the complex territorial development challenges in the study area (the subject) and to earn insight into the efficacy of the method in different decision-making context (the process).

The aim of geodesign workshops made with the Geodesign Hub platform is the collaboration among different stakeholders in the first phases of the strategic planning. The output of this collaboration is different plan alternatives made by stakeholder groups, each one with its interests and priorities. The following negotiation phase is specially designed to resolve the conflicts and leads to a shared final synthesis.

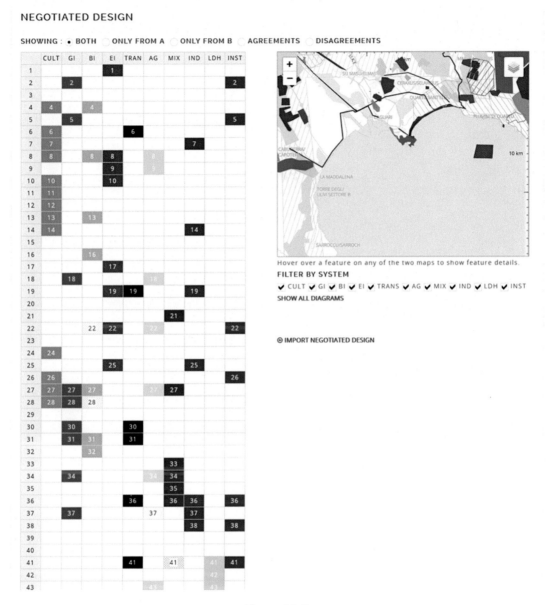

Figure 12.5
Geodesign Hub: negotiation interactive support tool.

The case of Oristano Gulf in 2019 was one of the first geodesign project in the world applied in real practice. This geodesign process led to a strategic plan shared among five municipalities and several private socioeconomical actors. It took 3—6 months/person in order to setting up all the assessment phase in Geodesign Hub (representation, process, and

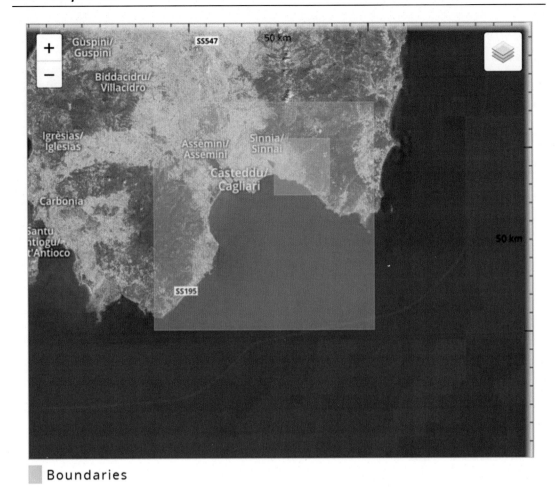

Boundaries

Figure 12.6
Metropolitan City of Cagliari case study areas in Geodesign Hub.

evaluation models of geodesign framework), and the workshop itself took just 2 days for the intervention phase (change, impact, and decision models) in order to acquire an enhancement in the cognitive processes of the actors about the territorial issues and to reach a consensus about options and alternatives, which are strategic and based on real projects developed on the different municipalities of the study area.

12.4 The case of Méridia Neighborhood in Nice, France

Méridia Neighborhood is an urban development zone (ZAC), nearby Var river, in the west part of Nice municipality in France. This area, which covers 24 hectares with a building capacity of 370,000 m², is the object of a national interest project (OIN), and several

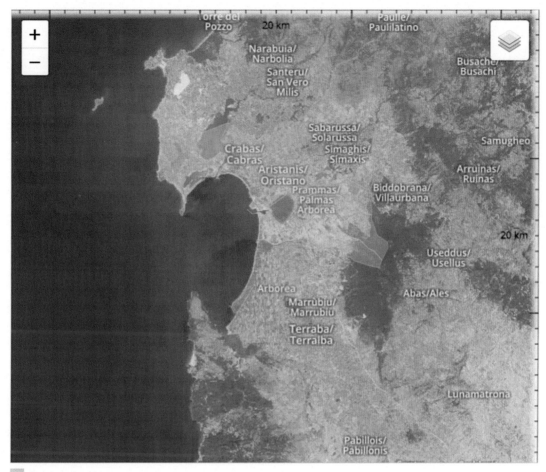

Boundaries

Figure 12.7
Oristano Gulf case study area in Geodesign Hub.

stakeholders are involved in its development. The National Planning and Development Authority (EPA), as project owner, announced for 2030 around 2500 collective dwellings and 5000 jobs spread over different activities such as private companies, retails and services, private and public research laboratories, and the new university campus of Nice. The general contractor Devillers & Associated is in charge to elaborate the site plan, which includes building volumes and their functions, and ensure the architectural and landscape coordination and its consistency. This plan is not fixed, but it constantly evolves during the time, which makes the full meaning of the geodesign process applied to this case study: it is important to be able to evaluate the environmental and social impacts of this strategic planning, considering different possible scenarios based on shapes and heights of buildings, new streets and squares, retail activities and services, current and future social context.

Geodesign activities took place at the Mediterranean Institute of Risk, Environment and Sustainable Development (IMREDD), member of University Côte d'Azur (UCA), in collaboration with three research units: *UMR ESPACE* (geography, spatial analysis, and planning), *UMR CEPAM* (history, culture, and environment), and *URE TransitionS* (media, knowledge, and territories). Geodesign was supported by two research projects in 2018, REACTIV'cities (in partnership with the French Electric Group EDF and its research center EIFER) and TransMetroMed (in partnership with the Architecture University of Genoa in Italy, and the Architecture, Urban Planning and Environment Council of Nice department). Two master programs at University Côte d'Azur were involved in the geodesign process, around 20 students during a week learning module: the second year of *GEOPRAD* (Geoprospective, Planning, and Spatial Sustainability) diploma and the *Engineers for Smart Cities* diploma. Other local authorities and private companies were invited, but unfortunately it did not result in a real participation of all stakeholders. People of the place were associated with this geodesign project and several surveys were carried out, in the Méridia Neighborhood, by ESPACE and TransitionS research units (Fig. 12.8).

In order to allow participants to manipulate and interact with geographical data, perform spatial analyses, measure or calculate spatial indices, build different scenarios of neighborhood development, and visualize them in a 3D virtual environment, an ESRI commercial software

Figure 12.8
Méridia neighborhood in the west part of Nice municipality in France.

called *CityEngine* was used (www.esri.com/cityengine). CityEngine is developed since 2008 by the ESRI R&D center in Zurich. It is a multiplatform software (Windows, MacOS, Linux), which has been chosen by Garsdale Design LTD (www.garsdaledesign.co.uk), pioneer in 3D spatial information services in the world, and applied in France in the Eiffage Phosphore project (www.eiffage-phosphore.com) for the 3D city model of Marseille. CityEngine allows users to import and manage volunteer open geographical data from OpenStreetMap (www.openstreetmap.org), like building footprints, their heights, road networks, digital elevation model, and satellite images of the study area. OpenStreetMap data is available for 2016 in Nice Méridia neighborhood, and this was useful to elaborate the preliminary scenario, showing the actual situation of the blocks in 2016 (Fig. 12.9).

CityEngine can manage multiple scenarios of the same area, and a new scenario has been built to receive 2D data from Devillers' site plan for 2030 and several expected buildings of the central block called Joia Méridia. In order to import 3D buildings in CityEngine with CityGML standard (LoD3) (Nagel et al., 2009), it has been necessary to create a 3DCityDB geodatabase (www.3dcitydb.org).

Concerning geodesign representation and change models, one of the advantages in using CityEngine software is the possibility to rapidly generate large 3D models in a parametric way. Actually, it is possible to produce 3D buildings or other geographical objects, at any Level of Detail (LoD), thanks to geometrical construction rules, which it is possible to interact with through their parameters. These rules are written in CGA (Computer Generated Architecture) language, which is a formal programming language that allows to build all 3D objects of the study area (buildings, streets, parks, furniture, etc.), and to calculate all Key Performance Indicators (KPIs) of the 3D model in the assessment and intervention phase. KPI values are visualized in a graphic dashboard and directly on the 3D

Figure 12.9
CityEngine: starting scenario of Nice Méridia neighborhood on 2016.

model to have a better understanding of the impacts of changes. CityEngine is a 3D geographic information system which performs easily network multiple centrality assessment (i.e., local and global Betweenness, Closeness, and Reach) and 3D visibility analyses (i.e., Viewshed, Dome 360 degrees, and View Corridor). It was unnecessary, for our purposes, to create rules from the scratch, but we adapted ESRI construction rules from the Redlands Redevelopment geodesign project in California (ESRI, 2014). For a better interactive experience, the final scenarios have been exported in a Virtual Reality platform, called *Unreal Engine Studio*, which allows user to explore the virtual world through an immersive technology (HTC VIVE).

This geodesign workshop falls within a research framework on possible evolutions of the functional structure of an urban area under construction. What was expected by participants is the anticipation of Méridia Neighborhood dynamics in the next two decades related to climate changes and its socioeconomic organization. Moreover, it was interesting to assess the energetic impacts of design on the study area through different future scenarios, evaluate air quality and noise pollution, and calculate the peak runoff due to the soil sealing, since the flood risk is extremely high in the region. In order to compare different scenarios made by participants, the following KPIs have been calculated in the CityEngine platform (Table 12.1). Using a semantic 3D model of Méridia neighborhood, it has been

Table 12.1: Key Performance Indicators calculated in CityEngine for Méridia Neighborhood.

Parcels/buildings/floors	Number of flats	Water runoff (m^3/h)	Slope (%)
Energy consumption (kWh/year)	Orientation (°N, °S, °E, °W)	**Green spaces**	Number of parking places
Heating consumption (kWh/year)	Window area (m^2)	Area (m^2)	Parking places area (m^2)
Water consumption (L/year)	Green area (m^2)	Construction costs (€)	Sidewalk area (m^2)
Domestic waste (kg/year)	Parcel area (m^2)	Number of trees	Number of bus stops
Graywater production (L/year)	Built-up area (%)	Tree cost (€)	Number of vehicles
Blackwater production (L/year)	Gross floor area (m^2)	CO_2 stocking (kg/year)	Number of pedestrians
Solar generated energy (kWh/year)	Building volume (m^3)	Water runoff (m^3/h)	Closeness
Construction costs (€)	Green space ratio (%)	**Streets**	Betweenness
Construction waste (kg)	People socioeconomic class	Construction costs (€)	Integration
Number of jobs	CO_2 production (kg/year)	Length (m)	Water runoff (m^3/h)

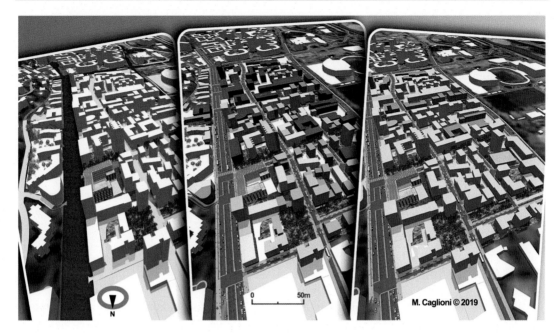

Figure 12.10
CityEngine: integration, energy consumption, land use.

simple to compute KPI on different geographic objects such as parcels, buildings, floors, public green spaces, and streets.

An example of visual KPI representation, resulting by geodesign activities in the Méridia neighborhood, is shown in Fig. 12.10.

The first 3D map on the left (Fig. 12.10) shows the integration assessment value (blue to red) of the Méridia street network, it gives information about the centrality of streets as a function of number and direction of the connected street segments. In the middle (Fig. 12.10), the energy consumption (green to brown) of the different buildings as a function of lifestyle of people that are supposed to live there. The last image on the right (Fig. 12.10) shows the land use of the neighborhood: street segments, green areas, built-up areas, apartments (yellow), offices (light blue), retail activities (dark blue), parking (gray), and university or learning centers (pink). In the platform, stakeholders are able to modify every single element, every geometry, every value and all indicators are recalculated on-the-fly.

This workshop has been a stimulating academic challenge: students learned how to perform a geodesign activity, involve stakeholders and discuss with them, and anticipate the future of the study area through a participatory change model approach. The assessment phase and its data have been prepared in advance by the students and the academic staff.

CityEngine has shown how 3D parametric modeling can be a major asset in the coconstruction of future scenarios for planning activities in Méridia neighborhood. Some relevant remarks are emerged from this geodesign process, for example, the need of public interfaces among the surrounding neighborhoods, or people demand to have a greater green connection between the hills of Nice and the Var river, or again a better consideration of traffic circulation, walkability and mobility demand, which can have an impact on the air and noise pollution in the block.

The activity carried out in the geodesign workshop on Méridia neighborhood did not end up on a master plan composition, since most of decisions on the study area were already taken, but it was beneficial, as we could expected, in showing a different approach for urban planning and design, and join local people expectations with professional points of view during a participatory activity.

12.5 Conclusions

In all experimented cases, several advantages were proven as evident workshop outcomes including the possibility to involve a great number of participants in collaborative design, achieving in a very short time a better understanding by all the involved actors about the complex territorial development challenges, and finding consensus on spatial development schemes, mediating conflictual values and priorities.

Design methods depend on the scale, and when planning at the regional scale, the number of actors involved tends to grow substantially. The geodesign workshop format demonstrated to be very successful in supporting collaborative design. With the workshop format, all actors have the possibility to contribute to the same level under a common language. This aspect was proven successful in education, research, and real-world practice settings. According to the feedback collected formally and informally in the various workshops, the participants were always favorably impressed by the level of interaction and collaboration achieved is such a short time as in the workshop. Other case studies documented in the literature confirm our conclusions.

While in the case of the workshop carried out in education settings, it was impressive observing that students with little experience in planning and knowledge on the territorial dynamic in the study area, in the workshop time grasped the main issues at hand and were able to develop meaningful design. In the case of real-world workshops, it was likewise possible to note as the understanding of the complexity of the development challenges evolved smoothly during the interaction of the different actors.

Geodesign activities are spreading out all over the world with hundreds of workshops based on Steinitz's framework. Like Geoprospective, Geodesign is an emerging research field, even though at the opposite of the former, it is not so diffused in France yet. Geodesign and

Geoprospective share several common points, such as the anticipation of the evolution of the study area, or the willing for public decision support. They show also some differences, for example, in their purposes. Geodesign is useful at the beginning of a strategic planning activity, in order to build, together with stakeholders, different plan alternatives and reach a consensual decision on projects. Geoprospective has no strategic goal, but it aims to recognize, understand, and anticipate the territorial development through the study of its spatial dynamics in the context of a volatile, uncertain, complex et ambiguous (VUCA) future. Nevertheless, their activities are strongly overlapping, and we are sure that each one can benefit from the other approach. Geodesign framework could be adapted and help to formalize Geoprospective, without to fall in a merely normative approach, and Geodesign could take advantage from the geoprospective approach to the change models in order to anticipate the future of the study area.

References

Batty, M., 2013. Defining geodesign (=GIS + design?). Environ. Plan. B Plan. Des. 40 (1), 1–2.

Benenson, I., Torrens, P.M., 2004. Geosimulation: Automata-Based Modeling of Urban Phenomena. John Wiley & Sons, Ltd., p. 287.

Calthorpe, P., 2012. Urbanism in the Age of Climate Change. Island Press, p. 225.

Campagna, M., 2014. Geodesign from theory to practice: from metaplanning to 2nd generation of planning support systems. TeMA J. Land Use Mobil. Environ. 211–221 [special issue].

Dangermond, J., 2010. Geodesign and GIS: designing our future. In: Buhmann, E., Pietsch, M., Kretzler, E. (Eds.), Proceedings of Digital Landscape Architecture, Wichmann. pp. 502–514.

Ervin, S., 2011. A System for Geodesign. Anhalt University of Applied Science, Berlin.

ESRI, 2014. CityEngine Example: Redlands Redevelopment. ESRI Press, p. 18.

Flaxman, M., 2010. Fundamentals of Geodesign. Anhalt University of Applied Science, Berlin, pp. 28–41.

Foody, G., See, L., Fritz, S., Mooney, P., Olteanu-Raimond, A.M., Costa Fonte, C., et al., 2017. Mapping and the Citizen Sensor. Ubiquity Press, p. 400.

Fusco, G., Caglioni, M., Emsellem, K., Merad, M., Moreno, D., Voiron-Canicio, C., 2017. Questions of uncertainty in geography. Environ. Plan. A Econ. Space 49, 2261–2280.

Gourmelon, F., Houet, T., Voiron-Canicio, C., Joliveau, T., 2012. La géoprospective, apport des approches spatiales à la prospective. L'Espace Géogr. 41 (2), 97–98.

Heppenstall, A.J., Crooks, A.T., See, L.M., Batty, M., 2012. Agent-Based Models of Geographical Systems. Springer, Dordrecht.

Hill, D., 2013. Smart citizens make smart cities. In: Hemment, D., Townsend, A. (Eds.), Smart Citizens. Future Everything Publications, pp. 87–90.

Nagel, C., Stadler, A., Kolbe, T., 2009. On the automatic reconstruction of building information models from uninterpreted 3D models. In: Geoweb 2009 Conference, Vancouver, Canada.

Steinitz, C., 2008. On scale and complexity and the need for spatial analysis. In: Specialist Meeting on Spatial Concepts in GIS and Design, December 15–16.

Steinitz, C., 2012. A Framework for Geodesign: Changing Geography by Design. ESRI Press, Redlands.

UN, 2018. World Population Prospects 2018 Revision. Department of Economic and Social Affairs, Population Division, United Nations.

Voiron-Canicio, C., 2006. L'espace dans la modélisation des interactions nature-société. Actes du Colloque International Interactions nature-société, analyses et modèles, La Baule.

Wiener, N., 1954. The Human Use of Human Beings: Cybernetics and Society. Houghton Mifflin, Boston, MA.

How do public policies respond to spatialized environmental issues? Feedback and perspectives

Christine Voiron-Canicio[1], Emmanuel Garbolino[2], Nathalie Cecutti[3], Carlo Lavalle[4] and José Juan Hernández Chávez[5]

[1]*Université Côte d'Azur, CNRS, Laboratory ESPACE, Nice, France,* [2]*Climpact Data Science, Nova Sophia—Regus Nova, Sophia Antipolis, France,* [3]*Ministry of the Ecological and Solidary Transition, Paris, France,* [4]*European Commission—Joint Research Center, Ispra, Italy,* [5]*Ministry of Environment and Natural Resources, Mexico*

Chapter Outline

13.1 Feedback on the Sustainable Territory 2030 prospective program carried out by the Prospective Mission of the French Ministry in charge of Sustainable Development and the Environment

Interview with Natahlie Cecutti, conducted on February 7, 2019

Natahlie Cecutti is State architect and urban planner in chief. After 20 years of territorial actions in connection with land development and urban planning in the private and para-public sectors, she led the Prospective Mission of the Ministry in charge of Sustainable Development and the Environment from 2011 to 2017. She is currently an expert to the Head of the Ministry of the Ecological and Solidary Transition's research department for

Ecosystem and Territorial Resilience.
DOI: https://doi.org/10.1016/B978-0-12-818215-4.00013-4

© 2021 Elsevier Inc. All rights reserved.

developing relations with non-State actors on research issues and she also coordinates the European research and innovation framework programs "Horizon 2020" and the future "Horizon Europe," on behalf of the ministry.

How was the spatial dimension integrated into your prospective works? Was it introduced implicitly?

N.C.: I will take the example of the work we carried out on the Sustainable Territory 2030 program (Box 13.1) when I was at the Prospective Mission, the MEDDE's[1] seminal task, initiated in part by Jacques Theys[2] on the principles, bearing in mind that it consisted in integrating the notion of sustainable development into the development of territories (Commissariat Général au Développement Durable, 2013). Indeed, from 2010 onwards, the main objective was to move to a second phase of land development, which until then was mostly devoted to organizing infrastructure and facilities. Jacques Theys and I realized that the global concern about the environment should be reflected in a different manner in the territories.

How to move from sustainable development to sustainable territory?

N.C.: Well. When he left the Prospective Mission, in the context of the prospective work that we were carrying out, we looked into the way of going from the idea of sustainable development to the territory. How to apply the concept to the territory? We thought that, in the main themes of Sustainable Territory 2030, we had to work in a concrete manner on some subdimensions, subprograms included in the fields covered by the Ministry of the Environment. That's how we decided to work on the biodiversity program for 2030 and on Aqua 2030 relating to continental waters (CGDD website).

The program Sustainable Territory 2030

The aim of the *program Sustainable Territory 2030*, conducted by the French Ministry of the Environment's Prospective Mission between 2010 and 2012, was threefold:

To prepare territories to the challenges of CC, social cohesion, biodiversity and green growth;

To set out the visions of a sustainable development for these territories;

To put forward strategic recommendations.

(Continued)

---

[1] MEDDE: Ministry of Ecology, Sustainable Development, and Energy (France).

[2] Jacques Theys was Head of prospective planning at the Ministries of the Environment and of Equipment, scientific director of the French Institute for the Environment, technical advisor for several ministers and co-president of one of the GIEC-IPCC's subgroups. He is vice-president of the Société Française de Prospective and of the Mediterranean Plan Bleu program.

(Continued)

The prospective exercise co-constructed by some thirty local stakeholders, was based on a systemic, holistic and multidimensional approach of the dynamics—economic, social, ecological, climate, energy and institutional—and on the scenario method. The prospective group paid much attention to the notions of abrupt change and crisis, as well as to changes depending on territorial contexts.

Two subprograms—Aqua 2030 and Biodiversity and Territory 2030—devised along the same principles, were conducted at the same time.

The subprogram *"Biodiversity and Territories 2030": a prospective approach structured in 6 steps*

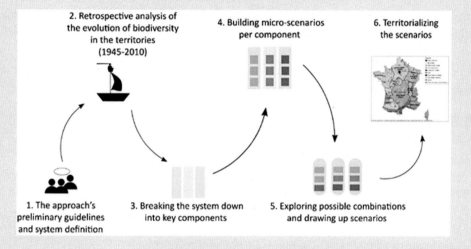

Five global scenarios: Sc. 1 Trend; Sc. 2 Community biodiversity; Sc. 3 Imposed biodiversity; Sc. 4 Ignored biodiversity; Sc. 5 Market biodiversity

Based on: Commissariat Général au Développement Durable, 2013, Biodiversité et Territoires 2030, cinq scénarios d'évolution, synthèse de l'exercice de prospective: volets 1 et 2, p.8. Collection Etudes & Documents de la Délégation au Développement Durable, n° 86, juin 2013.

By conducting the "Water, aquatic environments and sustainable territories 2030" prospective approach, called Aqua 2030, we raised interesting questions. We realized that there were questions of water quantity and quality that had to be addressed in the territories, but that these were not all in the same situation. The differentiation of territories has always been very important in my view. I have always maintained that the question of land development arises on differentiated territories. Each territory has its own components. I'll come back on this notion of components which cross the prospective dimension with the territorial dimension. The components that are specific to the territory, its characteristics, are important elements which are not addressed in the same manner when you conduct a prospective work on ways of life, for example.

Within the framework of Aqua 2030, we worked on the question of water governance, the volume of water, water quality, then on the impact of climate change on water supply, etc. I then realized that the water system had its own logic, that is, you are not in similar logics when you are on an upstream spring or in the forming of the first torrential valleys, when you are in an urbanized environment or in a basin of irrigated crops. There is already a logic of systems showing that you are forced to sequence territories. In themselves, these provide the ideas on the way to work on prospective spaces.

More precisely, the work on Aqua 2030 was carried out at the national level. The Ministry was meant to set the tone to territories, starting from a vision of national recommendations, without interfering in any way in regions, départements, and cities, except in the case of prerogative of major general interest. Therefore the difficulty was to find the happy medium between recommendations of a general nature, applicable in the entire national territory, and targeted recommendations for such and such territory, town, etc.

As such, the water environment provided the possibility to determine typologies on a more reduced scale. We thus were able to understand part of the territory with its own logics, via water, to have an influence on good water management, on the reparation of environments, on several components which determine the water system's values, contributions, benefits, or dysfunctions. It was then becoming relevant to try and find the kind of typologies in which the territories could recognize themselves. We defined seven types of water course named "model-systems with territorialized stakes".[3] A collaboration was entered into with Futuribles[4] and IRSTEA,[5] the aim being to produce prospective recommendations on these series of territories so that any local councilor could recognize his territory's type and reappropriate the recommendations (Hervieu and Jannès-Ober, 2017). At the same time, five global scenarios concerning water were worked out, integrating not only physical elements but also regulatory—European directives, for example—as well as economic, all elements that have an impact on the territory but are difficult to represent. Once these scenarios had been drawn up, we applied them on each of the seven model systems, representing their effects in the form of a block diagram.[6]

That's how water prospective became territorialized, in a form of innovative and subtle in-between, without any interference in the arrangements made by the communities,

---

[3] The seven model systems with territorialized stakes: "Intensive agriculture plain"; "Rural headwaters with high tourist component, Plain-Piedmont"; "Rural headwaters with high tourist component, Mountain"; "River metropolis"; "Coastal metropolis"; "Coastal wetlands"; and "Continental wetlands."

[4] An independent center for prospective studies and thinking on possible futures.

[5] National research institute in sciences and technologies for the environment and agriculture.

[6] cf. Figure 4.6 in Chapter 4.

because that's the role of the state, while leaving room for interpretation and progress to the communities or territories.

How was this implemented? How did you communicate on this modus operandi?

N.C.: We didn't do enough popularization or exploitation because it was a subproduct of the Sustainable Territory 2030 approach. We used more or less the same method for the work on biodiversity. France was divided into large biodiversity systems. Fundamentally, the large biodiversity systems are roughly spread over a geography which has not changed much. Although there have been transformations inside, and species have evolved, there are still territorial invariants. We wanted to work on large systems independently from the administrative division.

The two subprograms, Aqua 2030 and Biodiversity 2030, have fed the Sustainable Territory 2030 program. My original intention was to show prospective orientations on a representation of the territories. This problem is a matter of language. How to visually translate, on the map, prospective principles and orientations seriously, that is, with a somewhat more scientific content than a nice drawing? I wanted to push codification a little further and also use modeling. There were three entries: how to make known data speak (modeling), how to make representation tools speak, and how to insert prospective elements, to make them visible and legible by decision-makers in this instance because they are the ones who implement (Fig. 13.1).

I previously had some experience with choremes[7] on the territory of the Besançon Agglomeration to decipher the principle or prospective orientation of a SCOT,[8] with the aim of working on the symbol, including both text and image, to make complex notions understood by local authorities. With the Sustainable Territory 2030's "Geoprospective" study to which the ESPACE laboratory contributed, we have achieved an original prospective graphic representation.[9]

Until then, the choreme was only a means of making people understand the ongoing dynamics. You have invented the prospective choreme with ESPACE. What is nowadays your view on this geoprospective work? What operational benefits did you get from it?

N.C.: This work was first and foremost a scientific base. Nowadays, local authorities are very busy enhancing the image of their territory, other tools are being used—marketing, design thinking, which is fine. But if you want to go further and help to decipher the phenomena at work on territories, you require sufficiently stable sources, sufficiently updated and available. We are aware of all the stumbling blocks it entails, especially in a

---

[7] A choreme is a geographic model built using a combination of simple graphic forms.

[8] SCOT: Coherent Territorial Planning Scheme.

[9] In addition to the ESPACE laboratory, the consortium in charge of the Geoprospective study included four consulting firms: GAIAGO, ECOVIA, MOEBIO, and 3Liz.

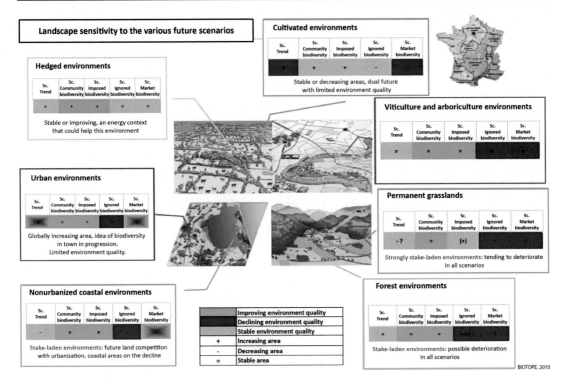

Figure 13.1

Sensitivity of each of the main landscape environments to the content of the five global scenarios. The " + ", " − ," and " = "signs indicate the variation of the landscape's area, and the colors (green, gray, and red) indicate the evolution of these landscapes' ecological quality, still according to the scenarios worked out. A color gradient was used in cases where the evolution of the landscape's quality was uncertain. *Source: Commissariat Général au Développement Durable, 2013, p.27.*

small community of municipalities. True, we have at our disposal the official manna of European websites and of INSEE (the National Institute for Statistical and Economic studies), which are rather well done, but nevertheless, it would be interesting to have more "indigenous" data banks. Indeed, a local representative who wants to carry out this work on his territory is not always on the right scale in relation to the available source data, and since he lacks these, he doesn't do it. Although he could have data that can help him to mature his project or make progress because he has long known his environment, and he has a technical department where people remember certain things even orally. That's vernacular planning.

When people were seeking my expert advice on such territory for a prospective study, I very often realized by consulting the specifications that the amount set aside for the diagnosis was nearly 50% of the total amount devoted to the study and that little was left for the stake and the project. A good diagnosis should account for 25% of the study.

The local community has practically everything it needs, it has collected numerous data, knows about evolutions. Somehow, it carries out a diagnosis every day. The consulting firm is there as a guide to monitor the process. 30%−35% should be allocated to the stakes, and the rest to the project itself, because that's what gives the orientation and above all, that's where questions of decision arise. Yet, in general it is the opposite: a lot on diagnoses, the stakes are passed over quickly and the questions are not really addressed. I think that the prospective exercise must really be rethought differently in its pace, especially in a territory, everything relating to stake and negotiation should account for 75% of the job. It's important to recalibrate the phases of the prospective process.

Lets come to the issue of spatializing the great principles. How, once the main territorial injunctions have been established, to help find more local specificities? How do these principles become spatialized in the actions chosen?

N.C.: Precisely, I see there the right cross between the spatial—a geographic reality on a given territory—and the "human intention," in any case the intention of the group, which is not only that of the institutional but also of the civil society. This is where we enter the operational phase. There is the territorial "concrete," the potential and the territorial capital. Béatrice Villari (2012), a researcher at the Milan Politecnico, has developed, in her book on territorial design, the concept of territorial capital, which is the wealth of a territory. This capital is at the crossroads of three spheres—that of entrepreneurs in the broad sense, the sphere of research, and the sphere of governance, of institutions in fact, and it is this matrix that brings them all together and enables them at a given moment to be inspiring for the future of a territory. There is also the notion of heritage and legacy. You draw on a territory for its potentials. The spatial evolves through the actions of the other dimensions of the territorial capital. When it is decided to set up a factory somewhere, it is not the territory that wants it. There are conditions, and the territory allows for these conditions to be met, but it is indeed the decision that makes that the factory will be set up in that place. Which means that the territory prepares the conditions for the human investment. The human and identity dimension is fundamental: to recover old traditions and rework them to plan for a different future, not exactly the same. What is somewhat passed over in silence is the intergenerational transmission of this heritage, the territorial lineage which imprints the collective unconscious, and which makes that sometimes people succeed in reviving their territory.

This reminds me of the project culture which has always guided me. A lecturer of the Nancy school of architecture used to tell us: "First thing when you arrive in a territory or a space for an architectural project, take an interest in the 'already there'." What are the limits, the constraints, how much room for improvement? The "already there" is inspiring. It's when you start from scratch that you are really lost. If you don't work on that principle you create dysfunction.

There are territorial logics. What I like in territories is that they impose their logics, at a given moment. They are intelligent enough to show us that human intervention, there, was an error and make us understand that it is necessary to repair and make amends.

What is missing as regards prospective in the territories? The needs in geoprospective, on the way of rooting the principles on the spaces so as they are spatially operational and, if possible, effective?

N.C.: A few years ago, I had the opportunity to attend a lecture by Peter Bishop, a futurist at the University of Houston, on the theme of going from prospective to strategy. His presentation on the way to go from idea to action, as the Americans do, was very interesting. You examine whether the action is effective, and if it's not, either you reformulate or you abandon the idea. You have no qualms. You test. I was rather fascinated by that concrete aspect.

I think that we should fairly quickly learn to prioritize prospective orientations that are mature in terms of decision-making, governance, and implementation on the territories, so that you could then field-test them, that is, find an experimentation space, a community, a territory. It is very difficult to apply on a territory orientation of coherent territorial planning scheme that is effective on a Local Urban Plan. For all that, why not do the opposite, while respecting the local or intercommunal town-planning schemes, of course? Namely, choose a territory where the local authorities are game and apply certain orientations on certain projects. Take two or three orientations—climate, biodiversity, for example—, and start, even if the prospective exercise has not been completed. Massify the intentions of projects and practice. If a dysfunction occurs, correct it, have the courage to correct. In an era of experimentation and demonstrators, it could be interesting to create a territorial geoprospective demonstrator, and ask oneself what it is like. We have French experimentation models, like Positive Energy Transition and Green Growth, but on the other hand, you see, we have a European demonstrator such as Nature-Based Solutions which has not always had the expected success in France.

Prospective of the stratospheric kind, brilliant but disconnected, is a preliminary. If you want to adapt it to the territory, you have to delimit it and show that such orientation, when finally applied, is indeed valuable because it is productive. On the other hand, you don't have the complete ecosystem, you don't have the complete dynamics of the prospective exercise. You decide to carry out a one-shot prospective exercise on 1 year, focusing on a specific objective, such as obtaining exceptional water quality within 30 years, for example. You keep only that aspect in a town and you get down to it; you define the interlocutors, and the framework for action allowing to obtain a pollution-free water quality. Big cities do it but it is more difficult for small territories.

Do you think that you are a lone voice? Or do you feel that other people think the same as you?

N.C.: In the days of Sustainable Territory 2030 and "Penser Autrement les Modes de Vie en 2030" (Think differently about ways of life in 2030), I was feeling very lonely, somewhat atypical, but afterwards I realized that I seemed to be followed within the ministry and in the professional networks that were awakening to these problems. All these questions of nature in the city are widely taken up nowadays. There is great progress. Everything relating to freeing public spaces in town and green requalification of traffic is starting to spread. It is gaining ground with the high authorities, and after that, we know that it percolates... On the whole, things are changing. A lot is happening around urban renaturation or restoring agriculture, especially in peri-urban areas, or technological forms of agriculture. In the United States, they are experimenting on *agrihoods*.[10] The European commission is also interested in the subject. In particular, it is working on the European green capital, focusing on sustainable territories. The European green capital is determined considering 12 fields, insisting notably on urban agriculture approaches in the field devoted to sustainable territory. It's an interesting factor that the Commission should now take into account to elect a green capital based on new emerging themes, as was not quite the case 3−4 years ago.

At the moment, there is an array of convergences between those involved in spatial development to include environmental and societal changes, which is promising for the sustainable development of territories.

13.2 Territorial planning in Europe: the contribution of European Commission Joint Research Centre

Dr. Carlo Lavalle (PhD) has over 25 years of experience in modeling and data analysis for policy applications. In 1990, he started to work with the Joint Research Centre (JRC) of the European Commission located in Ispra, Italy. He coordinates the development of the LUISA Territorial Modelling Platform and of the European Commission Knowledge Centre for Territorial Policies. The field of expertise of Carlo Lavalle deals with the evaluation of European policies with an integrated and prospective view by taking into account different aspects like economy, transports, and household. In his researches, Carlo Lavalle integrates all types of spaces and environments and all territorial thematic like infrastructures, tourism, transports, demography, and urbanization. The environmental parameters are integrated into

---

[10] Agrihoods, short for "agricultural neighborhoods" are housing developments centered around community farming with the objective to provide access to local food production and healthy living. Planned in suburban and urban spaces, they are an alternative neighborhood growth model.

the model in some specific cases, especially when these parameters can explain the dynamics of territory.

The integration of common European policies in these researches needs to interact strongly with the European Commission in order to have a better representation of the impacts and evolution such policies.

The approach developed into the LUISA Territorial Modelling Platform and of the European Commission Knowledge Centre for Territorial Policies can be applied at different spatial scales, from plots of 100 m^2 to the European scale. Carlo Lavalle considers that the change of spatial resolution is both a challenge and a way in order to identify the most relevant drivers of land-use changes according to the scale of the perception of the phenomena. In this frame, the modeling process applies statistical methods based on correlations and spatial analysis in order to understand the relationships between the drivers of land change.

In this case, we can say that the methodology can be considered as spatially explicit as mentioned into the technical report relating to the LUISA model "The final output of LUISA is in the form of a set of spatially explicit indicators that can be grouped according to specific themes, defined as 'territorial indicators'" (Lavalle et al., 2017; Jacobs-Crisioni et al., 2017). But some data cannot be considered as "spatially explicit" due to the type of data like the political orientations of the economy.

The integration of future environmental parameters estimated for 2050 and 2100 is also a challenge for the modeling process. Some data are coming from public institutions and organizations like the IPCC (Intergovernmental Panel on Climate Change), others are time series managed by the JRC like remote sensing data.

The Territorial Modelling Platform LUISA aims to evaluate the potential impacts of the European policies and trends on the European territory at local and regional scales. In this frame, LUISA contributed to establish the report on the European Territorial Trends, published in 2017 (Lavalle et al., 2017) that introduces the regional and urban diversities in Europe and their potential development toward 2030.

The spatial representation of the results depends on the European Nomenclature of territorial units for statistics (NUTS) that divides the European territory into subareas according to economic parameters. Three levels are proposed by this classification (Fig. 13.2):

- NUTS 1: major socioeconomic regions, with population thresholds between 3,000,000 and 7,000,000 inhabitants;
- NUTS 2: basic regions for the application of regional policies with population thresholds between 800,000 and 3,000,000 inhabitants;
- NUTS 3: small regions for specific diagnoses with population thresholds between 150,000 and 800,000 inhabitants.

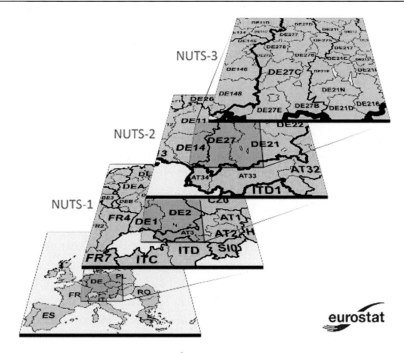

Figure 13.2
Maps of NUTS levels (European Commission Internet site).

In January 2018, the amount of NUTS gathered 104 regions for NUTS 1, 281 regions for NUTS 2, and 1.348 regions for NUTS 3. In order to have statistic data at the lower scale, Eurostat (European Statistical Office) has set up a system of local administrative units (LAUs) compatible with NUTS and called LAU 1 (groups of municipalities but this level is not applied for all European countries) and LAU 2 (municipality level). Some of data integrated by LUISA have been elaborated at the LAU levels. The results provided by LUISA are at the NUTS 2 and 3 levels.

Among the different data that The Territorial Modelling Platform integrates, some of them are related to the public health, the regulatory constraints, the economic drivers, etc. because they have a significant impact on the territorial dynamics. LUISA also incorporates time series in order to identify historical trends and propose current state and future projections of land-use changes. According to these data, LUISA apprehends the interaction between policy scenarios and social, economic, biotic, abiotic, and abiotic drivers in order to propose scenarios of territorial evolution. At the end of this process, LUISA provides results about the demand and the supply concerning the main resources and activities that induce land transformation like housing, infrastructures, transports, industries, energy, biotic, and abiotic resources. The scenarios can be defined as "business as usual" or they can focus on the integration of new policies with direct or indirect territorial impacts on the different scales of the study. The comparison between the results of these scenarios and the baseline situation procures a comprehensive frame to estimate the potential impacts of a particular policy.

LUISA is structured according to three main sets:

1. The Territorial Knowledge Base: it gathers data layers with the finest spatial resolution on the main territorial elements and driver like human and industrial settlements, economic indicators, infrastructures, climatic data, and energy;
2. The advanced analytical and modeling modules: it is the core of the modeling platform. It simulates the spatiotemporal dynamics of the main socioeconomic variables at the different scales (NUTS 2 and 3 levels) in order to assess the dynamics of population, production systems, and services;
3. The production and visualization of territorial indicators: LUISA is able to produce spatially explicit indicators grouped in specific themes that are defined as territorial indicators. Among these indicators, we can cite population dynamics, education, health, energy, environment and climate, urban development, social issues, transport and accessibility, employment, etc. Their evolution is computed within 2060.

It is also possible coupling LUISA with other models that are more specific for the assessment of a particular issue. For example, LUISA has been coupled with the CBM model (Carbon Budget Model of the Canadian Forest Sector, Kurz et al., 2009) in order to assess the potential impacts of future regulation, land dynamics, and climate change on forests, especially in terms of Afforestation/Reforestation, Deforestation, and Forest Management activities (Pilli et al., 2016). The final aim is to assess the potential emissions and removals of CO_2 resulting from forest management and use.

Linking LUISA and CBM allowed assessing how much forest biomass could be available given the area for harvest and given the management of forests according to current and future periods.

Fig. 13.3A shows the aggregated data at the scale of Europe concerning the level of conversion of land to forest areas from 2010 to 2030 and Fig. 13.3B, the deforestation for the same time scale.

These results have shown the trends of afforestation and deforestation at the scale of European countries until 2030.

Fig. 13.4 shows a specific example with Germany in order to demonstrate the spatial resolution of the model outputs.

(A)

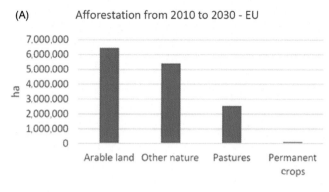

(B)

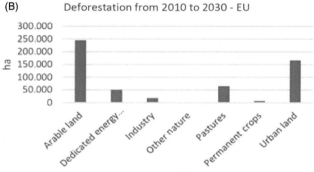

Figure 13.3

Afforestation and deforestation from 2010 to 2030. In EU, (A) Conversion of land to forest land, (B) Conversion of forest land to other land.

Statistics of deforestation and afforestation are also extracted from the selected areas (Fig. 13.5).

Other applications of LUISA have been developed in different topics like energy demand, tourism, agriculture, and water management. This modeling platform represents a generalized approach attesting the need of European Commission to develop models and tools transposable at different scales of perception.

13.3 Environmental planning in the Mexican territory

José Juan Hernández Chávez has a Bachelor degree in Biology, of the National School of Biological Sciences of Instituto Politécnico Nacional, México. He is currently studying a master degree in Animal Biology at Sciences Faculty, Universidad Nacional Autónoma de México. He is also a fellow of cohort 20 from LEAD (Leadership for Environment and Development) program México, Colegio de México. His current position is Director for Envionmental Policy in the Secretaría de Medio Ambiente y Recursos Naturales, México where he is in charge of technical counseling and institutional support in the

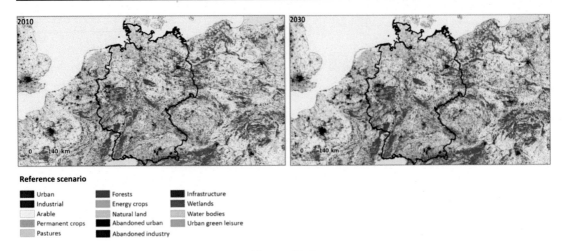

Reference scenario

Urban	Forests	Infrastructure
Industrial	Energy crops	Wetlands
Arable	Natural land	Water bodies
Permanent crops	Abandoned urban	Urban green leisure
Pastures	Abandoned industry	

Figure 13.4
Land use observed (2010) and predicted (2030) by the LUISA model.

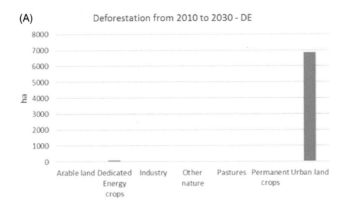

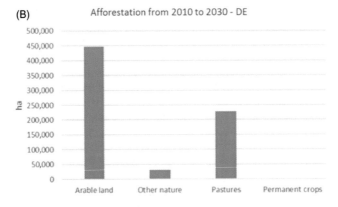

Figure 13.5
Afforestation (A) and deforestation (B) surfaces in Germany, from 2010 to 2030.

processes of environmental planning (ordenamiento ecológico del territorio) in the Mexican Territory.

According to the interview in January 2020, José Juan Hernandez Chávez explained his vision of ecological land planning in Mexico that requires a prospective approach in order to integrate the potential evolution of the territory. He shares with us his opinion on how a geoprospective approach should play a role in territorial planning in Mexico, according to the challenge of biodiversity and nature protection.

Do you know the concept of geoprospective?

Geoprospective can be understood as a conceptual and methodological framework used as a guide to perform an alternative scenario of a future environmental state for a territory, in order to use them as tools, to figure out how to make a management of natural resources with a time reference. In the geoprospective research, you should have to follow some general rules:

- The future is unpredictable, but we can make an educated assumption (hypothesis) of trends of the future based on the discovery and interpretation of historical patterns.
- The reality is complex, interrelated, and nonlinear, so you must bear in mind those issues when you give an interpretation of the data.
- Keep in mind that high correlations in patterns not necessarily mean causality among them.
- Always try to make a cross-check between geographic and tabular data available.

How the spatial dimension is taken into account in your activities in prospective, especially concerning the evolution of the territorial structures?

For the environmental planning in México (known as *Ordenamiento ecológico del territorio* in the Mexican law), the spatial dimension is a strategic matter. The main issue is modeling the change through time of spatial variables like the land uses (including native and anthropogenic ecosystems), the identification of biodiversity spots (high number of species or distribution of endangered species in our country), soil erosion, carbon sequestration, and water captured to the aquifers. The rates of modeled change are used to develop thresholds in the use of natural resources.

On what topic is addressed your activity about prospective? That is, evolution of an ecosystem or many ecosystems? Evolution of the spatial distribution of species, etc.?

The main topics of prospective are the evaluation landscape changes, changes of income of people due to a change in ecosystem services (erosion, pollution, water quality, loss of biodiversity, etc.) which is a real challenge due to the complexity and interconnectivity of these dimensions that characterize our socioecosystems.

Table 13.1: Methodology for land suitability assessment integrating stakeholders' participation.

Steps	Activities	People involved	Means
1	Definition of objectives and goals	Stakeholders	Lectures
2	Methods for suitability analysis	Interdisciplinary team	Lectures
3	Sectorial objectives and goals	State and municipal authorities and stakeholders representatives	Lectures
4	Description of activities	Stakeholders representatives	Small group involvement
5	Identification of land-use criteria	Stakeholders representatives	Small group involvement
6	Identification of environmental conflicts	Stakeholders representatives and State and municipal authorities	Small group involvement
7	Identification of compromises for settlement of conflicts	Stakeholders representatives and State and municipal authorities	Small group involvement

Based on Bojórquez-Tapia, L.A., Díaz Mondragón S., Ezcurra E., 2001. GIS-based approach for participatory decision making and land suitability assessment. Int. J. Geogr. Inf. Sci. 15 (2), 129—151.

On what kind of geographical scales is based your approach? (Microstation, municipality, watershed, regional, national, international)

In Mexico, the environmental law establishes a hierarchy of planning, from national to municipality levels, including the marine territory. In each level, there is a set of criteria mixing geomorphology, edaphology, basins, sub and microbasin and vegetation types used to create a nested arrangement of polygons. In general terms, for the national and marine level the working scale is 1:1,000,000, for the regional level 1:250,000, and for the local level is 1:50,000. But, sometimes when it is relevant, we use multiscale information as complement for specific purposes.

Does your methodology can be considered as systemic? If yes, what are the main methods you use?

Yes, indeed. The methodology for the environmental planning (*Ordenamiento ecológico del territorio*) in México was created from an academic research,[11] the main concepts of the methodology were included in a regulatory law[12] and finally the specific methods were published in the « *Manual del Proceso de Ordenamiento Ecológico del Territorio* ».

Table 13.1 shows the different steps proposed by Bojórquez-Tapia et al. (2001) in order to integrate stakeholders' participation in ecological planning of the territory.

---

[11] Bojórquez Tapia, L. A., Díaz Mondragóny, S., Saunier R., 1997. Ordenamiento Ecológico de la costa norte de Nayarit. Instituto de Ecología, UNAM, Organización de los Estados Americanos Departamento de Desarrollo Regional y Medio Ambiente.

[12] Reglamento en materia de ordenamiento ecológico de la Ley General del Equilibrio Ecológico y la Protección al Ambiente de México.

After the integration of such data, the methodology uses a multivariate statistical procedure for classifying land units into land suitability groups, according to the different sectors. This methodology of land suitability assessment based on multicriteria analysis is described by Bojórquez-Tapia et al. (1994).

Do you analyze the spatial interactions between some parameters of the territory (water, animals, floral, climate, etc.)?

The core of the methodology is the ≪ *Análisis de Aptitud Sectorial* ≫ ; it consists in a geographic exploration of the interactions in the territory conducted by the stakeholders (farmers, cattle ranchers, mining owner, industrial developers, etc.) and translated into a Geographic Information System (GIS) in order to achieve cartography of aptitude gradients in the territory for each interest sector. Also, there are methods called ≪ *conflict analysis* ≫ between stakeholder and ≪ *Area identification for conservation and compensation of negative environmental impacts* ≫.

All this cartographic information is the input for the prospective (*sensu stricto*) that includes modeling and simulations of social-environmental systems with software (STELLA or others) in order to describe and quantify future scenarios.

According to your experience, what are the limits of the methods that you use and what kind of approaches and results should be relevant in order to help you?

The limits of the methodology of ≪ *Ordenamiento Ecológico del Territorio* ≫ are the following:

- Public participation is low because there are few time and resources in order to involve the citizen opinion and vision.
- It doesn't perform a landscape analysis that includes issues like fragmentation, corridors, matrix permeability, relaxation effect, and border effect. Those elements have a strong impact on nature, ecosystems services, and biodiversity conservation.
- It doesn't perform an evaluation of the effect of climate change in the vegetation and the ≪ *aptitud sectorial* ≫, which is a significant challenge in territorial planning due to our context of climate forcing.
- It didn't define the maximum amount of native vegetation change, which is also a significant parameter to take into account in territorial planning.
- It doesn't include the socioeconomic issues in the environmental planning program, which is another major dimension according to the dynamics of socioecosystems.

Does your institution seek to establish projections in order to adapt the public policies according to global changes?

Yes, definitely. The formulation process of the *Ordenamiento Ecológico del Territorio* have a set of analysis to predict paths for some variables like rates in landscape change, water consumption, and population grow. This analysis is elaborated with tabular information and GIS. The main reason is to establish thresholds in the use of natural resources.

Does the concept of resilience is integrated into your activities and the reflection of territorial planning currently? If yes, how? If not, is it considered as implicit?

The resilience is a topic not included in the environmental planning process, but it could be considered as an issue in the climate change adaptation.

Acknowledgments

Christine Voiron and Emmanuel Garbolino want to express their gratefulness to Nathalie Cecutti, Carlo Lavalle, and José Juan Hernández Chávez for the time they dedicated to these interviews and for sharing all of these interesting information and data.

References

Bojórquez-Tapia, L.A., Ongay-Delhumeau, E., Ezcurra, E., 1994. Multivariate approach for suitability assessment and environmental conflict resolution. J. Environ. Manag. 14, 187–198.

Bojórquez-Tapia, L.A., Díaz Mondragón, S., Ezcurra, E., 2001. GIS-based approach for participatory decision making and land suitability assessment. Int. J. Geogr. Inf. Sci. 15 (2), 129–151.

CGDD, Mission prospective, Le programme Territoire 2030. <http://www.territoire-durable-2030. developpement-durable.gouv.fr/index.php/td2030/programme/?id = aqua> (in French).

Commissariat Général au Développement Durable, 2013, Biodiversité et Territoires 2030, cinq scénarios d'évolution, synthèse de l'exercice de prospective : volets 1 et 2, Collection Etudes & Documents de la Délégation au Développement Durable, n° 86, juin 2013 (in French).

Hervieu, H., Jannès-Ober, E., 2017. L'exercice Aqua 2030: comment imaginer les politiques de demain sur l'eau et les milieux aquatiques à la fois dans ses dimensions nationale et territoriale? Sci. Eaux Territ. 22, 62–67 (in French).

Jacobs-Crisioni, C., Diogo, V., Perpiña Castillo, C., Baranzelli, C., Batista e Silva, F., Rosina, K., et al., 2017. The LUISA Territorial Reference Scenario 2017: A Technical Description. Publications Office of the European Union, Luxembourg, ISBN 978-92-79-73866-1, doi:10.2760/902121, JRC108163. 42p.

Kurz, W.A., Dymond, C.C., White, T.M., Stinson, G., Shaw, C.H., Rampley, G.J., et al., 2009. CBM-CFS3: a model of carbon-dynamics in forestry and land-use change implementing IPCC standards. Ecol. Model. 220, 480–504.

Lavalle, C., Pontarollo, N., Batista E Silva, F., Baranzelli, C., Jacobs, C., Kavalov, B., et al., 2017. European Territorial Trends. Facts and Prospects for Cities and Regions Ed. 2017, EUR 28771 EN. Publications Office of the European Union, Luxembourg, ISBN 978-92-79–73428-1, doi:10.2760/148283, JRC107391. 62p.

Pilli, R., Fiorese, G., Abad Viñas, R., Rossi, S., Priwitzer, T., Hiederer, R., et al., 2016. LULUCF Contribution to the 2030 EU Climate and Energy Policy; EUR 28025. Publications Office of the European Union., Luxembourg, JRC102498; doi:10.2788/01911.- 690p.

Villari, B., 2012. Design per il territorio. Un Approccio Community Centred. Ed. Franco Angeli, Milan (in Italian).

Index

Note: Page numbers followed by "*f*" and "*t*" refer to figures and tables, respectively.